HUMAN ANATOMY AND PHYSIOLOGY

A COMPREHENSIVE REVIEW FOR THE HEALTHCARE STUDENT®

BY

KELLY DRAGON, M.D.

ISBN

Paperback: 979-8-3306-9359-7

Table of Contents

About the Author

Kelly Dragon, M.D., graduated from Pace University in Pleasantville, New York, with a Bachelor of Science in Biology and a minor in Chemistry. He completed his medical education at Ross University School of Medicine. Dr. Dragon is a professor at Bowie State University in Bowie, Maryland. He also teaches Anatomy and Physiology at Prince George's Community College (PGCC) in Largo, Maryland.

Acknowledgments

This volume consolidates several years of teaching human anatomy and physiology in the classroom. It also represents the contribution of many people in my life, beginning with my parents and grade school teachers, who early on had a profound effect on my education.

I would like to thank Mark Hubley, Ph. D., professor and former chair of the natural sciences and engineering department at Prince George's Community College (PGCC), for his mentorship, guidance, and wisdom throughout my years there.

To my students, I thank you for constantly pushing me to improve in the classroom and making me a better teacher over the years. This is for you!

To my family,

thank you for your

unwavering support. Without you,

this would not be

possible.

Introduction

Comprehensive Review Guide: Anatomy & Physiology

This essential manual provides a thorough and accessible review of anatomy and physiology, featuring HD illustrations to support students at every academic level—whether in high school, college, university, or professional programs. Designed to simplify complex concepts, this guide is an invaluable resource for mastering the foundational knowledge of the human body.

Covering the body's major systems—skeletal, muscular, cardiovascular, respiratory, digestive, nervous, and endocrine—each chapter explores the anatomy (structure) and physiology (function) of organs, tissues, and cells, demonstrating how these systems work together to maintain homeostasis. With a strong focus on clinical applications, the guide connects theoretical learning to practical healthcare settings.

Detailed explanations of vital processes such as cell metabolism, nerve conduction, muscle contraction, and cardiovascular function are enhanced with flowcharts, diagrams, and mnemonics, making challenging material easier to grasp. Key terms, definitions, and review questions further aid in exam preparation, ensuring students are well-equipped for success.

Ideal for students in health sciences, nursing, and other medical fields, this comprehensive guide provides a solid foundation for advanced study and professional practice.

Part I
Human Anatomy and Physiology I

Chapter 1: Introduction to the Human Body

I. Scientific Overview of Anatomy and Physiology

Human **anatomy** is the scientific study of the structures and their locations in the body. These structures vary in size. Some are so small that they can only be seen and analyzed using a microscope (**microscopic anatomy**). On the other hand, observing larger structures that can readily be studied without the assistance of a microscope is referred to as **gross anatomy** or **macroscopic anatomy**. There are many ways, however, to view detailed structures inside the body with precision using a variety of imaging techniques such as Functional Magnetic Resonance Imaging (fMRI) and Computed Tomography (CT) scans as well as ultrasonography (U/S). CT scan is an imaging technique that uses multiple angles of X-rays at once to produce a more detailed image and, as such, exposes patients to high levels of radiation. And so does X-ray imaging, which is best for hard structures such as bones, to analyze bone fractures, for example.

It should also be noted that there are two general approaches to studying body structures: regional and systemic. Regional **anatomy** is the study of the interrelationships between the structures in a specific body region to serve that specific region, such as the head (cephalic region), or the abdomen (abdominal region). In contrast, **systemic anatomy** studies the structures that make up a specific body system (i.e., the cardiovascular system), with those structures working together to perform a unique body function. For example, the respiratory system - one of the eleven body systems containing multiple structures working together to allow the individual organism to breathe.

Human **physiology** is the scientific study of how structures work together to support the functions of life. For example, a cardiovascular physiologist studies how the heart regulates blood pressure and flow in the vessels (hemodynamics), among other things.

Microscopic anatomy includes **cytology**, the study of cells, and histology, the study of tissues. Biology, being the field of study of all living organisms, has other branches in addition to the ones already mentioned that students must be familiar with. These are **embryology**, the study of the embryo from conception to birth (development); **pathology**, the study of how disease affects the body; and **endocrinology**, the study of hormonal levels in the body.

II. Levels of Organization of the Human Body

The various levels at which the body is organized, from the simplest to the most complex level are very important to truly understand anatomy.

a. **Chemical level**. The most basic level of organization includes atoms and molecules. Examples are calcium, glucose, and phospholipids.
b. **Cellular level**: At this level, various molecules aggregate into larger structures that make up a cell, the smallest independently functioning unit of a living organism. In this level, cells with similar functions aggregate to form a larger structure. Examples include biceps brachii muscle cells, and heart muscle cells.
c. **Tissue level**: Cells associate to form tissues, a group of many cell types that work together to perform a specific function. Examples include skeletal muscle tissue, cardiac muscle tissue, and skin tissue.
d. **Organ level**: An organ is a structure composed of two or more tissue types; each organ performs one or more specific physiological functions. Examples include the heart, the lungs, or any other organ of the body.
e. **Organ system level**: A group of organs that work together to perform major functions or meet the body's physiological needs. An example is the respiratory system.
f. **Organismal level**. An entire organism (the whole body).

Levels of organization of the human body. Complete this chart by Identifying each level.

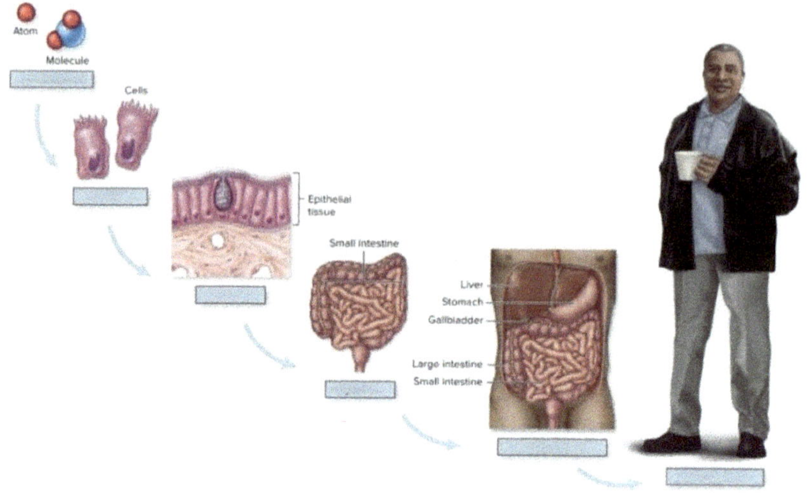

III. Homeostasis: Keeping Internal Conditions Stable

Much of the study of physiology centers around the body's tendency toward homeostasis.

Homeostasis is the state of dynamic stability of the body's internal conditions. It is the body's ability to maintain a relatively constant internal environment despite changes in the external and internal environments that tend to put stress on the body and move it away from homeostasis.

Maintaining homeostasis requires the body to continuously monitor its internal conditions. This maintenance requires control mechanisms (**homeostatic system**) that detect and respond to changes. The components of the homeostatic system are: (1) **sensor** (receptor). Receptors monitor stimuli in the internal and external environments; (2) **control center** monitors information from receptors and compares this information to set points for the body; and (3) **effectors** which are used to bring about changes in the body to maintain homeostasis.

Consider the following example: Thermoreceptors (receptors) in your body may detect that your body temperature is rising excessively and send signals to your brain. Neurons in your brain (control center) detect that your body temperature is in danger of rising well above the 98.6 degrees set point. The brain sends signals to your sweat glands (effectors) to increase perspiration. Perspiration cools the body to help maintain the proper temperature.

Homeostasis is typically maintained by negative feedback systems. A negative feedback system minimizes deviations of a variable from a sct value. For example, when glucose levels rise above the set value, the pancreas releases insulin. Insulin then lowers the amount of glucose in the blood. Similarly, when glucose levels in the blood drop below the set value, the pancreas releases glucagon. Glucagon, in return, raises the amount of glucose in the blood. Both are examples of negative feedback. (see diagram below)

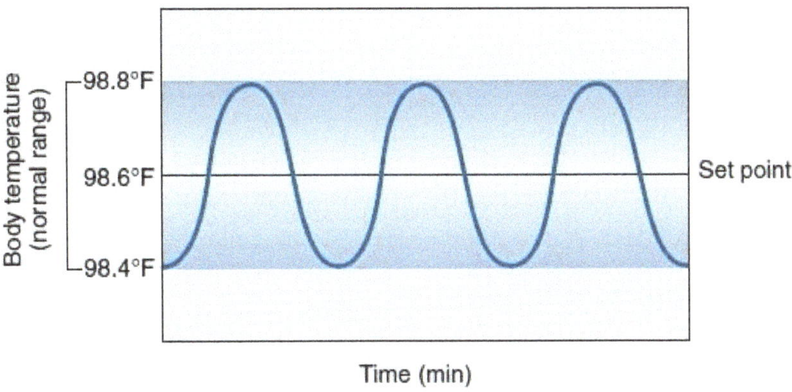

Homeostasis

Homeostasis is the maintenance of a variable, such as body temperature, around an ideal normal value, or set point. The value of the variable fluctuates around the set point to establish a normal range of values.

Positive feedback systems increase a variable's deviation from its normal value. In other words, a positive feedback system intensifies a change in the body's physiological condition rather than reversing it. An example of a positive feedback system is the role of the hormone oxytocin in childbirth. During childbirth, dilation of the cervix stimulates the release of oxytocin, which increases uterine contractions. Increased contraction of the uterus will push the fetus farther through the cervix and increase cervical dilation. Increased cervical dilation causes the release of more oxytocin until the fetus (baby) exits through the birth canal.

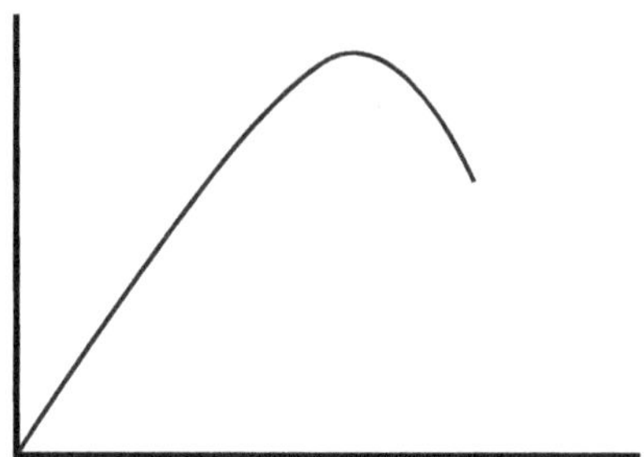

Positive feedback mechanism graph above. This mechanism is involved in childbirth, and lactation among other processes for example, and here the line upstroke represents the increased intensity of uterine contractions (i.e., one contraction brings on another stronger contraction and so on) until the climactic event at the peak of the upstroke, in this case, the baby passing through the birth canal. Eventually, the downstroke follows and represents the body returning to its normal state (i.e., uterine relaxation).

IV. The Eleven Body Systems

a. *Integumentary system*
   - Creates a barrier that protects the body from pathogens and fluid loss
   - Sensory reception

b. *Skeletal system*
   - Supports and protects the body

c. *Muscular system*
   - Creates the movement of the body
   - Contributes to body temperature homeostasis

d. *Nervous system*
   - Acts as the sensor for homeostasis
   - Connects the brain to every part of the body

e. *Endocrine system*
   - Secretes hormones that regulate many bodily processes

f. *Cardiovascular system*
   - Delivers oxygen, nutrients, hormones, and waste products throughout the body
   - Contributes to temperature regulation

g. *Lymphatic/Immune system*
   - Regulates fluid balance in the body
   - Houses some of the body's immune cells that defend the body from pathogens

h. *Respiratory system*
   - Exchanges air with the atmosphere
   - Provides surface area for the diffusion of oxygen and carbon dioxide with the blood

i. *Digestive system*
   - Breaks down food and absorbs nutrients into the body

j. *Urinary system (Excretory)*
   - Contributes to blood pressure and pH homeostasis
   - Removes waste products from the body

k. *Reproductive system*
   - Produces and exchanges gametes
   - Houses the fetus until birth
   - Lactation

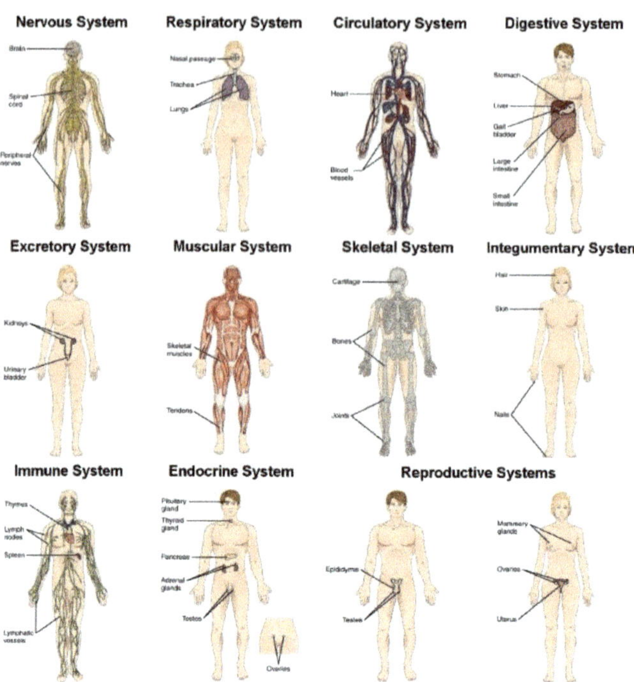
Nervous System
Brain
Spinal cord
Peripheral nerves
Respiratory System
Nasal passage
Trachea
Lungs
Circulatory System
Heart
Blood vessels
Digestive System
Stomach
Liver
Gall bladder
Large intestine
Small intestine
Excretory System
Kidneys
Urinary bladder
Muscular System
Skeletal muscles
Tendons
Skeletal System
Cartilage
Bones
Joints
Integumentary System
Hair
Skin
Nails
Immune System
Lymph nodes
Spleen
Lymphatic vessels
Endocrine System
Pituitary gland
Thyroid gland
Pancreas
Adrenal glands
Testes
Ovaries
Reproductive Systems
Testes
Ovaries
Uterus

Chapter 2: Terminology of the Human Body

I. Anatomical Position

Recall that Anatomy is the “study of structures in the human body," and Physiology is “how body parts function together.” To learn human anatomy, locating specific body structures using points of reference is crucial. So, to determine the location of various body parts, the standard frame of reference is the **“Anatomical position.”** In this position, a person stands upright with the arms on the sides of the body, the palms of the hands facing forward, and looking straight ahead. (See below)

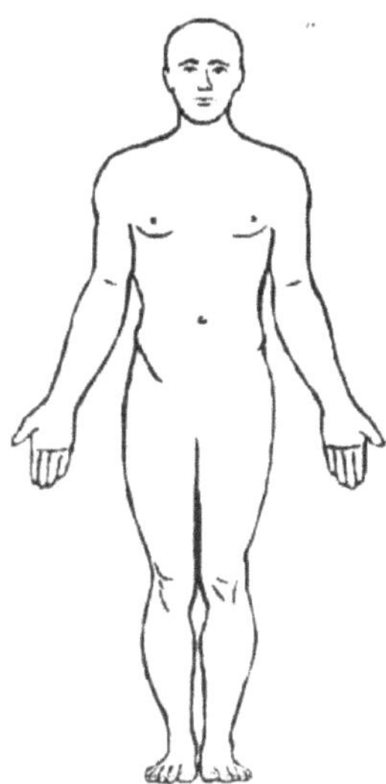

Other body positions to consider are **1) supine position** - when a person is lying on their back and **2) prone position** - when a person is lying on their stomach.

II. Identifying Body Parts

The human body has a right, left, front, and rear/back side. The body is not cylindrical in shape, so it is important to use proper terminology when referring to body surfaces to locate or describe specific structures precisely. Hence, instead of 'front,' the correct anatomical term is **anterior**, and for the rear, the proper anatomical term is **posterior**. Another similar term with subtle differences used for anterior is **ventral**, which refers to the belly portion of an animal, and for posterior, the similar term is **dorsal**, meaning the surface of the back. Remember that structures are termed in this manner based on their anatomical position and by comparing their location to other nearby structures. In addition to the former terms, paired **directional** terms to know are the following (see illustration below):

a. **Medial** describes the location of a structure close to the midline (a line drawn straight down the middle of the body separating the body into right and left halves) of the body. **Lateral** describes the location of a structure that is farther away from the midline. For example, the little toe is lateral to the great toe, which makes the great toe medial to the little toe. In the same manner, the thumb is lateral compared to the pinky, and therefore, the pinky is medial to the thumb.
b. **Superior** and **inferior** denote the location of a structure within a body part that is above another structure within that same specific body part. For example, the eyebrows are superior to the eyes. The mouth is inferior to the nose; therefore, the nose is superior to the mouth.
c. **Superficial** and **deep** can be used when comparing structures closer to the body's surface vs. those farther inward. For example, the skin is superficial relative to the pectoral muscles of the chest; the pectoral muscles are deeper than the skin.
d. **Proximal** and **distal** refer to the locations of structures on the extremities/appendages (i.e., arms, legs, hands, and feet) depending on their proximity to the torso. For example, the shoulder joint is proximal compared to the elbow, and therefore, the elbow is distal to the shoulder. Another example is the ankle, which is distal to the knee joint, which makes the knee joint proximal to the ankle.
e. **Ipsilateral** and **contralateral** describe the location of a structure on the same side of the body. For example, the right arm is ipsilateral to the right leg. Contralateral is the location

of a structure on the opposite side of the body. For example, the right arm is contralateral to the left leg.

Directional terminology

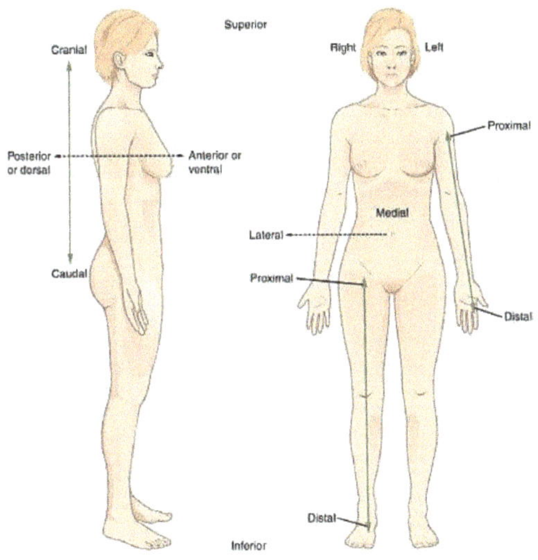

Directional Terms

- **Ipsilateral** – on the same side of the body
  - The right arm and right leg are ipsilateral.
- **Contralateral** – on the opposite side of the body
  - The right arm and left leg are cotralateral.

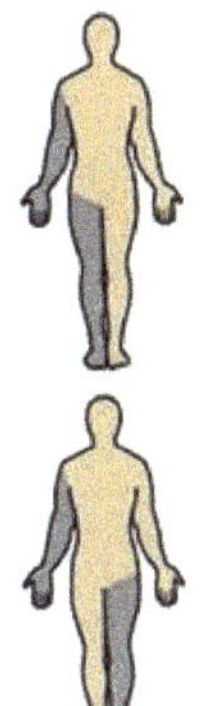

III. Body Sections/planes

To view the internal structures or internal organs of the body, it must be cut through, and as such, there are two primary directional cuts:

a. **Longitudinal** and **transverse.** A longitudinal cut runs along the long axis of the body from top to bottom, separating the body into a right portion and a left portion. So, any cut made along the midline or parallel to the midline is considered longitudinal. A transverse cut, however, goes across the long axis of the body, separating the body into an upper portion and a lower portion. A transverse section can also be called a cross-section. Note that viewing the internal structures from these cuts would be looking at the organs from 2 different angles.

b. In addition to the longitudinal and transverse sections, there are other ways to cut through individual organs using the two previous directional cuts (i.e., longitudinal and transverse). So, there are two ways to make a longitudinal cut:
   i. A **frontal** section, also known as a **coronal** section, is a cut through the head that separates the front from the back of the head. A **sagittal** section made at the midline is a **midsagittal** section, and if the cut or plane is off the midline either to the right or the left of the midline, then it is a **parasagittal** section.

Parasagittal

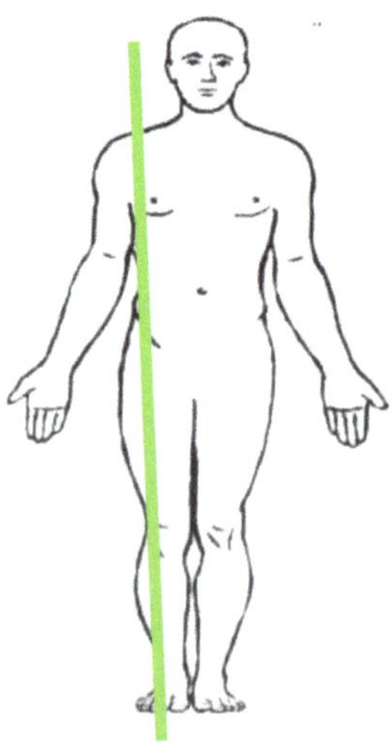

Midline/Median/Midsagittal including medial and lateral forearm

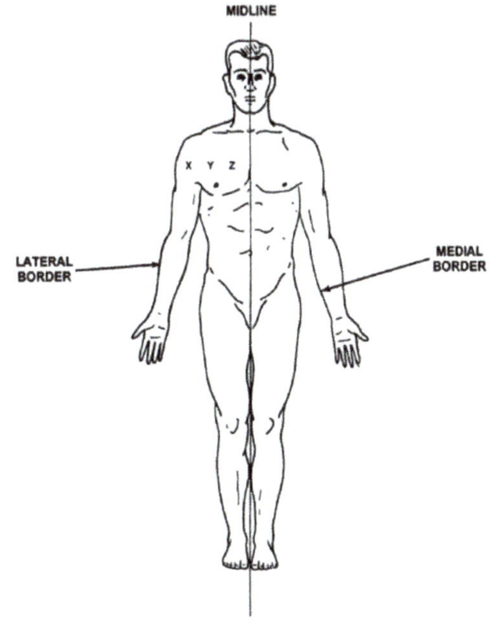

Body Planes

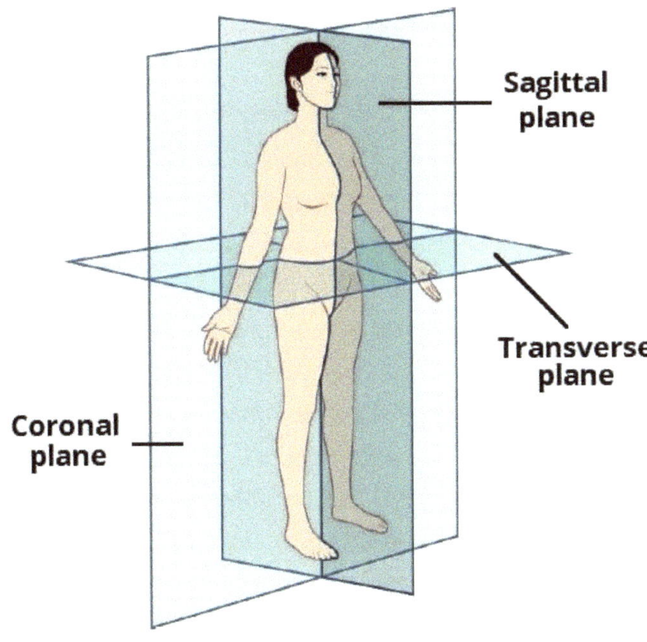

Body planes (continued)

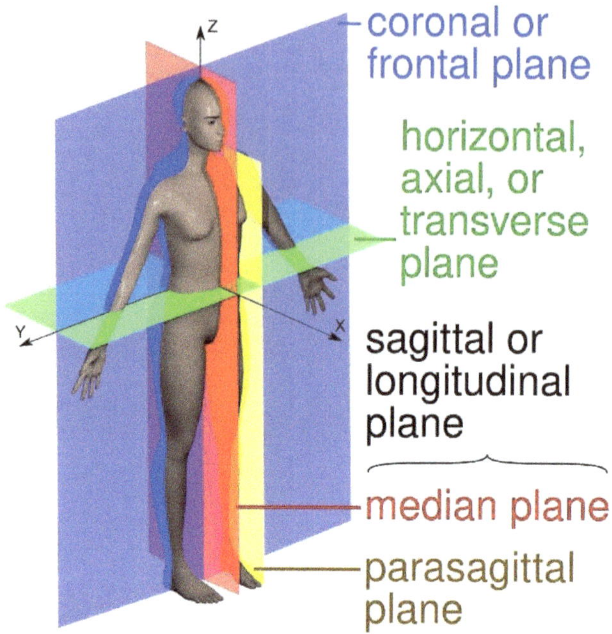

IV. Major Body Cavities and Regions

- There are two major cavities within the human body: the **dorsal body cavity** and the **ventral body cavity**. (See image below).
- The dorsal body cavity lies within the skull and the spine as one continuous space. However, it is divided into two parts: the **cranial cavity**, the space inside the skull that contains the brain, and the space where the spinal cord resides, the **vertebral canal**.
- The ventral body cavity is much larger than the dorsal body cavity and contains more organs. The superior portion of the ventral body cavity is separated from the inferior portion by the diaphragm. The entire superior portion is called the thoracic cavity, and it is further subdivided into a medial portion, called the mediastinum, and two lateral portions, called the **pleural cavities**. The mediastinum contains the trachea and the esophagus. It also contains another cavity,

called the pericardial cavity, which surrounds the heart. Each of the two pleural cavities surrounds one of the lungs.

- The inferior portion of the ventral body cavity is called the abdominopelvic cavity. This cavity is also further subdivided into a superior portion called the abdominal cavity and an inferior portion called the pelvic cavity. The abdominal cavity contains many organs, including the stomach, liver, spleen, pancreas, and intestines. The pelvic cavity contains the urinary bladder, rectum, and, in females, the uterus and ovaries.

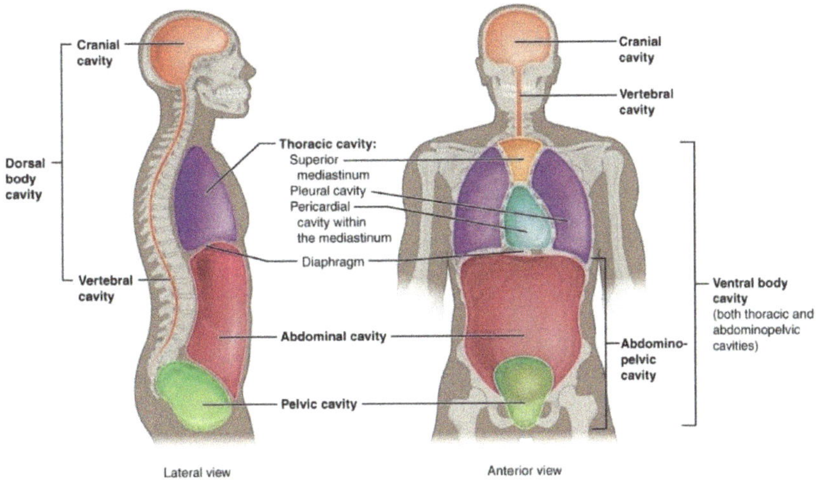

V. Membranes lining the major body cavities

- Organs (brain and spinal cord) in the dorsal body cavities are wrapped in membranes called the **meninges**. Organs (stomach, spleen, pancreas, intestines, liver) are wrapped in a thin layer of membrane referred to as the **visceral serosa**. The inside of the body cavity wall is lined with a layer of membrane referred to as the parietal serosa. Between the visceral serosa and the parietal serosa is **serous fluid**, which lubricates the membranes.
- The serous membranes are given specific names according to the specific parts of the ventral body cavity in which they are found. The membranes in the pericardial cavity are the

visceral pericardium and **parietal pericardium**. The membranes in the pleural cavities are the **visceral pleura** and the **parietal pleura**. The membranes in the abdominopelvic cavity are the **visceral peritoneum** and the **parietal peritoneum**. Additional membranes, called **mesenteries**, hold the abdominopelvic cavity in place, attaching them to each other and to the inside of the wall. (See images below).

Cranial meninges

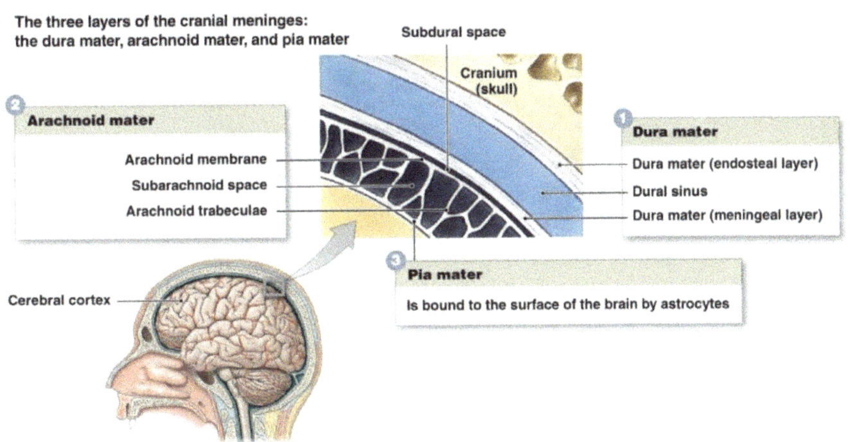

Body Cavities and Regions

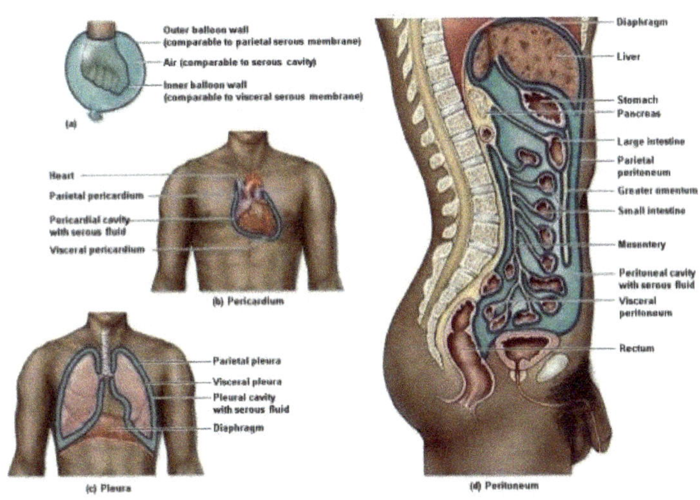

The following diagram illustrates the body regions.

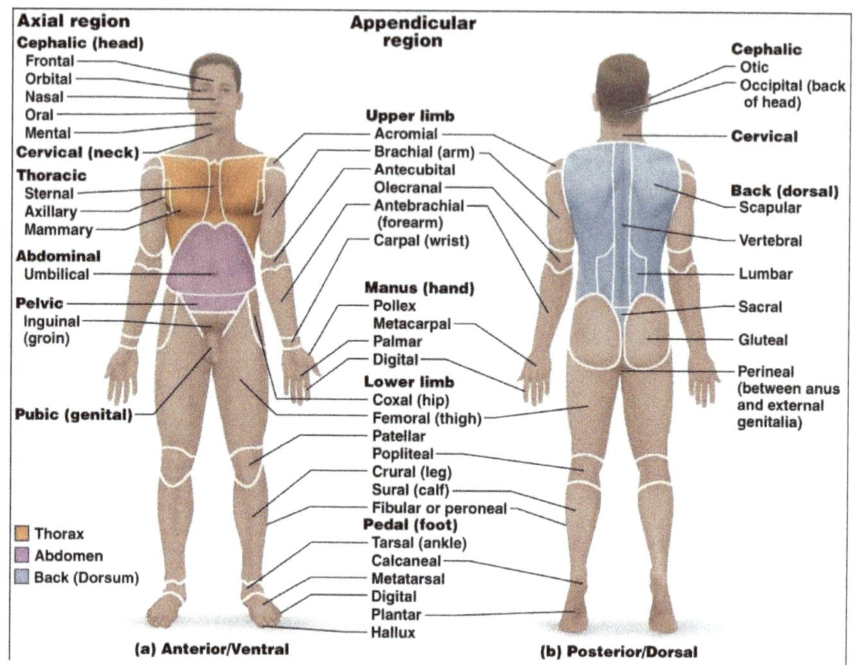

Below is a transverse section of the abdomen showing visceral and parietal peritonea. Imagine this is your patient getting a CT (Computed Tomography) scan of the abdomen and their head is coming out of the page towards you, can you identify their body position?

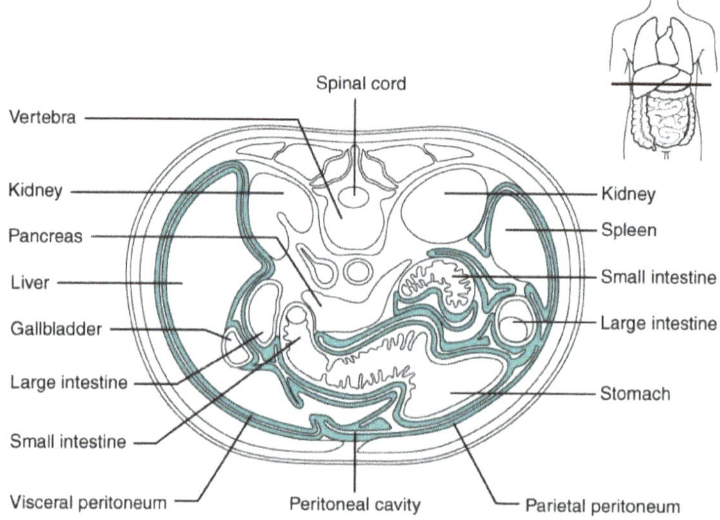

What section or plane is displayed here in the heart below?

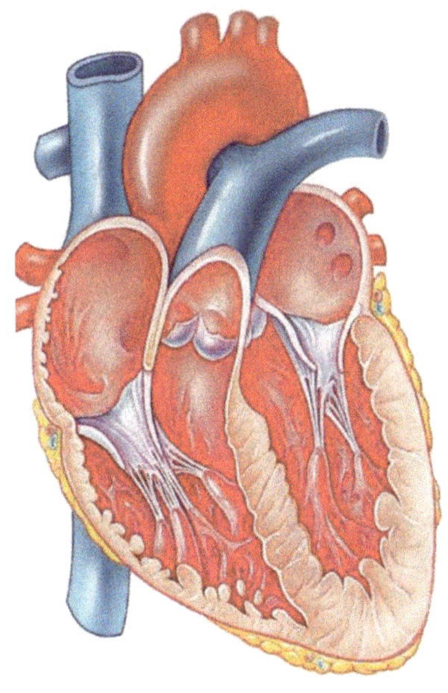

Label the diagram of the body cavities below.

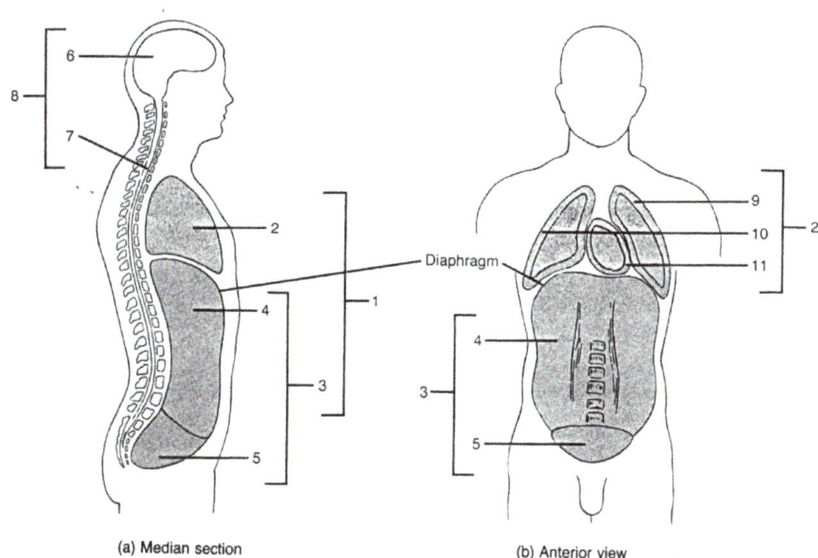

Chapter 3: Body Tissues

There are four general types of body tissues: **epithelial**, **connective**, **nervous**, and **muscle**.

Epithelial tissues: Simple and Pseudostratified Epithelia

Cover surfaces of the body, both external and internal. For example, both the surface of your skin and the inner lining of your stomach are made of epithelial tissues. All epithelial tissues have certain properties in common: (1) Each epithelium has an **apical surface**, which is exposed to the outside of the body or the lumen of an internal cavity. (2) The cells of an epithelium are held together with specialized contacts, including **tight junctions** and **desmosomes**. (3) A **basement membrane** connects the basal surface of the epithelium to an underlying layer of connective tissue. (4) Epithelial tissues are **avascular**.

The various epithelial tissues are classified according to the number of layers of cells in the epithelium and the shape of the cells in the epithelium. An epithelium made of one layer of cells is a **simple** epithelium. In contrast, an epithelium made of two or more layers of cells is a **stratified** epithelium. Simple epithelia are generally found where substances are exchanged across the surface. Stratified epithelia typically provide strength and protection. Epithelial cells are **squamous** if they are shaped like pancakes, **cuboidal** if they are shaped like cubes, and **columnar** if they are elongated. Most epithelial tissues are named for the number of layers in combination with the shape of the cells. For example, an epithelium made of one layer of cube-shaped cells is called "simple cuboidal." An epithelium that contains many layers of pancake-shaped cells is called "stratified squamous."

A. Simple squamous epithelium

The thinnest of the epithelial tissues is often found where the exchange of materials occurs. In the lungs, structures called **alveoli** are made of simple squamous epithelium, allowing the exchange of oxygen and carbon dioxide between the air and the blood. In the kidneys, structures called **glomerular capsules** are made of simple squamous epithelium used to filter the blood (See image below).

Glomerular capsule diagram and histology of the kidney below

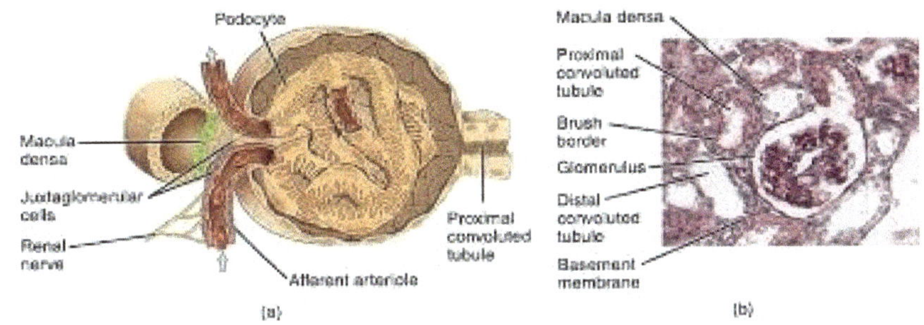

Capillaries of the cardiovascular and lymphatic systems are made of simple squamous epithelium, as are the inner linings of the heart, blood vessels, and lymphatic vessels. In these locations, the simple squamous tissue is referred to as **endothelium**. The serous membranes described earlier (visceral pericardium, parietal pleura, etc.) are also made of simple squamous epithelium. These tissues are referred to as **mesothelium**, and the mesothelium cells produce serous fluid.

B. Simple Cuboidal Epithelium

Can be found in various places in the body, including the kidney tubules and multiple glands. The most important functions of this tissue are generally secretion and absorption of materials into or from a tube or cavity. In the kidneys, the fluid that is filtered from the blood into the glomerular capsule is called "filtrate," and it is on its way to becoming urine. The glomerular capsule is the beginning of a structure called the renal tubule, and although the glomerular capsule is made of simple squamous epithelium, the rest of the renal tubule is made of cuboidal epithelium. As the filtrate passes through the renal tubule, cells of the simple cuboidal epithelium reabsorb materials that your body wishes to keep and secrete materials that your body wishes to eliminate. Thus, the filtrate starts with a composition similar to blood plasma and ends up as urine.

C. Simple Columnar Epithelium

Lines the digestive tract, ducts of some glands, and portions of the respiratory and female reproductive tracts. Specialized goblet cells secrete mucus, which may lubricate the apical surface of the epithelium or trap harmful particles. In the digestive tract, the columnar cells have microvilli on their apical surfaces to increase the area for nutrient absorption. The columnar cells in the respiratory and reproductive tracts have cilia on their apical surfaces to remove debris trapped in mucus.

D. Pseudostratified Columnar Epithelium

Forms the lining of the respiratory tract, portions of the male reproductive tract, and ducts of some glands. You may already know that "pseudo" means false. This tissue is called pseudostratified because it appears stratified, but it really is not. Like columnar epithelia, pseudostratified columnar epithelia may contain goblet cells and cilia. Smoking can damage this epithelium, resulting in a "smoker's cough."

E. Stratified Epithelia and Glands

Simple epithelia generally have the function of exchange, allowing materials to move in or out of the body by diffusion, filtration, or secretion. Stratified epithelia typically perform the functions of protection and secretion. The most obvious example of an epithelium with the function of protection is the stratified squamous epithelium of the skin, which protects the body from abrasion, water loss, and UV radiation. Stratified epithelia are also found in various glands, such as sweat and salivary glands, where they form ducts for secretion.

Glands are found throughout the body, secreting various substances for many purposes. Glands are divided into two general categories, exocrine and endocrine (See images below). An **exocrine gland** secretes a substance onto a surface of the body, which may be an external surface (such as sweat being released onto the skin) or an internal surface (such as mucus being released into the small intestine). Exocrine glands may be unicellular or multicellular. The most common unicellular exocrine gland is the goblet cell, which may be found in simple and pseudostratified columnar epithelia. Multicellular exocrine glands include sweat glands, salivary glands, mammary glands, the prostate gland, etc. Multicellular exocrine glands have one or more ducts that deliver the secretion to the epithelial surface.

An **endocrine gland** is a ductless gland that secretes a substance directly into the bloodstream, which then distributes it throughout the body. The substances secreted by endocrine glands are called **hormones**. Examples of hormones are the various steroids (such as estrogen and testosterone), epinephrine and norepinephrine, insulin, glucagon, and growth hormone. One example is the pancreas, which functions as both an endocrine and an exocrine gland.

Exocrine glands:	**Endocrine glands**:
. have ducts . carry substance to membrane surface. **Example**: Sweat produced by gland acinus and released by duct onto the surface of the skin.	. no ducts . secrete hormones into the blood and eventually reach the target cell/organ. **Example**: Pancreas releases insulin into the blood by diffusion; thyroid hormones are secreted directly into the blood.

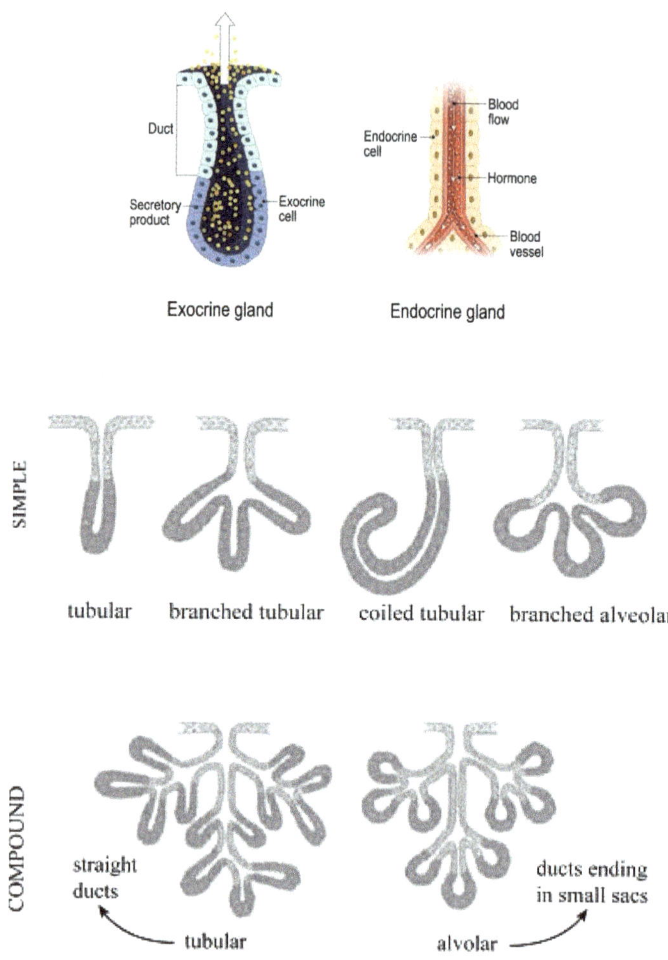

F. Stratified Squamous Epithelium

Forms the surface of your skin. It also lines the inside of the oral cavity, esophagus, vagina, anal canal, and much of the pharynx. In all of these places, the many layers of cells protect the underlying tissues from abrasion. Although this tissue is called the stratified squamous epithelium, the tissue has cells with more than one shape: the deeper cells are typically cuboidal, whereas the superficial cells are squamous. The convention for naming a stratified epithelium is to use the shape of the cells in the apical layer as part of the name.

The stratified squamous epithelium of the skin is known as the **epidermis**. The many layers of cells in the epidermis are grouped into four or named layers of epidermal cells. From deep to superficial, they are the **stratum basale, stratum spinosum, stratum granulosum**, **stratum lucidum**, and **stratum corneum** (see illustrations below). The thin, hairy skin that covers the rest of the body has four layers; there is no stratum lucidum. Deep in the epidermis is a layer of connective tissue called the dermis.

Most epithelial cells in the epidermis are called **keratinocytes**, and they produce a protein called **keratin**. Keratin provides some strength to the skin, but its primary function is to make it resistant to water loss. The stratum corneum also contains cells called **melanocytes**, which produce a pigment called **melanin**. Melanin makes skin dark and protects the underlying cells from ultraviolet (UV) radiation.

G. Transitional Epithelium

Forms the inner lining of the urinary bladder, ureters, and part of the urethra. It is also referred to as **urothelium**. Transitional epithelium may not have obvious protective or secretory functions, but it does protect the body by containing urine in the urinary bladder until the urine is eliminated from the body. The key to the ability of transitional epithelium to maintain its protective barrier is that it can stretch as the bladder fills with urine and expands. As the bladder stretches, cells of the transitional epithelium change from a round, cuboidal-like shape to a flattened, squamous shape. Hence the name "transitional' epithelium. These epithelial cells are held together with tight junctions to help prevent urine from leaking out of the lumen of the bladder

H. Stratified Cuboidal Epithelium

Forms the ducts of most exocrine glands, including sweat glands of the skin, mammary glands, and salivary glands. Generally, there are two or just a few layers of cells.

I. Connective Tissues

Introduction

The category 'connective tissue' includes a wide variety of tissues, many of which appear to have completely different properties. You would likely think that bone, fat, and blood have nothing in common, yet all three tissues are considered connective tissues. Since many kinds of connective tissues have various properties, it is convenient to categorize them further. Accordingly, connective tissues are divided into three main classes: (1) connective tissue proper, (2) supporting connective tissue, and (3) fluid connective tissue. So, why are all of these tissues grouped together as connective tissues? There are two key features that all connective tissues have in common: All connective tissues have a common origin during embryonic development, and they all contain **an extracellular matrix**. As you examine connective tissues, you should notice two ways in which connective tissues can generally be distinguished from epithelial tissues: (1) Epithelial tissues are always found on body surfaces; connective tissues are located internally. (2) In epithelial tissues, the cells are packed tightly together; in connective tissues, the cells are generally separated from each other, with extracellular matrix occupying the space between cells.

Epithelium, connective tissues, and basement membrane

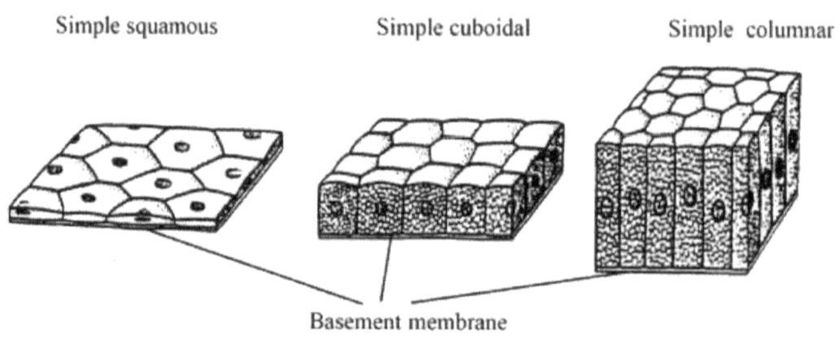

Types of Epithelial Tissues

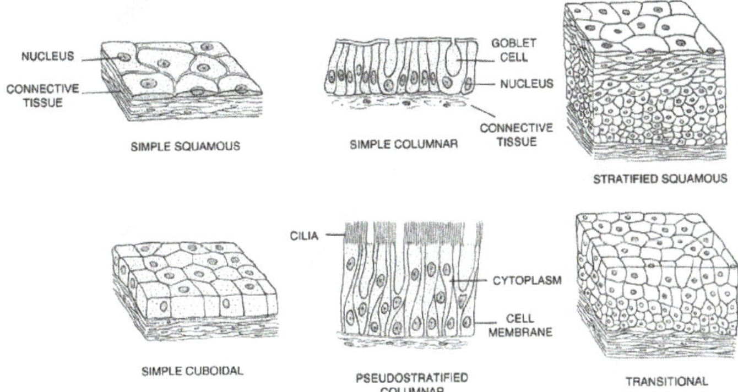

Below is the histology of the pseudostratified ciliated columnar epithelium that forms the lining of the trachea. Use this illustration to identify these structures and their function: a) *cilia*, b) *cell nuclei*, c) *goblet cell*, and d) *basement membrane*.

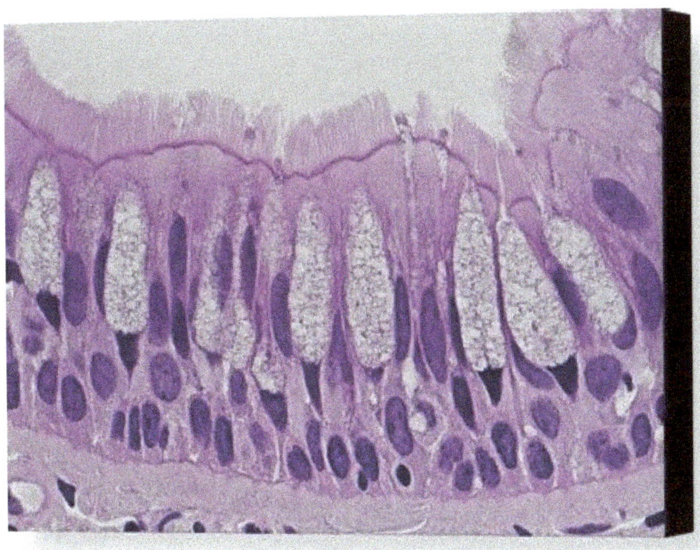

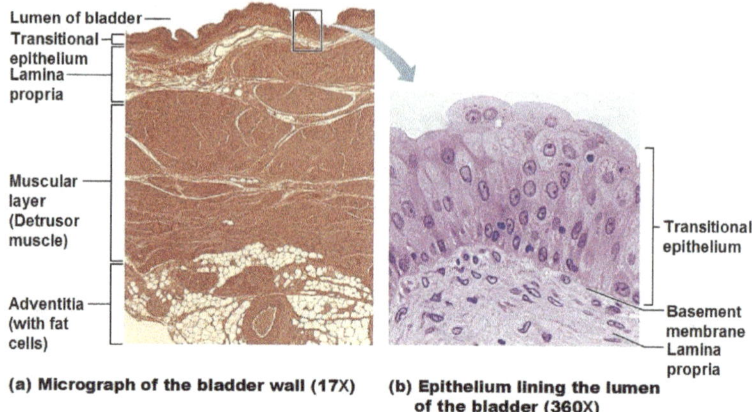

(a) Micrograph of the bladder wall (17X) (b) Epithelium lining the lumen of the bladder (360X)

Note picture (b) above shows the stratified transitional epithelium lining the bladder in its relaxed state. Take special note of the bulbous, or rounded appearance of the cells at the surface when the bladder is empty; these cells flatten and elongate when the bladder is distended (fills with urine).

There are four types of connective tissue proper: (a) areolar connective tissue. (b) adipose connective tissue, (c) dense regular connective tissue, and (d) dense irregular connective tissue.

Connective Tissue Proper

A. Areolar Connective Tissue

It is found throughout the body, underlying epithelial tissues and surrounding organs, and wrapped around blood vessels and nerves. You may think of it as the body's packing material. The predominant cell type in areolar connective tissue is the **fibroblast**, which secretes fibers into the extracellular matrix. Macrophages (a type of white blood cells whose primary function is "phagocytosis," which means "cell eating" of harmful materials) and mast cells (a type of immune system cells that cause inflammation upon foreign particle invasion of the body) are also present, and they act as part of the immune system to protect the body. Three types of fibers can be found in the matrix of areolar connective tissue: **collagen fibers** provide strength, **elastic fibers** can stretch and recoil, and **reticular fibers** may help connect the tissue to organs or other types of tissue. (See image below).

Areolar Connective Tissue with lots of ground substance (extracellular matrix-ECM). Note that this tissue can be found in the papillary (upper) layer of the dermis of the skin as well as the lamina propria of the small intestine and other organs.

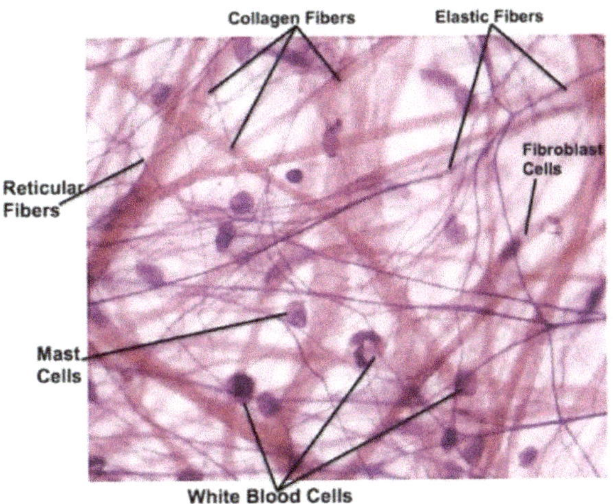

B. Adipose Tissue

Also known as fatty tissue is found throughout the body including the bones. The primary cell of adipose tissue is the **adipocyte**, and its main function is to store lipids (as a source of energy storage), insulation, and serves as a cushion for organ (i.e., protection). Inside each adipocyte is a **lipid droplet**, which fills with fat and pushes the cell's nucleus to the edge of the cell. As a person gains or loses weight, the number of adipocytes in the person's body does not change. Rather, existing adipocytes swell or shrink as the amount of fat increases or decreases. Adipose tissue typically contains a lot of blood vessels, which transport lipids to and from the adipocytes.

Histology of Adipose (Fat) Tissue below

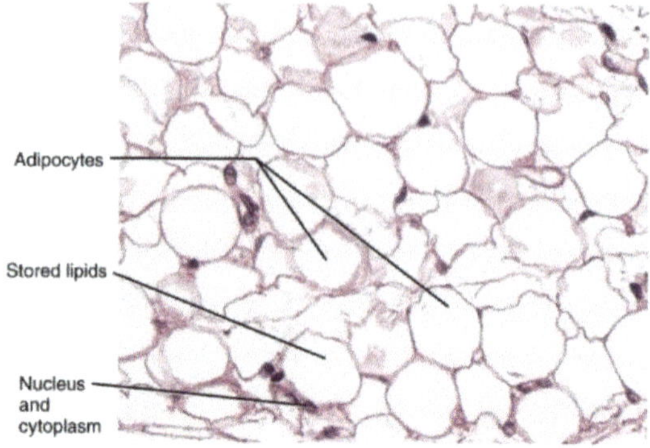

Why is the nucleus of an adipocyte not found in the center of the cell?

C. Dense Regular Connective Tissue

This tissue is termed 'dense' because it is composed of tightly packed collagen fibers, with little ground substance and few cells. It is called 'regular' because the collagen fibers are aligned in the same direction. Thus, this tissue provides great strength when it is pulled along the direction in which the fibers are oriented. Dense regular connective tissue is the main ingredient of **tendons**, which attach muscles to bones, and **ligaments**, which connect bones to bones (see image below).

Tendons and ligaments composed of dense regular connective tissue.

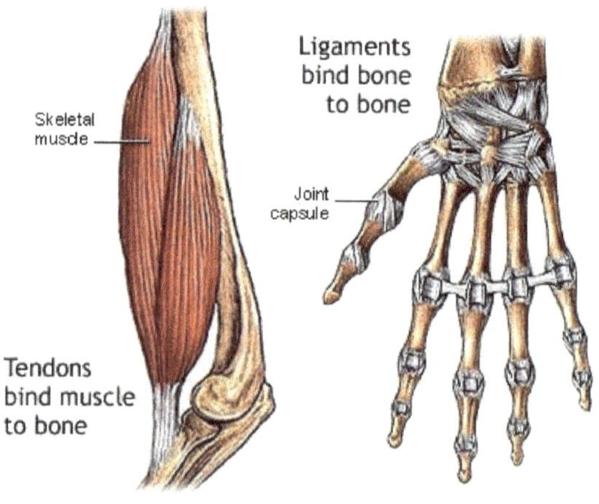

Connective tissues have varying degrees of vascularity. For example, adipose tissue has a rich blood supply as blood is necessary to transport lipids to and from the adipocytes. On the other hand, dense regular connective tissue has a poor blood supply. With few cells and mostly non-living collagen fibers, this tissue has little demand for oxygen or nutrients. Unfortunately, the lack of vascularization makes it difficult for this tissue to heal after an injury. It can take a long time to recover from a damaged tendon or ligament, such as a torn anterior cruciate ligament.

Histology slide of **dense regular connective tissue** (notice the regular arrangement of the parallel fibers).

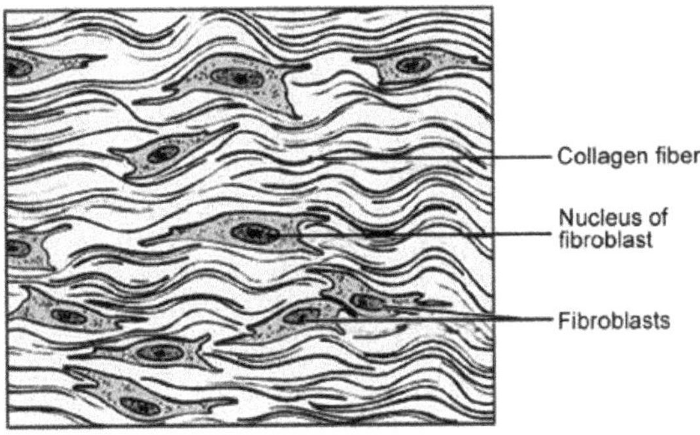

D. Dense Irregular Connective Tissue

This tissue is similar in structure to dense regular connective tissue, but as the name implies, the collagen fibers are arranged in a seemingly random and irregular manner. This tissue, therefore, provides considerable strength and support in all directions. Dense irregular connective tissue has a more extensive blood supply than dense regular connective tissue and contains more ground substance. Dense irregular connective tissue is found in the dermis of the skin, giving the skin its strength. It also is the main ingredient of the epimysium, which wraps around the muscles, the epineurium, which wraps around the nerves, and the periosteum, which wraps around bones (See image below).

Histology of **dense irregular connective tissue** displaying the irregular arrangement of the collagen fibers to one another as opposed to those of the dense regular connective tissue. This tissue is found abundantly in the reticular layer of the dermis providing tensile strength and thus, can be pulled in all directions.

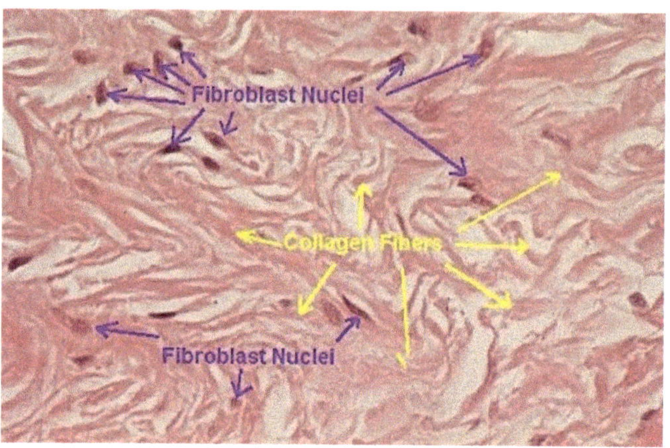

What type of tissue forms each of the following layers of the skin?

Stratum corneum

Stratum basale

Papillary layer

Reticular layer

How does the structure of dense irregular connective tissue relate to its function?

Supporting Connective Tissue

Introduction

This tissue includes **cartilage** and **bone**. In both types of tissue, the extracellular matrix is dense and firm, and it contains collagen fibers. The presence of varying amounts of elastic fibers in cartilage gives cartilage varying degrees of flexibility, while calcium phosphate and other chemicals make bone more solid. In addition to providing "support," cartilage and bone are used to protect, connect, assist with movement, transmit sound waves, and more.

There are three types of cartilage in the body: **hyaline cartilage**, **elastic cartilage**, and **fibrocartilage**. All three types of cartilage contain cells called **chondrocytes** (*chondros*-cartilage, *cyte*-cell), and each chondrocyte occupies a space in the dense extracellular matrix called a lacuna. (See images below).

A. Hyaline Cartilage

Forms the articular cartilage on the ends of long bones, the costal cartilages that attach ribs to the sternum, and the rings that support the trachea. Hyaline cartilage contains a large amount of collagen fibers, and it can provide firm support and flexibility.

B. Elastic Cartilage

Elastic cartilage as the name suggests, is the most flexible type of cartilage. This results from the tissue having a relatively high proportion of elastic fibers in the matrix. Elastic cartilage is found in the ear's outer structure and the epiglottis, which blocks the entrance to the larynx during swallowing.

C. Fibrocartilage

Fibrocartilage contains densely packed collagen fibers that contribute to its function of support. Intervertebral discs, the pubic symphysis (the anterior joint between your two coxal bones), and the menisci in your knees are made of fibrocartilage.

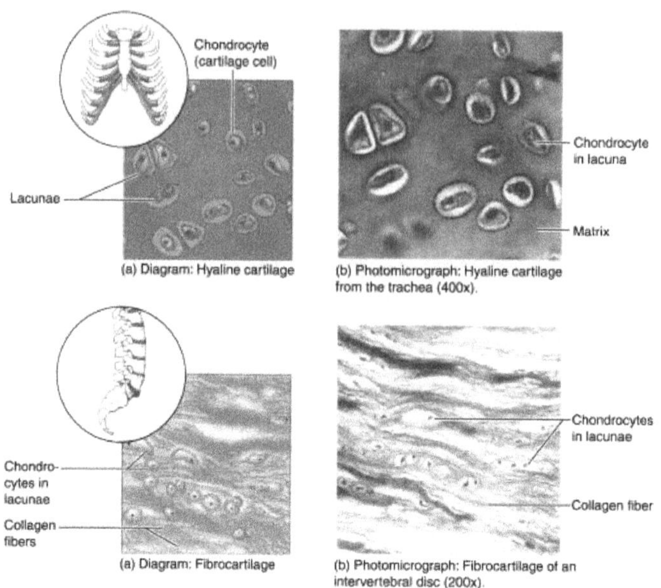

(a) Diagram: Hyaline cartilage

(b) Photomicrograph: Hyaline cartilage from the trachea (400x).

(a) Diagram: Fibrocartilage

(b) Photomicrograph: Fibrocartilage of an intervertebral disc (200x).

There are two types of bone tissue: **compact bone**, and **spongy bone.**

The most numerous cell in bone is the osteocyte (*osteo*= bone). Osteocytes are responsible for maintaining a healthy bone matrix. One of the key ingredients of the extracellular matrix in bone is collagen, which is particularly important in resisting pulling and stretching forces that act on a bone. Another key ingredient of bone matrix is **calcium phosphate**, which hardens bone and is important in resisting compressive forces. Since the bone matrix is hard, osteocytes are unable to move about the tissue. Like chondrocytes, osteocytes reside in lacunae. The hard consistency of the matrix also makes it very difficult for nutrients and waste products to diffuse through the tissue. To compensate for this, osteocytes connect to each other with tentacle-like extensions. These extensions pass through tunnels in the matrix, called **canaliculi**, which connect the lacunae. Gap junctions are found at the ends of the extensions, where two cells join, allowing nutrients and other small molecules to pass from one osteocyte to another.

Everything mentioned in the previous paragraph applies to both types of bone tissue, compact and spongy bone. Also, in both types of tissue, the matrix is formed in thin layers, called **lamellae**. The big difference between compact bone and spongy bone is the manner in which the lamellae are organized. In compact bone, the lamellae form a distinctive pattern of concentric circles around a **central canal**. In spongy bone, the lamellae have no apparent pattern.

Diagram of compact bone with osteocyte within lacuna below

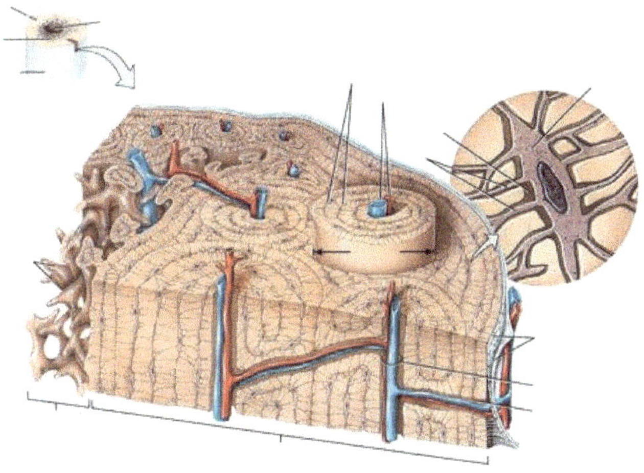

Transverse section of an osteon below. Note the concentric lamellae.

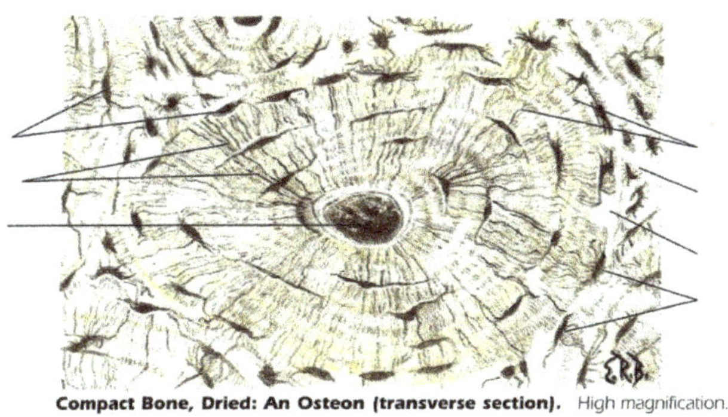

Compact Bone, Dried: An Osteon (transverse section). High magnification.

When identifying bones, look for features visible from the surface of the bone. Most of the time, a visible structure that protrudes from the surface of a bone (known as projection) typically has one of two functions: to articulate (make a joint) with another bone or to serve as an attachment site for a muscle. If the articulation (joint) is able to move, then the articulating surface will be smooth. A smooth articulating protrusion (projection) typically has one of the following names:

1. A **condyle** is a rounded projection
2. A **head** is the rounded end of a bone and normally used for articulations (joint formation)

Immovable joints, like the joint between the hip and spine, are generally roughened. However, there are many projections for muscle attachment. A projection for muscle attachment generally has one of the following names:

1. A **tubercle** is a small, roughened projection
2. A **tuberosity** is a larger, roughened projection
3. A **trochanter** is a very large, roughened* projection found only on the femur for skeletal muscle attachment.
4. A **spine** is a slender ridge of bone.
5. A **line** is a long, narrow ridge.
6. A **crest** is a ridge that usually forms the border of the bone.
7. An **epicondyle** is a roughened projection superior to a condyle.
8. A **process** may have just about any shape.

***All roughened skeletal projections provide increased surface area that makes it ideal for a stronger fusion of skeletal muscle attachment because more space on the bone makes for a greater number of muscle fibers to fuse with the bone.**

Depressions and holes in a bone have a wide variety of functions, the list follows:

1. A **facet** is a shallow, smooth indentation for articulation.
2. A f**ossa** is a deeper indentation, often for articulation.
3. A **meatus** is a tube-shaped cavity.
4. A **foramen** is a hole that goes through a bone, often serving as a passageway for nerves and blood vessels.
5. An **alveolus** is a socket for a tooth found in the maxillae and mandible (facial bones).

6. A **sinus** is a cavity within a bone.

Why are surfaces for moveable articulations smooth?

Why are surfaces for muscle attachment rough?

The nervous and muscle tissues are discussed in detail in the last two chapters.

Classification of Connective Tissue

Type	Structure of Matrix	Example of Location
FIBROUS CONNECTIVE TISSUE		
Loose (areolar) connective tissue	Collagen, elastin, and reticular fibers	Between tissues and organs
Adipose tissue	Fibroblasts enlarge and store fat; very little matrix	Beneath skin; around organs
Dense connective tissue		
Regular	Bundles of parallel collagen fibers	Tendons; ligaments; aponeuroses
Irregular	Bundles of nonparallel collagen fibers	Dermis of skin; joint capsules
Reticular connective tissue	Reticular fibers	Lymphatic organs and liver
CARTILAGE		
Hyaline cartilage	Fine collagen fibers	Ends of long bones; rib cartilages; nose
Elastic cartilage	Many elastin fibers	External ear
Fibrocartilage	Strong collagen fibers	Between vertebrae of spine
BONE		
Compact	Collagen plus calcium salts; arranged in *osteons*	Skeleton
Spongy	Collagen plus calcium salts; arranged in *trabeculae*	Ends of long bones
BLOOD	Plasma plus red and white blood cells, platelets	Inside blood vessels

Chapter 4: Biology of the Cell

I. Introduction to Cells

This chapter's primary focus will be the plasma membrane and the movement of materials across it. You should already be familiar with the basic structures and roles of various cellular organelles (including the nucleus) and the process of cell division. Review these topics (cell structure and organelle functions) on your own as necessary.

Everything we know as life occurs within an aqueous environment. This is obvious when we consider fish and other aquatic organisms, but it is also true of land-dwelling organisms. All of the chemical processes required for life occur within the aqueous environments of the cells and surrounding fluids, which are contained within the body. To survive, a human must maintain normal volumes and compositions of the **extracellular fluids** (*e.g.*, interstitial fluid, blood plasma, lymph, and cerebrospinal fluid) and the **intracellular fluids** (*i.e.*, cytoplasm).

Eukaryotic cell diagram below

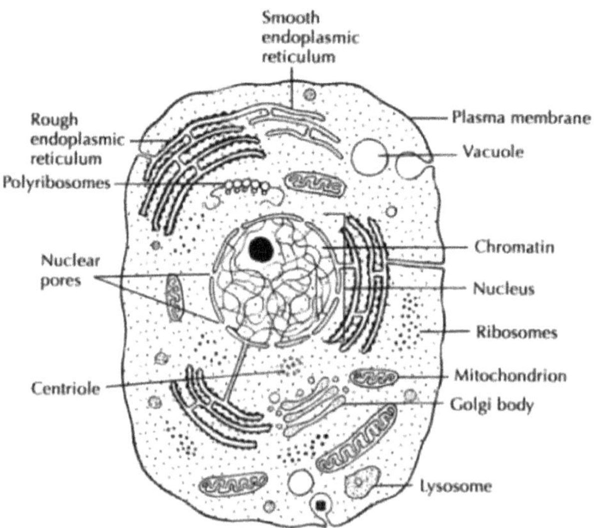

Define the following terms:

Intracellular fluid–

Extracellular fluid–

Interstitial fluid–

II. Chemical Structure of the Plasma Membrane

Every cell is surrounded by a **plasma membrane**. The plasma membrane forms a boundary between the cell and its surroundings. The fluid enclosed by the plasma membrane is the intracellular fluid, also called the **cytoplasm**.

Diagram of the phospholipid bilayer of the cell membrane below

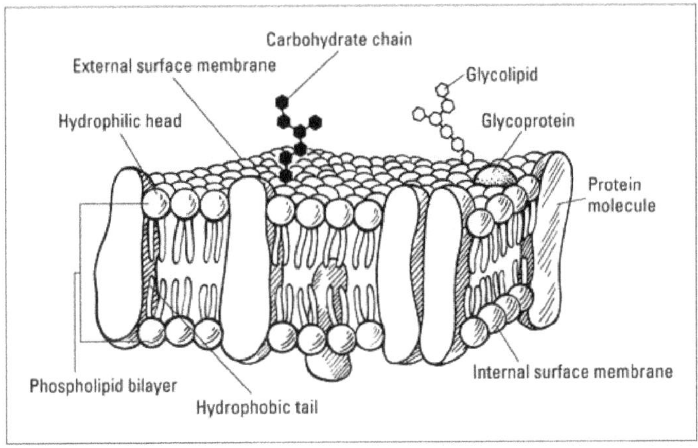

Various organelles are wrapped in their own lipid bilayer membranes inside a eukaryotic cell List some of these:

The primary structure of a cell membrane is a bilayer of lipid molecules, most of which (about 80%) are **phospholipids** (hence the term **phospholipid bilayer**). While phospholipids are the main constituent of the lipid bilayer, most of the remaining 20% of the lipids in the cell membrane are **cholesterol** molecules. Cholesterol stabilizes neighboring phospholipids and decreases the flexibility of the cell membrane.

Proteins are associated with the membrane either by insertion into the membrane or by association with one of the surfaces. Generally, about half the mass of a cellular membrane is protein. Two major classes of membrane-associated proteins exist: (1) **Integral proteins** are tightly associated with the membrane. Part or all of the protein extends into the lipid bilayer, and they have hydrophobic amino acids that are held in the membrane by hydrophobic interactions. (2) **Peripheral proteins** are bound to one surface of the membrane. They are generally held in place with weak, non-covalent bonds.

The outer surface of the plasma membrane is coated with **glycoproteins** and **glycolipids**, which form a "fuzzy" and sticky layer called the **glycocalyx**. Every type of cell in the body has its own pattern of carbohydrates in the glycocalyx, allowing cells within the body to recognize each other.

III. Membrane Transport

As defined by your text, **diffusion** is "the random movement of molecules or particles down their concentration gradient." In any aqueous solution, molecules/particles will be in motion and thus have kinetic energy. Try to grasp the concept that if there is a greater concentration of particles near point A in a solution and a lesser concentration of particles near point B (a **concentration gradient**), then the chance of a particle moving from A toward B is greater than the chance of a particle moving from B toward A. Thus, over time, there will be **net movement** of particles from point A toward point B until the concentrations at points A and B become equal.

Pay close attention to the phrase "net movement." In a solution with no concentration gradient, particles will still move about. However, *diffusion requires a concentration gradient* and <u>net</u> movement down the gradient.

For example, if you put a drop of dye into a beaker of water, there will be a high concentration of dye at the drop and a low concentration elsewhere. Given enough time, the dye will diffuse throughout the water until it is evenly distributed and there is no longer a concentration gradient.

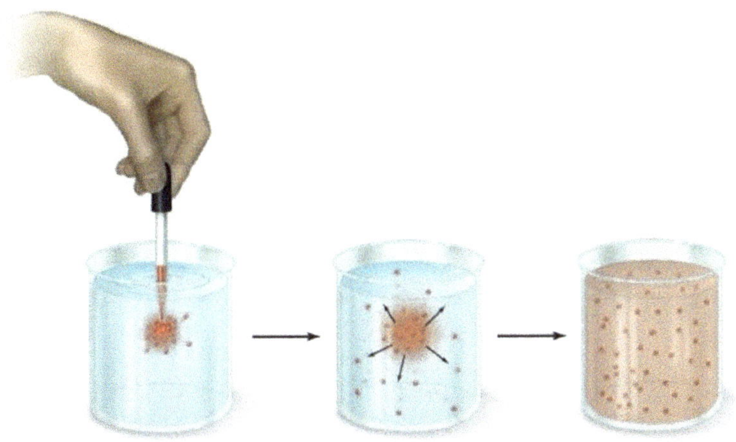

The importance of diffusion to this class is that the human body is essentially a large container of intracellular and extracellular fluids. Although cellular membranes compartmentalize these fluids, dissolved particles still tend to diffuse and become evenly distributed. For any living thing to survive, dissolved particles must be distributed throughout the body in an orderly fashion, not randomly.

Cellular membranes restrict and regulate the movement of some particles between compartments within the body. Some particles can move freely across cell membranes. Thus, cellular membranes (including the plasma membrane) are said to be "**semi-permeable**." Whether free or regulated, there are various ways that particles may cross cell membranes:

Simple diffusion

Simple diffusion is the movement of a molecule directly through the lipid bilayer. Only certain molecules can pass directly through the bilayer; these molecules are small and nonpolar (*e.g.,* oxygen, carbon dioxide, and methane). Some larger, nonpolar

Simple diffusion (does not require energy), is a passive process

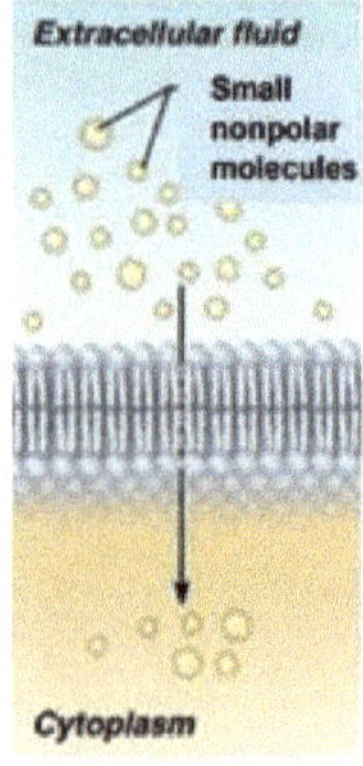

(a) **Simple diffusion** of small nonpolar molecules through the phospholipid bilayer

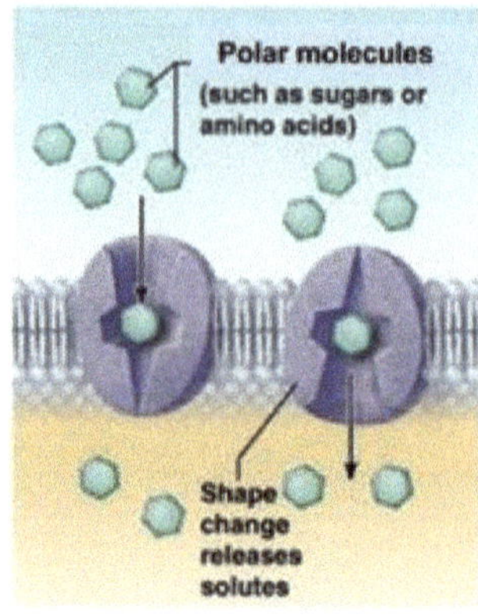

(b) **Carrier-mediated facilitated diffusion** via protein carrier specific for one chemical; binding of substrate causes transport protein to change shape

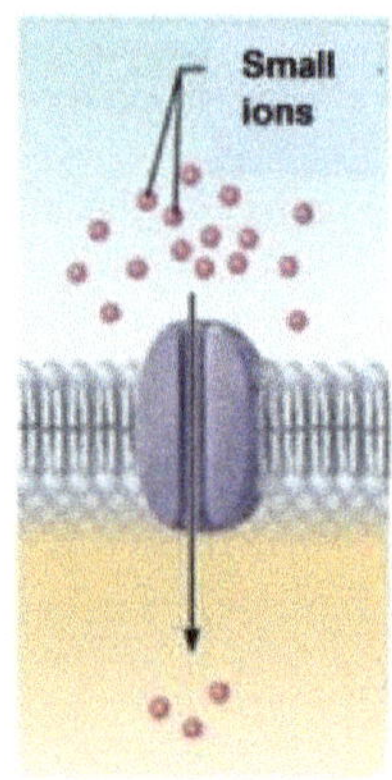

(c) **Channel-mediated facilitated diffusion** through a channel protein; mostly ions selected on basis of size and charge

molecules (such as steroid hormones and certain vitamins) may also pass directly through cellular membranes. Water, although polar, will also pass cell membranes as it is small, and there is simply so much of it in a cell.

Simple diffusion is referred to as a "passive" process because simple diffusion across a membrane requires no expenditure of energy by the cell. Simple diffusion always occurs from a high concentration to a low concentration.

Facilitated diffusion

While some molecules can cross membranes by simple diffusion, many molecules in your cells cannot. Some molecules, such as glucose and other sugars, are polar and too large to pass directly through the lipid bilayer. Electrically charged atoms and molecules, such as Na^+, Cl^-, and amino acids, are unable to pass through the lipid bilayer by simple diffusion, regardless of their size.

These particles need assistance from channels or carrier proteins to cross cell membranes. A **channel** comprises one or more proteins that form a hole through the lipid bilayer. **Leak channels** form holes in the cell membrane that are always open. **Gated channels** open and close in response to some stimulus. A **carrier protein** more or less grabs a specific molecule and shuttles it across the membrane.

Many (if not all) cells have **aquaporins** in their plasma membranes. These channels allow water to cross the membrane. Thus, water crosses plasma membranes by both simple and facilitated diffusion.

Like simple diffusion, facilitated diffusion is passive, allowing particles to move across the membrane only from high to low concentrations.

Osmosis

Because water is at such a high concentration in cells and is so important to the life of a cell, the diffusion of water across membranes (osmosis) is given special consideration. Osmosis arises from the fact that having different amounts of dissolved particles (**solutes**) in a **water solution** changes the effective concentration of water in the solution. Water will diffuse from a solution with fewer solutes (and a higher water concentration) to a solution with more solutes (and a lower water concentration).

The **osmotic pressure** of a solution is an indirect measure of the number of solutes in it; a solution with more solutes tends to have a higher osmotic pressure than a solution with fewer solutes. **Tonicity** is the tendency of water to move into a cell. Review and understand the meanings of the terms **isotonic**, **hypertonic**, and **hypotonic**.

Effect of osmotic pressure on red blood cell

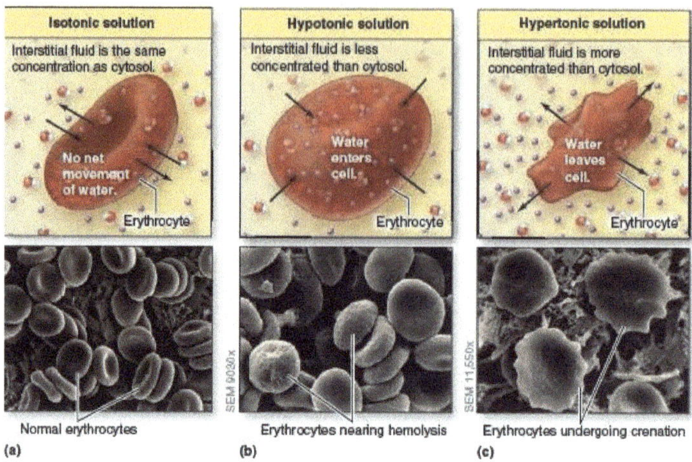

What will happen to a cell bathed in an isotonic solution? A hypertonic solution? A hypotonic solution?

Osmosis is another passive process in which water moves from a high concentration to a low concentration. Note that osmosis is just water movement across a membrane by simple and facilitated diffusion.

IV. Factors that affect the rate of diffusion

When considering simple or facilitated diffusion across a cell membrane, five factors generally affect the rate at which particles cross the membrane:

1. *Size of the particle.* Smaller particles tend to cross the membrane at faster rates than larger particles.
2. *Temperature.* As temperature increases, the kinetic energy of particles increases, and the rate of diffusion increases.
3. *Charge.* Nonpolar molecules tend to move easily across lipid bilayers. Polar molecules do not move easily across lipid bilayers. Fully charged molecules basically cannot move directly through lipid bilayers.
4. *Concentration gradient.* The greater the difference in concentration across a membrane, the greater the diffusion

rate across the membrane. What is the diffusion rate if the concentrations are equal on both sides of the membrane?

5. *The number of carrier proteins or channels.* As noted earlier, many particles need carriers or channels to diffuse across a membrane. The greater the number of carriers or channels, the greater the diffusion rate. What is the glucose diffusion rate across a membrane if there are no carriers or channels for glucose?

V. Active transport

Active transport involves the movement of a particle *against* its concentration gradient. This requires an energy input from the cell, generally in the form of ATP. An example of active transport is carried out by the Na^+/K^+ ATPase (*i.e.,* the sodium-potassium pump), which actively pumps two potassium ions and three sodium ions for every ATP hydrolyzed.

Obviously, active transport is an active process that requires the cell to expend energy. *It is not diffusion!*

Passive and active transport diagram below

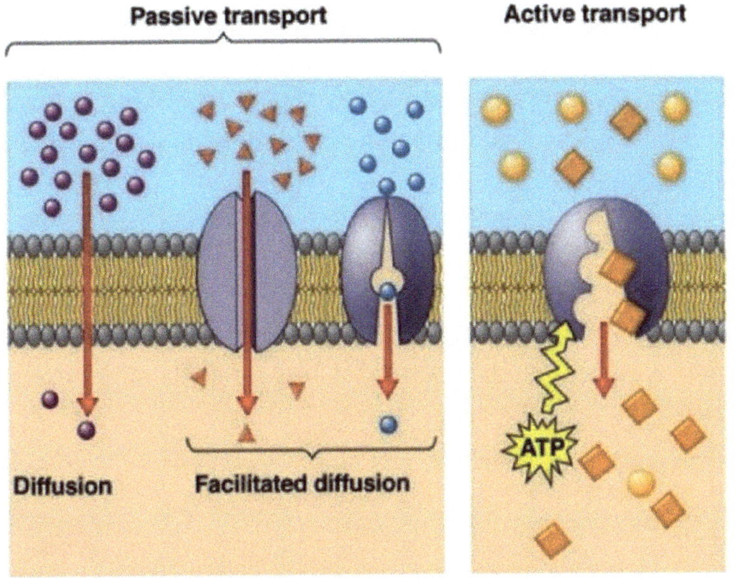

Vesicular transport

Some molecules are simply too large to pass through a cell membrane, even with the help of carrier proteins. Consider that sometimes cells want to move proteins themselves across a membrane. White blood cells may even ingest other cells! The movement of such large molecules across a cell membrane is accomplished by vesicular transport. The two main categories of vesicular transport, **endocytosis**, and **exocytosis**, occur in a similar manner; the difference is that endocytosis involves bringing things into a cell, whereas exocytosis involves exporting particles to the outside of a cell.

Endocytosis can take various forms, depending on what is brought into the cell and how. **Pinocytosis** occurs when the plasma membrane folds inward, bringing with it multiple substances. Eventually, the invaginated region of the membrane pinches off from the plasma membrane and enters the cell. This results in the formation of a vesicle inside the cell. The vesicle contains whatever particles were drawn in from the outside. The term **phagocytosis** is used when a cell engulfs something particularly large, such as another cell. An example of this process is when a white blood cell ingests a bacterium. **Receptor-mediated endocytosis** occurs when small particles attach to receptors on the external surface of the cell membrane. The receptors and particles aggregate together at a region of the cell membrane, and this region of the membrane invaginates to form a vesicle.

Exocytosis is simply the process going in reverse. A vesicle containing certain materials fuses to the inner surface of the plasma membrane, and the contents are dumped to the outside. During the process, the vesicle is incorporated into the plasma membrane.

Phagocytosis (cell-eating)

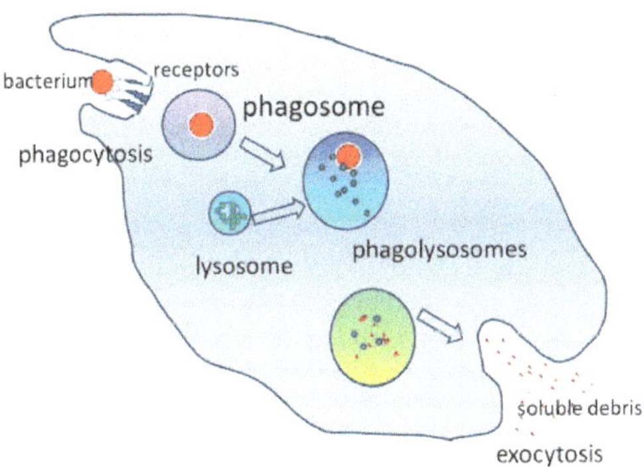

VI. Cellular Structures

Most cells in the body (blood cells are an exception) are attached to other cells. Cells are held together in part by the adhesion of the glycocalyx of one cell with those of neighboring cells. However, three types of specialized "membrane junctions" can also help hold cells together: **tight junctions**, **desmosomes**, and **gap junctions**. Review the structures and functions of these membrane junctions and briefly describe each of these junctions:

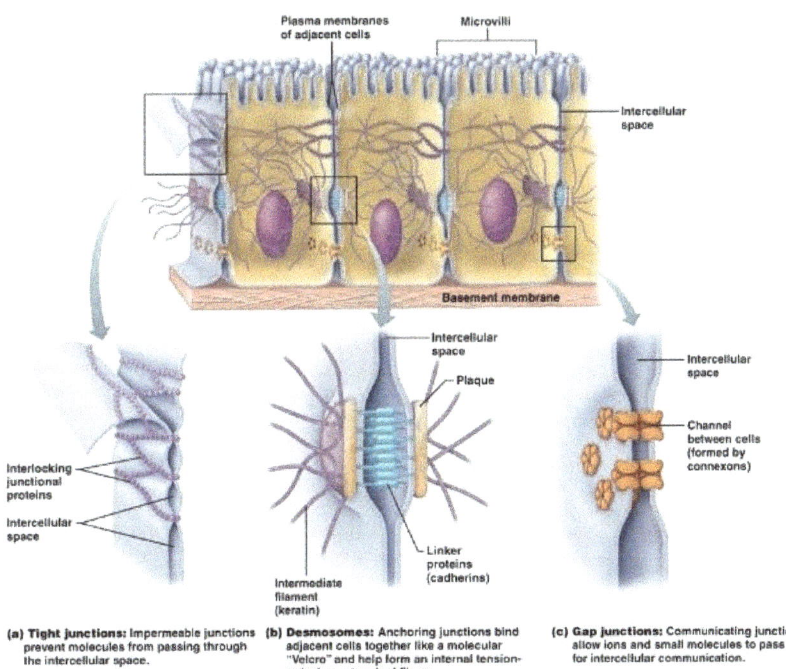

(a) Tight junctions: Impermeable junctions prevent molecules from passing through the intercellular space.

(b) Desmosomes: Anchoring junctions bind adjacent cells together like a molecular "Velcro" and help form an internal tension-reducing network of fibers.

(c) Gap junctions: Communicating junctions allow ions and small molecules to pass for intercellular communication.

Tight junction -

Desmosome -

Gap junction -

Chapter 5: Integumentary System

I. Integumentary System: Structure and Functions

a. Contains the **skin** (also called **integument** or **cutaneous membrane**) and its **derivatives** (sweat glands, sebaceous glands, hair, nails)
b. The skin (largest organ -covers the greatest surface area of the body) has two layers: **epidermis** and **dermis**.
c. Epidermis - **superficial** layer (part of skin we see), composed of **stratified squamous epithelial** tissue.
d. Dermis - **deep** layer (deep to the epidermis) primarily composed of **areolar connective tissue** and **dense irregular connective tissue**.
e. It protects the body from ultraviolet radiation, dehydration, pathogens (harmful microorganisms like bacteria, viruses, protozoa, etc.), and physical and chemical injury.
f. Skin blood vessels - site of vitamin D production and dissipation of body heat.
g. It allows us to be aware of our environment in terms of temperature and the sensation of touch.
h. Deep to the dermis is the **hypodermis** - not part of the skin. The hypodermis (also known as the superficial fascia) is primarily composed of adipose (fatty) tissue. Adipose tissue's functions are insulation, energy storage, and binding skin to underlying skeletal muscles and other structures.

II. Epidermis

a. Stratified squamous epithelium - prevents pathogen entry and water loss (dehydration).
b. Most of the cells in the epidermis are keratinocytes. They are so named because they produce copious amounts of keratin (Greek: *Keras*, meaning *horn*), the tough, waterproof, bacteria-resistant protein that gives skin its protective properties.

c. Other epidermal cells include melanocytes, tactile cells, and epidermal dendritic cells.
d. Melanocytes produce the pigment melanin (Greek: *Melanos*, meaning *black*), which provides protection from the damaging effects of UV radiation. Melanocytes have multiple cellular extensions that snake between the keratinocytes. This allows the melanocytes to transfer melanin to keratinocytes. The granules of melanin cluster on the apical side of the keratinocytes provide a barrier between the sun's rays and the cells' nuclei. This protects the DNA in the nuclei from mutations that may potentially cause skin cancer from UV radiation.
e. Tactile cells are found in the deep epidermis and, when compressed, stimulate nearby nerve cells. This provides sensory information about what is touching the skin.
f. Epidermal dendritic cells can perform phagocytosis (cell-eating and destruction of invading pathogens). They also take the remains of the pathogens to the lymph nodes and present pathogen fragments to white blood cells to prompt an immune response.
g. The epidermis is divided into several named layers (each of which can consist of multiple layers of keratinocytes).
h. There are five epidermal layers in thick skin (found on the ventral surface of the hands **[palms]** and the plantar surface of the feet **[soles**]).
i. There are four epidermal layers in thin skin (found everywhere else).
j. Thick skin has thicker layers and an extra layer (stratum lucidum), which helps the palms and soles withstand more abrasion and friction than other areas of the skin.
k. From most superficial to deepest, the five layers of thick skin epidermis are 1) stratum corneum, 2) stratum lucidum, 3) stratum granulosum, 4) stratum spinosum, and 5) stratum basale. A good acronym to use is the following:

Come	stratum corneum
Let's	stratum lucidum
Get	stratum granulosum
Sun	stratum spinosum
Burn	stratum basale

l. The **stratum corneum**, the most superficial layer (Greek. Cornu *meaning* horn) - consists of dead flattened keratinocytes filled with keratin. Its contents (keratin and lipids) provide the skin strength and waterproofing qualities. It can be up to 30 cell layers thick in thick skin. The top cell layers of the stratum corneum constantly slough off and are replaced by new cells produced from the stratum basale, a process known as cornification, which takes about 2 to 4 weeks.
m. The **stratum lucidum** is superficial to the S. granulosum but is only found in thick skin. The cells of this thin layer appear clear (Latin. Lucidum/ lux, *meaning* "light") in the microscope.
n. The **stratum granulosum** is superficial to S. spinosum and is a thin layer where cell keratinization occurs. The keratinocytes here are beginning to die and are accumulating granules of keratin precursors, which results in a dark and spotty appearance.
o. The **stratum spinosum** is superficial to S. basale and contains several layers of keratinocytes (the thickest of all five epidermal layers). The desmosomes mechanically linking the keratinocytes are quite visible in slides of the epidermis and are responsible for this layer's name. These keratinocytes are still alive and have not yet become filled with keratin. Epidermal dendritic cells are common in this layer.

p. The **stratum basale** is the deepest layer and is superficial to the basement membrane and the underlying areolar connective tissue. This layer is a single row of cells in constant mitosis, with new cells continually pushed upwards to the surface. Additionally, melanocytes are commonly found in the stratum basale.

III. Dermis

a. Deep to the epidermis, it has two layers: the **papillary** dermis and the **reticular** dermis.
b. **Papillary dermis**—The upper ⅕ of the dermis and primary tissue is loose areolar CT. It supports the epidermis and contains abundant blood vessels and nerves. Upward projections of the papillary dermis (dermal papillae) interdigitate with downward projections of the epidermis (epidermal ridges) to anchor the two layers to one another.
c. **Reticular dermis**: The lower ⅘ of the dermis and primary tissue is dense, irregular CT. This provides the skin with structural integrity and tensile strength to resist tension damage.
d. The dermis contains many nerves and blood vessels. In response to a rise in body T°, dermal blood vessels vasodilate. This increases blood flow to the skin and allows heat to radiate away from the body. In response to a drop in body T°, dermal vessel vasoconstriction occurs as a means of heat retention.
e. Dermal blood vessels are also the site of the initiation of vitamin D production. UV radiation acts on blood chemicals, which are then modified by the liver and kidney to yield active vitamin D. Vitamin D has several physiological roles including the maintenance of proper blood calcium levels, which is crucial for bone health.

I. **Skin Derivatives**: The skin includes several structures that are derived from the epidermis. These skin appendages include sweat glands, sebaceous glands, nails, hair follicles, and hair.

a. The skin contains two main types of sweat glands: **merocrine** and **apocrine**. Each is a coiled tube with the secretory portion in the dermis and a duct leading through the epidermis to an opening at the surface (in a merocrine sweat gland) or into a hair follicle (in an apocrine sweat gland).
b. There are roughly 3-4 million merocrine sweat glands. They release sweat via exocytosis onto the epidermal surface. This sweat is 99% water with only a few solutes (Na^+, Cl^-, lactic acid, antibodies, urea, and ammonia). The primary function of merocrine sweat glands is thermoregulation. The evaporation of sweat on the skin's surface removes heat from the skin, thus cooling the body. The acidity of sweat is bacteriostatic (i.e., it reduces bacterial growth).
c. Apocrine sweat glands release their sweat into hair follicles in the axillary regions, pubic region, anal region, and around the nipples. They release their sweat via exocytosis. This sweat is thick and contains lipids and proteins utilized by resident bacteria, which produce a distinctive odor. These glands become active around puberty.
d. Modified apocrine sweat glands called ceruminous glands are found in the external ear canal lining. Their secretion adds to sebum to yield cerumen (a.k.a. earwax), which blocks pathogen entry into the ear canal.
e. Sebaceous glands secrete sebum, an oily bactericidal (kills bacteria) substance that moisturizes the skin. They are typically found branching from hair follicles and are absent on the palms and soles.
f. Hair covers much of the human body and is abundant on the surface of the head, axillae, and pubic regions. It plays a role in sensation and provides some warmth and

protection from UV radiation. Hair consists of a hair bulb , a root, and a shaft. The bulb consists of swollen epithelial cells that are deep within the dermis. Deep to the bulb is a collection of connective tissue, blood vessels, and nerves known as the hair papilla. The hair papilla nourishes the bulb and allows distortion of the hair to convey information. The root and shaft of the hair are both composed of concentric layers of dead keratinized cells. The shaft is the visible portion of the hair, whereas the root is within the epidermis and dermis. Surrounding the lower shaft and the root is the hair follicle. Attached to the follicle is a small muscular bundle called arrector pili, a smooth muscle tissue. It contracts and causes the hair to stand up in response to cold temperatures and fright. There is no functional value to this in humans. In small furry mammals, it helps them look larger to protect them from predators and can trap warm air around the skin.

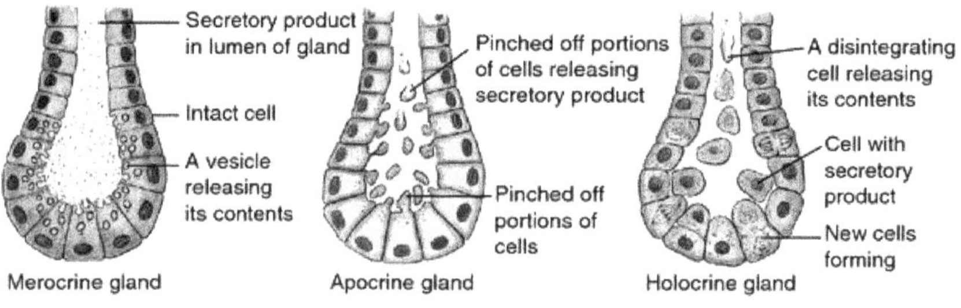

Anatomy of the epidermis and dermis and subcutaneous tissue below

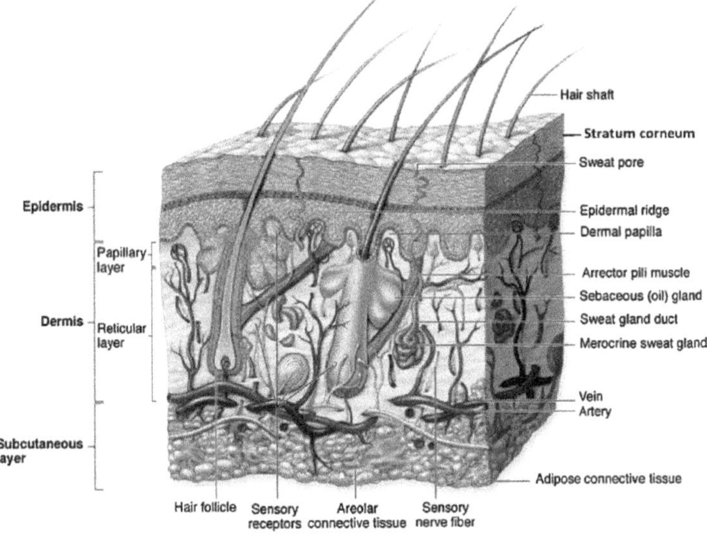

Anatomy of the epidermis

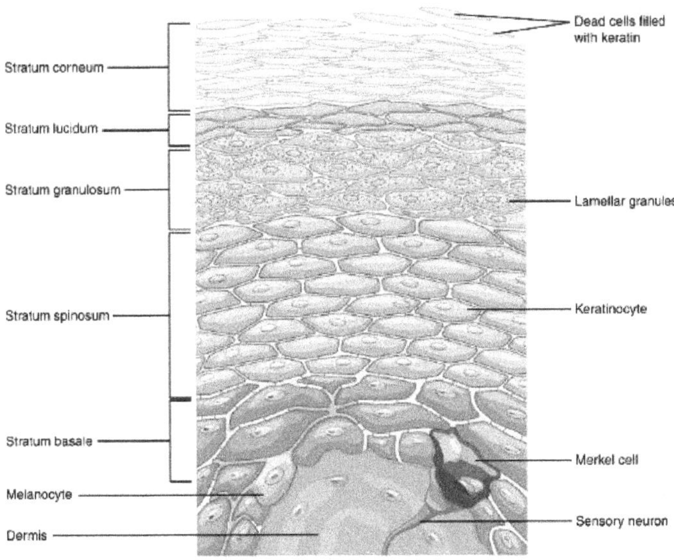

Chapter 6: Skeletal System

I. Introduction

The human skeleton consists of 206 named bones, each bone is considered an organ and performs various functions such as:

1. active support of body weight.
2. Organ protection like the brain, spinal cord and the lungs.
3. They create body movement when pulled upon by skeletal muscles during muscle contraction.
4. Serve as storage of minerals such as calcium (Ca^{2+}) and phosphate (PO_4^{2-}).
5. Site for hematopoiesis or hemopoiesis (production of blood cells in bone marrow).
6. Triglycerides (lipid/fat) storage to create energy for the body cells.

Bones have an exterior composed of compact bone and an interior that contains spongy bone (a.k.a. cancellous or trabecular bone).

- **Compact bone** appears solid to the naked eye.
- **Spongy bone** contains many spaces found between needle-like struts of bone called **trabeculae**.

II. There are four basic types of bone: long bones, short bones, flat bones, and irregular bones (see chart below).

1. Long bones are longer than they are wide with the midshaft being cylindrical in shape. Hence, the phalanges of the fingers and toes albeit small are designated as 'long' bones. Therefore, all the bones of the extremities or appendages (except patellae, carpals, and tarsals) are long.
2. Short bones are roughly cube-shaped. Carpals and tarsals are short bones.
3. Flat bones are thin, flattened, and usually a bit curved. The sternum, scapulae, ribs, and most skull bones, especially the cranial bones are flat.

4. Irregular bones have irregular shapes that don't fit the other designations. The vertebrae and hip bones are examples of irregular bones.

Bone Types	Appearance	Function	Picture	Example(s)
Long Bones	Longer Than They Are Wide	Mechanical Strength		Femur Tibia Fibula Humerus Ulna Radius
Short Bones	Cube-shaped	Multi-directional Motion		Carpal Bones (Of The Hands/Wrists) And The Tarsal Bones (Of The Feet/Ankles).
Flat Bones	Thin And Flat Has Large Surfaces For Muscle Attachments	Mechanical Protection to Soft Tissues Beneath		·Cranial Bones ·Sternum ·Ribs ·Scapulae
Irregular Bones	Complicated Shapes that cannot be Classified as "Long", "Short" or "Flat".	Provides Major Mechanical Support for the Body Vertebra Protects the Spinal Cord		·Vertebrae ·Hyoid Bone ·Sphenoid Bone ·Facial Bones.
Sesamoid Bones	Most Sesamoid Bones Are Un-named.	Protects From Additional Friction And Use - can form in Palms And Soles		Only One Type Of Sesamoid Bone Is Present In All Normal Human Skeletons So It Has A Name; The Patella.

III. Long Bone Structure

1. A long bone consists of a shaft (diaphysis), a pair of expanded ends (epiphyses), and a pair of membranes.
2. The diaphysis forms the long axis of the bone and consists of a collar of compact bone surrounding a central medullary cavity.
3. In adults, the medullary cavity typically contains yellow bone marrowrich in triglycerides.
4. Epiphyses are the expanded, rounded ends of the long bone that articulate (form a joint) with other bones.
5. Epiphyses have an exterior of compact bone and an interior of spongy bone.
6. The joint surfaces of the epiphyses are covered by a layer of hyaline cartilage (articular cartilage).
7. In children, the epiphysis contains a disc of hyaline cartilage (the epiphyseal plate), allowing growth along the long axis.
8. In adult long bones, only a remnant (the epiphyseal line) of the epiphyseal plate remains.

9. Long bones in the appendages have epiphyses designated as proximal or distal.
10. The exterior of the bone (except joint surfaces) is covered by a double-layered membrane known as the periosteum.
11. The outer portion of the periosteum is composed of dense irregular connective tissue and is, therefore, the fibrous layer.
12. The inner portion of the periosteum is made of bone-forming cells (osteoblasts) and is thus known as the osteogenic layer.
13. The periosteum is the anchor point for attaching ligaments and tendons to bone.
14. The periosteum is tightly attached to the compact bone by clusters of collagen fibers known as perforating fibers.
15. The periosteum is quite vascular and contains a lot of nervous tissue. The blood and nerve supply to the bone extends from the periosteum and enters the bone tissue through holes known as nutrient foramina.
16. A membrane known as the endosteum lines the medullary cavity, surfaces of the trabeculae, and interiors of the canals within compact bone.
17. The endosteum contains bone-forming cells (**osteoblasts**) and bone-destroying cells (**osteoclasts**).

Diagram of adult long bone structure below. Note the distinguishing feature of yellow marrow (adipose tissue) in the marrow cavity from that of the red marrow in the long bone of a child (prepuberty).

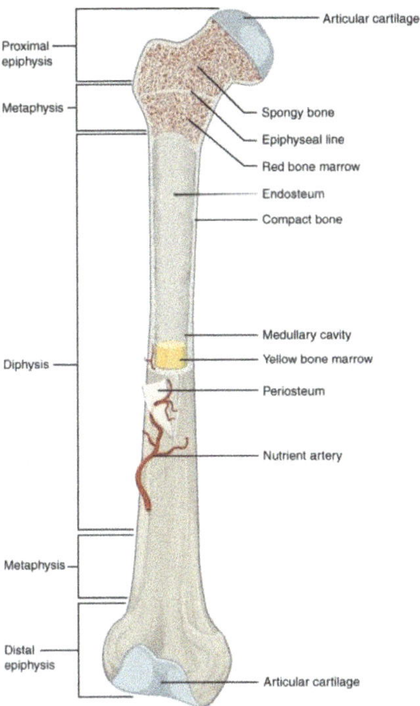

IV. The compact bone in the diaphysis of long bones consists of cylinders known as osteons.

1) An osteon is a weight-bearing tube oriented along the long axis of the bone.
2) It has a series of concentric layers of bone tissue (lamellae) surrounding a lumen (central canal).
3) Lamellae are composed of bone matrix containing lacunae that house mature bone cells (osteocytes).
4) The central canal contains the blood vessels and nerves serving the cells of the osteon.
5) Central canals are linked to one another and to nutrient foramina by perforating canals, which run perpendicular to the long axis of the bone.
6) Osteocytes are spider-like cells with long extensions that spread through the matrix via small channels called canaliculi

. The extensions of one osteocyte are linked to the extensions of other osteocytes via gap junctions, allowing nutrients and signals to pass from cell to cell within the bone.

7) Spongy bone does not contain osteons.

V. Flat bones consist of a pair of plates of compact bone separated by an interior of spongy bone (the diploë). Red bone marrow, containing hematopoietic (blood-forming) tissue, is found in the spaces of the spongy bone.

Flat bone and diploe (spongy portion sandwiched between two layers of flat compact bones)

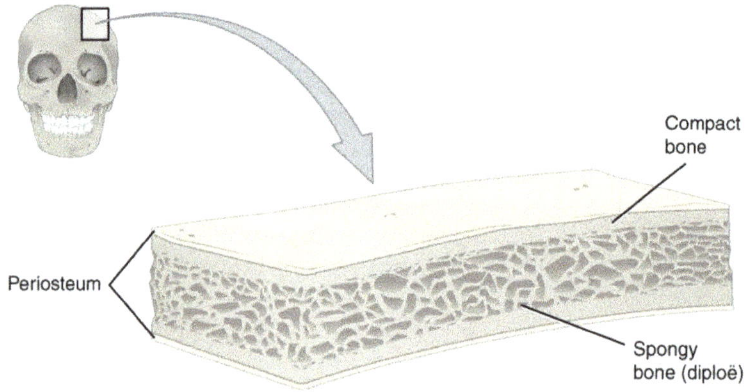

VI. Bone matrix surrounds osteocytes in both compact and spongy bone.

1. The bone matrix is strong, flexible, and dynamic.
2. ⅓ of the matrix is composed of osteoid, which consists of ground substance and collagen fibers produced by osteoblasts. Collagen provides the bone with flexibility as well as incredible tensile strength.
3. ⅔ of the bone matrix is composed of minerals – primarily calcium phosphate ($CaPO_4$). The calcium phosphate combines with calcium hydroxide ($Ca(OH)_2$) to yield hydroxyapatite crystals that deposit around the collagen fibers and give bone its hardness and ability to resist compression.

Compact bone matrix and its structures

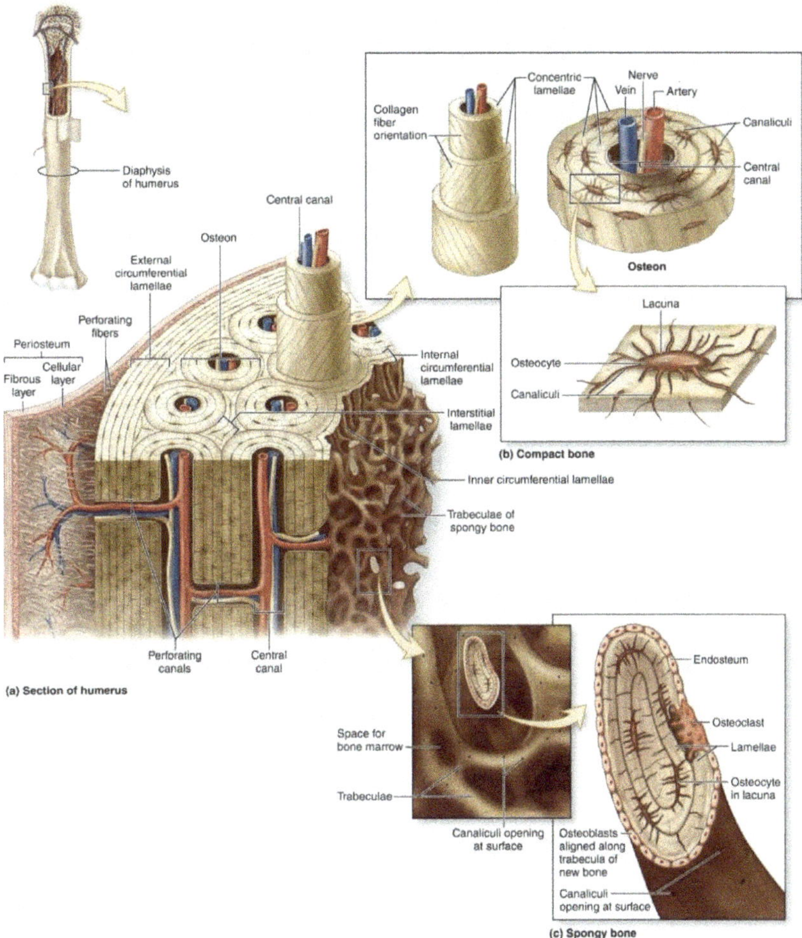

VII. Long bone growth in length (known as interstitial growth) occurs at epiphyseal plates.

1. Within the plate, chondrocytes divide, producing new cells. These new cells push older chondrocytes towards the diaphysis, causing the bone to lengthen. The older chondrocytes near the diaphysis then grow, causing their lacunae to enlarge. Then, the surrounding matrix calcifies, which diminishes the ability of nutrients to diffuse to the chondrocytes. These chondrocytes then die. This leaves an area next to the diaphysis consisting of spaces separated by

calcified cartilage. Osteoblasts and osteoclasts move in and transform the calcified spicules into new spongy bone. As this occurs, the epiphyseal plate does not change in size since its growth rate at its epiphyseal end is equal to its transformation rate into bone at its diaphyseal end.

2. During childhood, epiphyseal plate activity is stimulated by growth hormone released by the pituitary gland.
3. At puberty, blood levels of sex hormones (testosterone or estrogens) increase. This causes two things to occur. It causes an increase in the conversion of the cartilage to bone (resulting in increased bone length), but it also decreases the rate of production of new cartilage in the plate (ultimately ending the ability of the long bone to grow)

VIII. Growth in width (known as appositional growth) is a product of the activity of the osteoblasts in the osteogenic layer of the periosteum. The deposition of bone on the outer surface is balanced by osteoclasts' destruction of bone adjacent to the marrow cavity within the endosteum.

Bone is a dynamic tissue. It constantly remodels and reshapes in response to the stresses and demands placed on it.

1) Bone remodeling consists of both bone deposition or osteogenesis (production of new bone matrix by osteoblasts) and bone resorption or osteolysis (destruction of bone matrix by osteoclasts).
2) The external stress (or lack thereof) placed on the bone often determines where these events occur.

IX. Additionally, blood calcium levels control bone deposition and resorption. Calcium is essential to physiological functions, including muscle contraction, nerve signaling, and blood clotting.

1. A decrease in blood calcium levels prompts the parathyroid glands (found on the posterior of the trachea) to secrete parathyroid hormone (PTH). PTH increases blood calcium

by acting on both skeletal and non-skeletal tissues. It stimulates the activity of osteoclasts. The resulting breakdown of the inorganic bone matrix releases calcium into the blood. PTH also increases the absorption of dietary calcium in the intestines and decreases calcium output in the urine.

2. An increase in blood calcium prompts the C cells of the thyroid gland to release the hormone calcitonin, which decreases osteoclast activity and increases urinary calcium excretion.
3. The absorption of dietary calcium depends on vitamin D, produced within the dermis's blood vessels upon exposure to ultraviolet radiation.
4. Vitamin C promotes new bone production by increasing osteoblast activity and decreasing osteoclast activity.

Osteogenesis in long bones (Bone formation from cartilage)

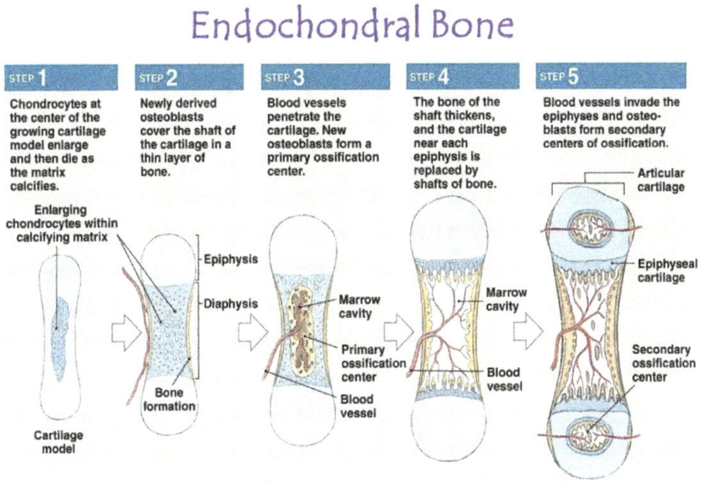

Skeletal System: Articulations

Although it is important to know the forms and functions of individual bones, it is also important to remember that, with one exception (the hyoid bone), every bone in the body articulates with at least one other bone. Furthermore, the skeleton is a dynamic structure in which many bones are able to move at their **articulations** (or **joints**). Some joints are designed for strength, some are designed for mobility, and others reflect compromises of strength and mobility.

Classification of Joints

Joints can be classified according to their structures or according to their functions.

With regards to structure, joints are either **fibrous**, **cartilaginous**, or **synovial**.

With regard to function, joints are either **immovable**, **slightly moveable**, or **freely moveable**.

Fibrous Joints

Fibrous joints are held together with dense fibrous connective tissue. Most are immovable, but some are slightly moveable. **Sutures** are immoveable joints found in the skull. The bones involved in these joints interlock and are held together by dense fibrous connective tissue. During middle age, the bones of the skull tend to completely fuse.

The joint between the radius and ulna and the joint between the tibia and fibula are held together by broad sheets of dense regular connective tissue, called **interosseous ligaments**. These joints are considered slightly moveable.

Fibrous joints diagrams

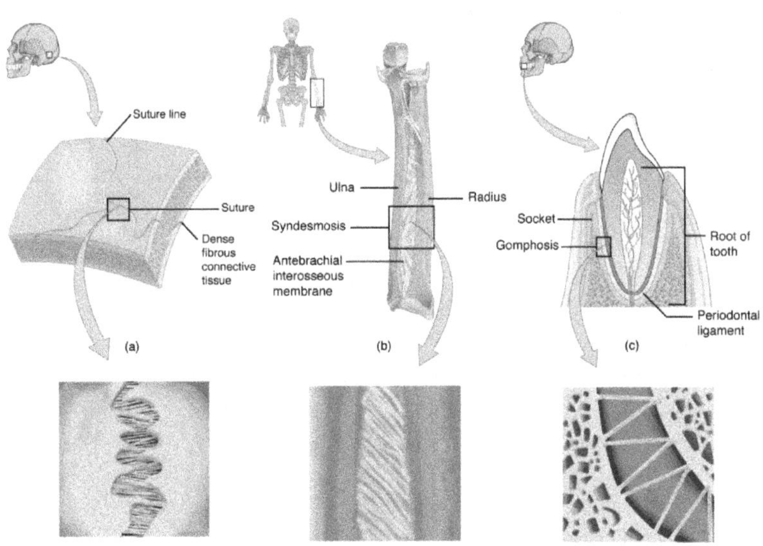

Cartilaginous Joints

Cartilaginous joints are held together with cartilage. As with fibrous joints, movement is generally limited.

Epiphyseal plates are immoveable joints made of hyaline cartilage.

A **symphysis** is a slightly moveable joint held together with fibrocartilage. Examples are the intervertebral discs and the pubic symphysis. Under what condition might the pubic symphysis allow some movement?

Cartilaginous joints

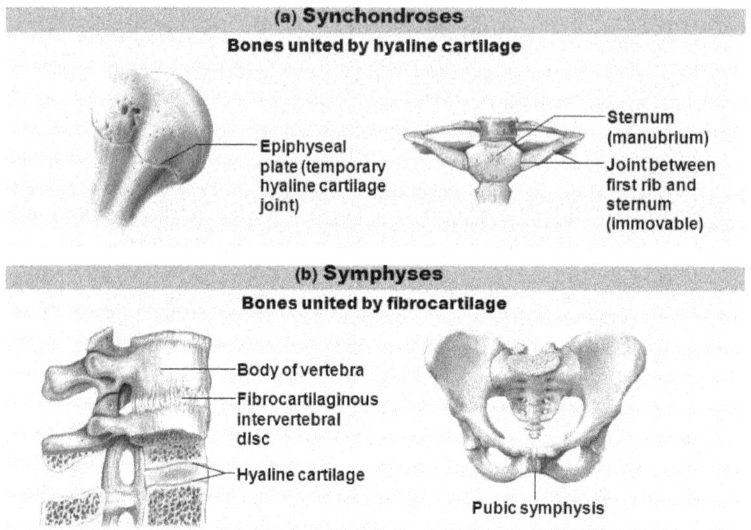

Synovial Joints

Synovial joints are characterized by a fluid-filled cavity between the two articulating bones. Although many do not allow a very wide range of movement, all synovial joints are nonetheless considered freely moveable.

Distinguishing features and anatomy of synovial joints

Synovial joints have five characteristic features:

A. The articular surfaces of the bones are covered with **articular cartilage**.

B. There is a "space" called the **joint cavity** between the two articulating bones. The joint cavity is filled with **synovial fluid**. What is the function of synovial fluid?

C. A membranous **articular capsule** surrounds the synovial joint. There is an outer **fibrous layer** composed of dense connective tissue, and this is continuous with the periosteum. There is an inner **synovial membrane** composed of loose connective tissue.

D. Synovial joints are stabilized by **reinforcing ligaments**.

E. Synovial joints contain nerves and blood vessels. Some of these nerves and their associated receptors help your brain monitor the position of your body.

Synovial joints

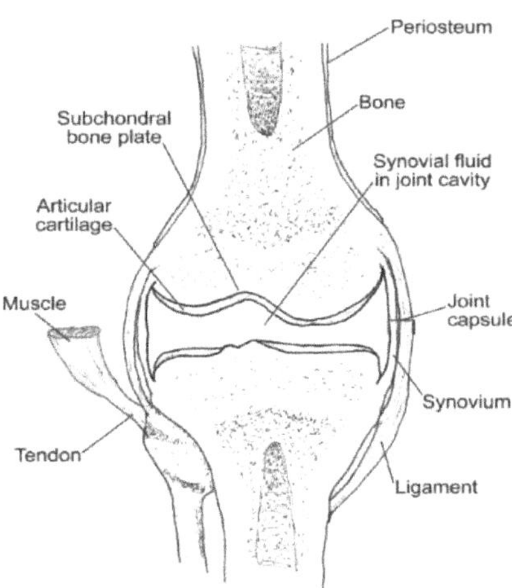

Classification of synovial joints

Synovial joints can be classified on the basis of their structures:

A. **Plane joints** have flattened or only slightly curved faces, and they allow gliding movements. As mentioned previously, examples include the femoropatellar and intercarpal joints.

B. **Hinge joints** permit angular motion in a single plane. Examples are the elbow, knee, and ankle.

C. **Pivot joints** allow rotation about a single axis. An example is the joint between the atlas and the axis.

D. **Condylar joints** allow angular motion in two planes and circumduction. Note the way the wrist moves on the radius from side to side or front to back, but the wrist cannot rotate on the radius.

E. **Saddle joints** also allow angular motion in two planes and circumduction, but not rotation. An example is the carpometacarpal joint at the base of the thumb.

F. **Ball-and-socket** joints allow various combinations of angular and rotational movement. Examples are the shoulder and hip joints.

Types of Synovial Joints	Models of Joint Motion	Examples
Gliding joint 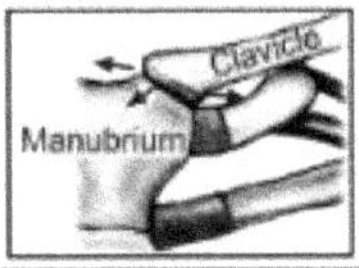	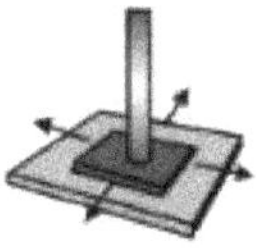	• Acromioclavicular and sternoclavicular joints
Hinge joint 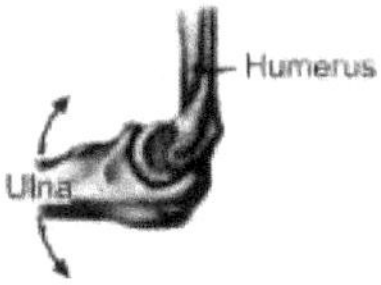	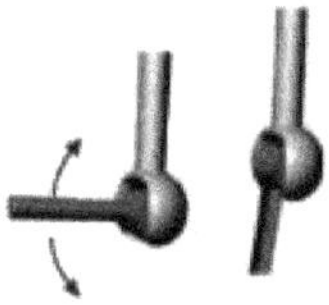	• Elbow joints • Knee joints • Ankle joints • Interphalangeal joints
Pivot joint 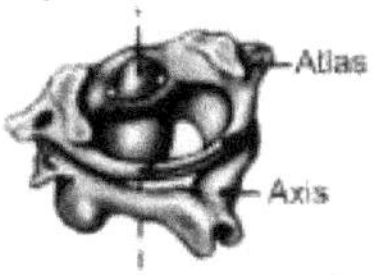	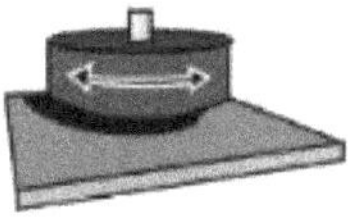	• Atlanto-axial joint • Radio-ulnar joints
Condyloid joint 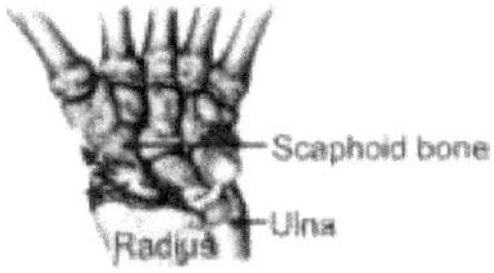	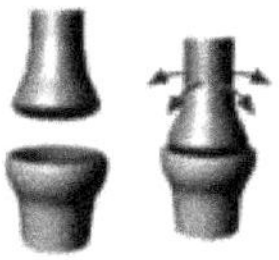	• Wrist joints • Metacarpophalangeal joints • Metatarsophalangeal joints
Saddle joint 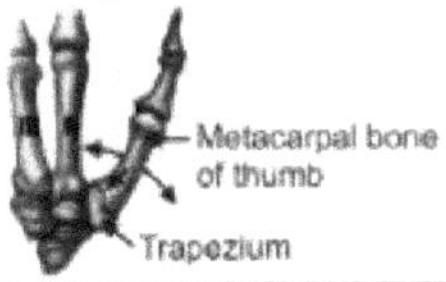	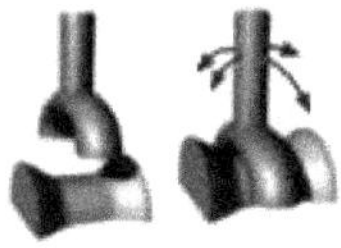	• Carpometacarpal joints
Types of Synovial Joints	**Models of Joint Motion**	**Examples**
Ball and socket joint 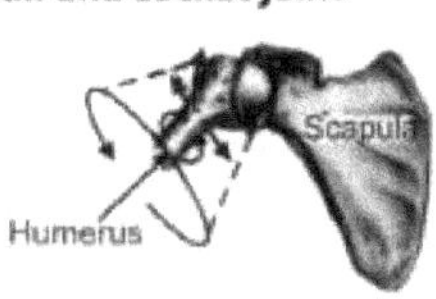	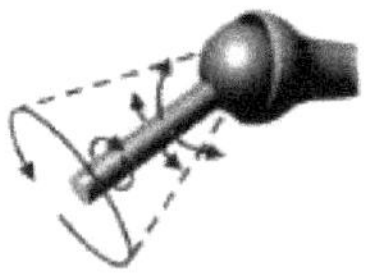	• Shoulder joints • Hip joints

Different types of synovial joints

The Movements of Synovial Joints

Different joints are designed to allow different types of motion. The different motions can generally be described as gliding, angular motion, or rotation.

Gliding occurs when two surfaces slide past each other. A good example is the femoropatellar joint, in which the patella glides across the patellar surface of the femur. Gliding joints can also be seen at the intercarpal joints.

Angular motion takes a variety of forms, and it is important to keep these terms in their proper frames of reference based on a body in the anatomical position. **Flexion** is movement in the anterior/posterior plane that reduces the angle between the articulating elements. **Extension** occurs in the same plane but increases the angle between articulating elements. Note that when a person is in the anatomical position, all major joints (except the ankle) are considered to be at full extension. Extension beyond the anatomical position is considered **hyperextension**. **Abduction** is movement away from the longitudinal axis of the body (midline). **Adduction** is movement toward the longitudinal axis of the body (i.e., midline or median plane). The terms abduction and adduction apply only to the appendicular skeleton. **Circumduction** is the movement of the arm or leg in a loop independent of actually rotating the limb in its socket. (See images below for all joint movements).

Rotational movements involve the rotation of a portion of the body on an axis. Head rotations are considered left or right, and limb rotations are either medial or lateral. **Pronation** is the rotation of the forearm such that the palm faces posteriorly, and **supination** is the rotation of the forearm such that the palm faces anteriorly. The forearms are supinated in an anatomic position.

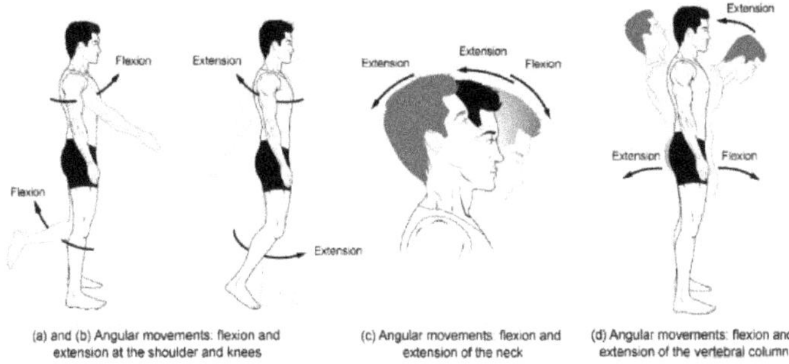

(a) and (b) Angular movements: flexion and extension at the shoulder and knees

(c) Angular movements: flexion and extension of the neck

(d) Angular movements: flexion and extension of the vertebral column

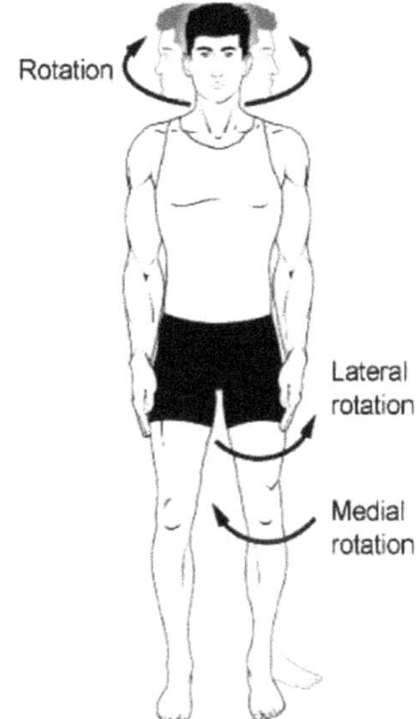

Rotation of the head, neck, and lower limb

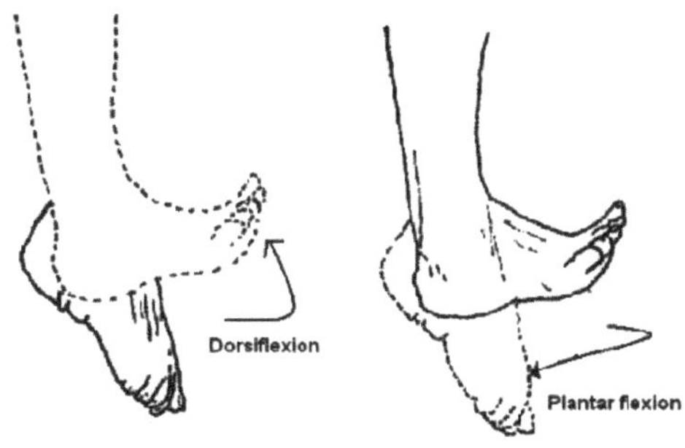

Hip movements below

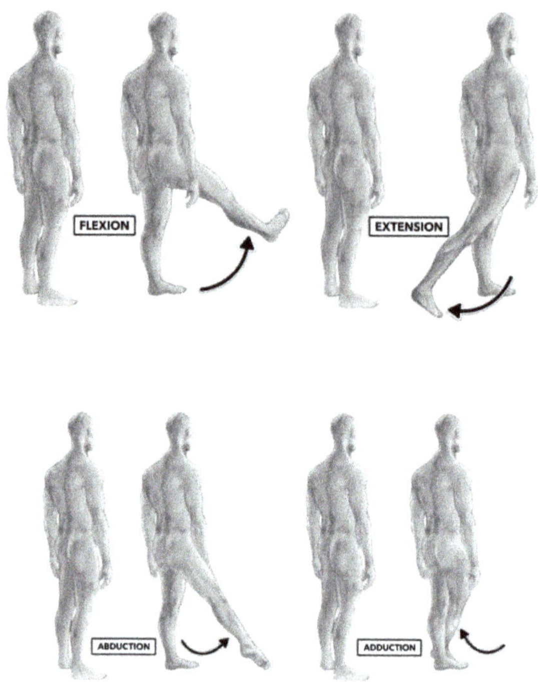

The Human Skeleton

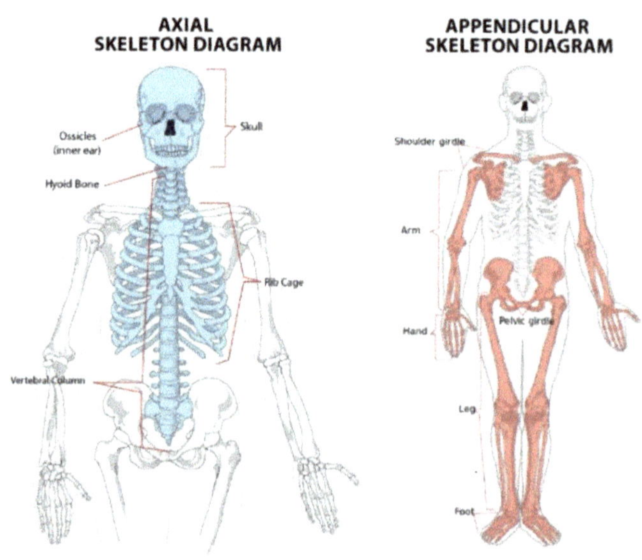

Pectoral girdle includes the clavicle and scapula

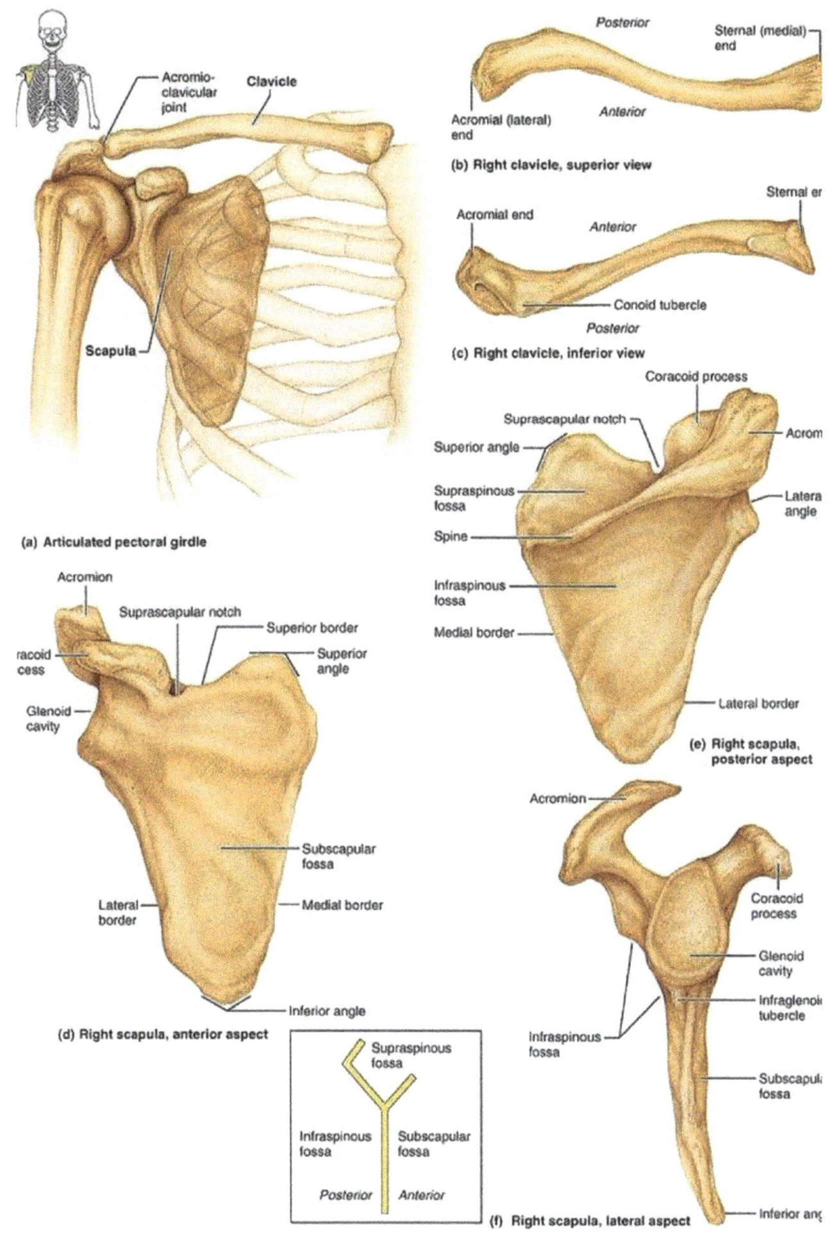

Anterior and posterior views of humerus and elbow joint

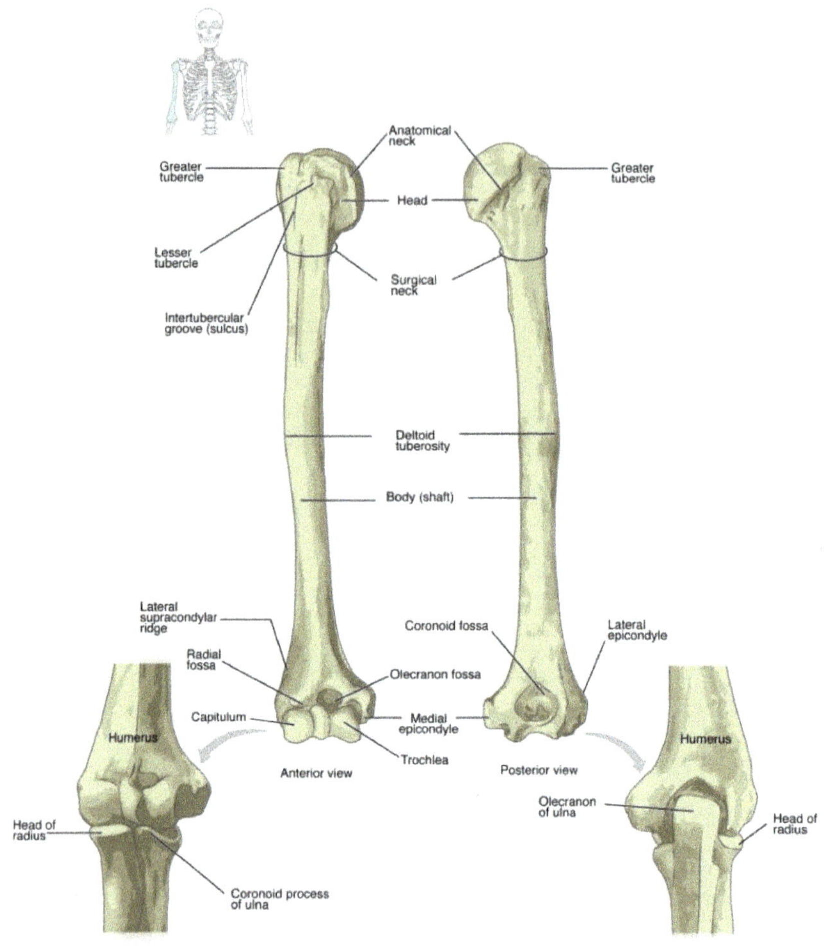

Anatomy of the radius and ulna (anterior and posterior views)

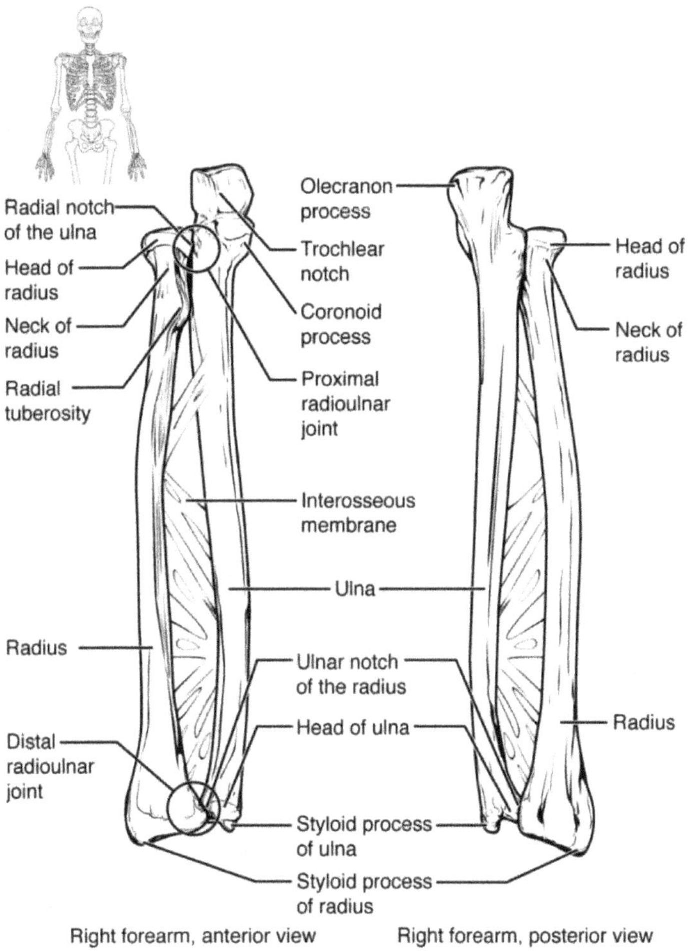

Bones of the hands and wrists

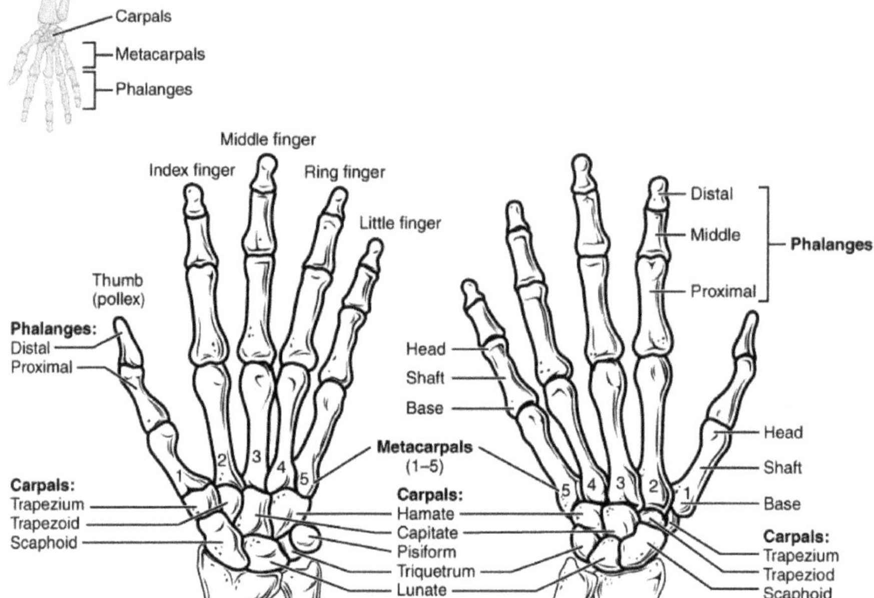

The pelvic girdle and the lower appendage

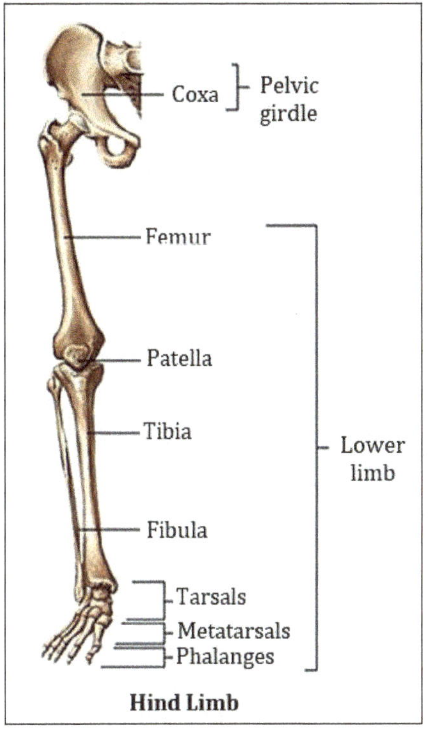

Hind Limb

Anterior view of the pelvis

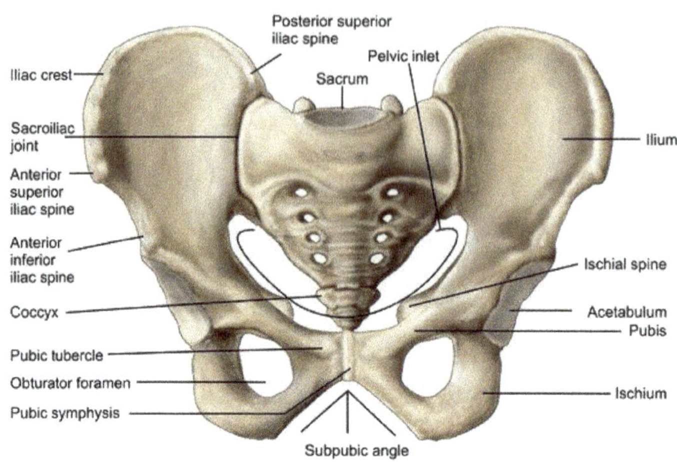

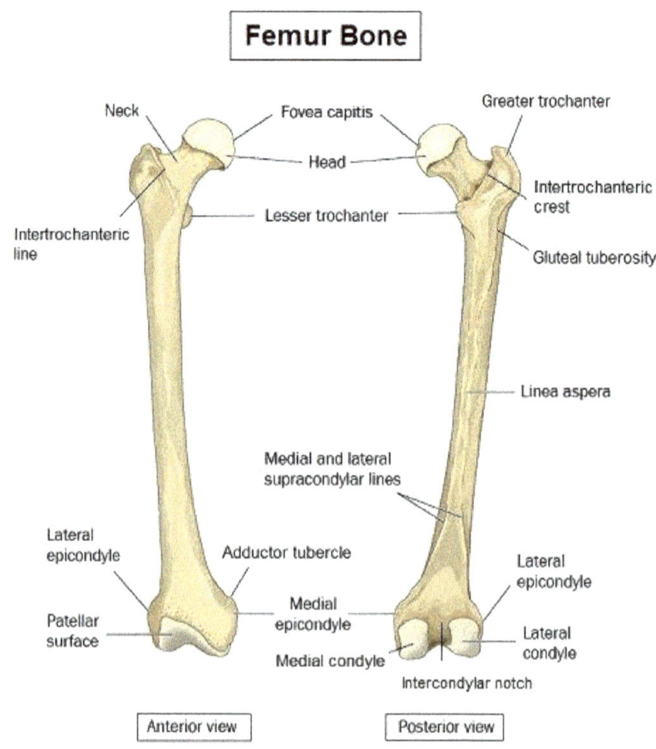

Anatomy of the tibia and the fibula (anterior and posterior views)

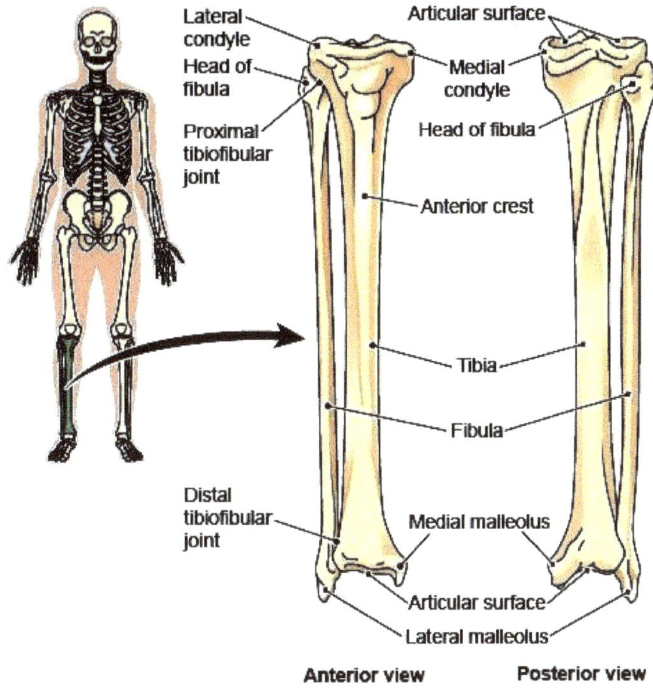

The ankle (7 tarsal bones) and the foot (5 metatarsals and 14 phalanges)

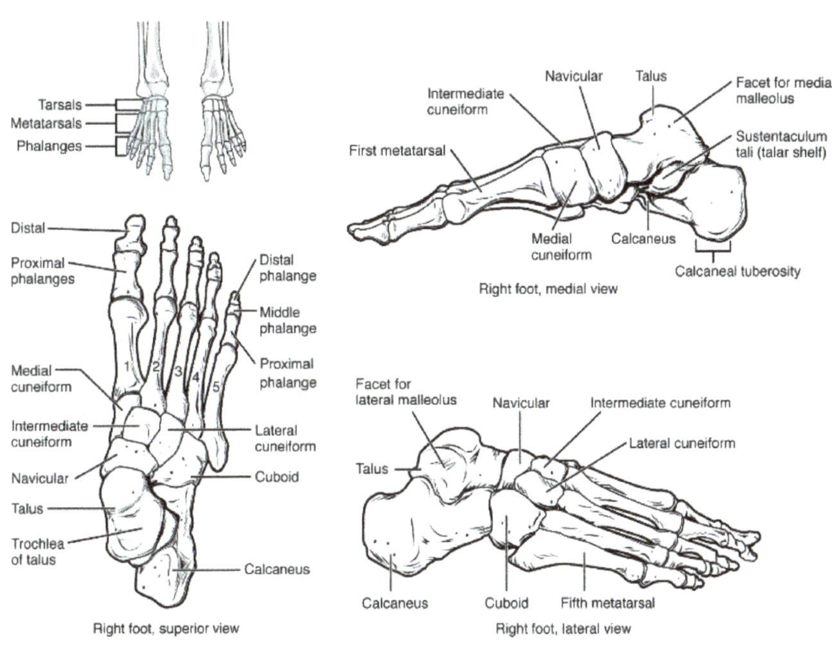

The skull (8 cranial and 14 facial bones – lateral, frontal, posterior views)

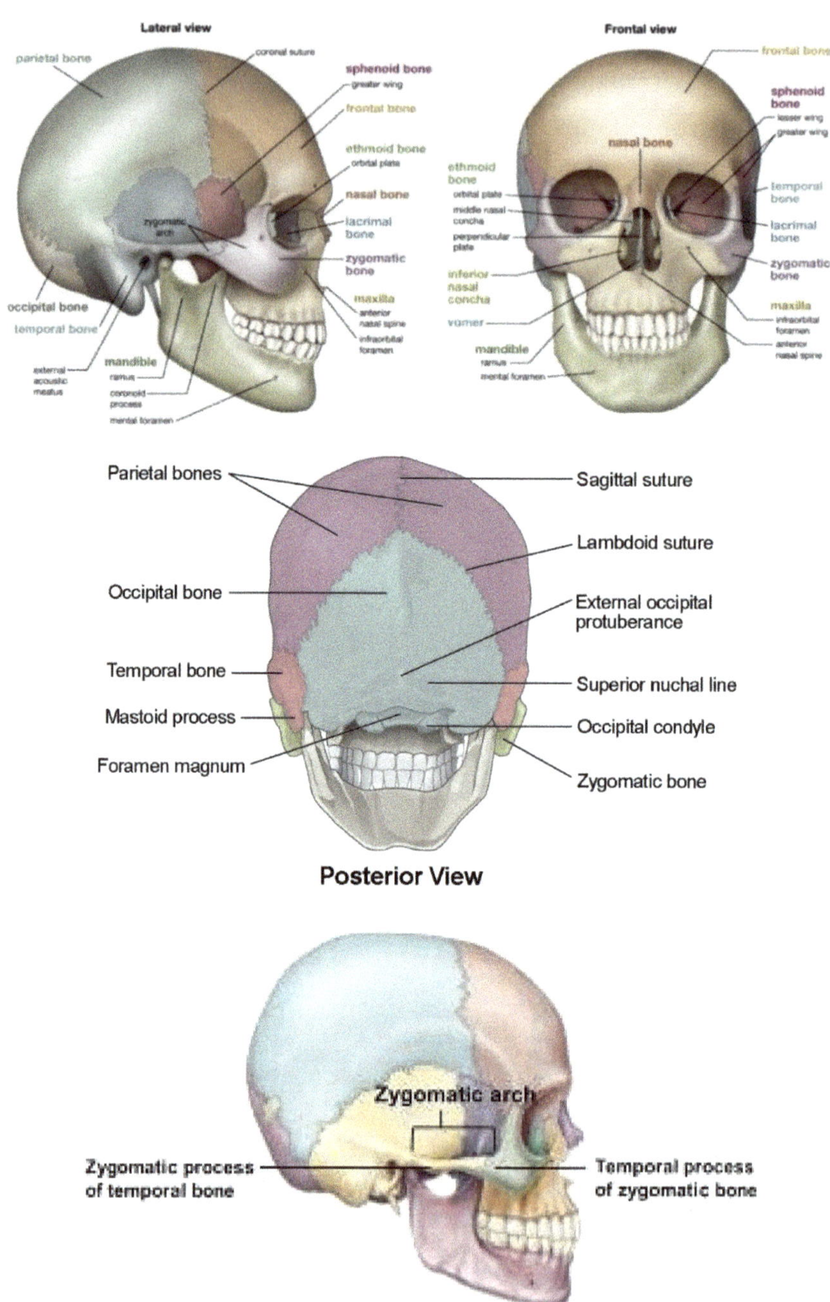

Temporal bone

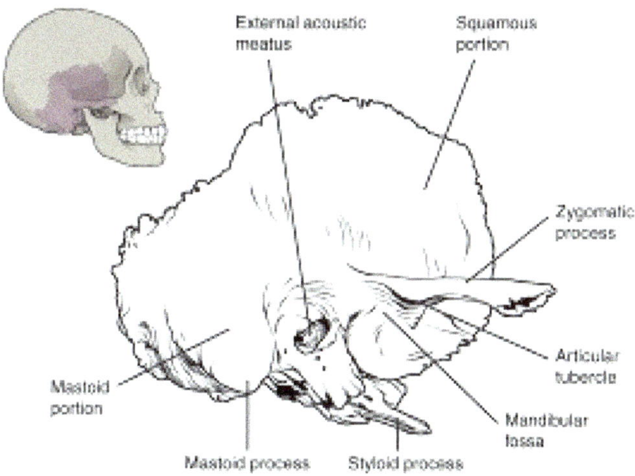

Floor of the cranium

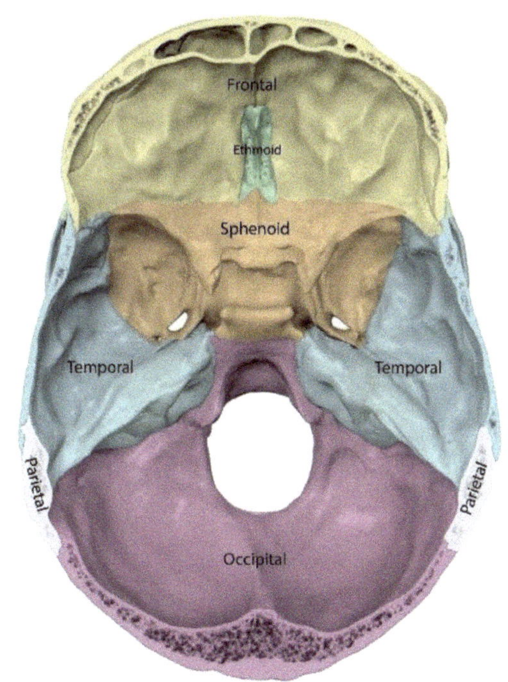

Inferior view of the skull without the mandible

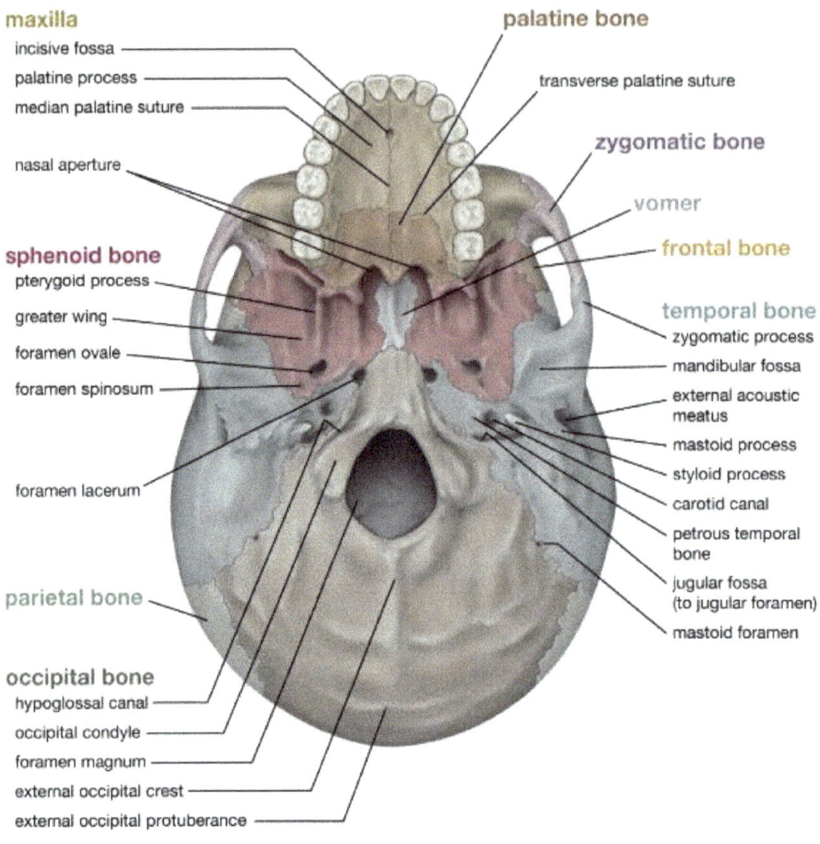

Sphenoid bone: superior and posterior views

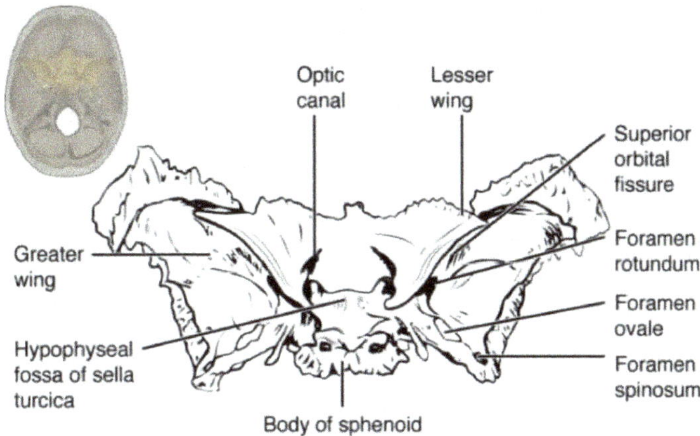

(a) Superior view

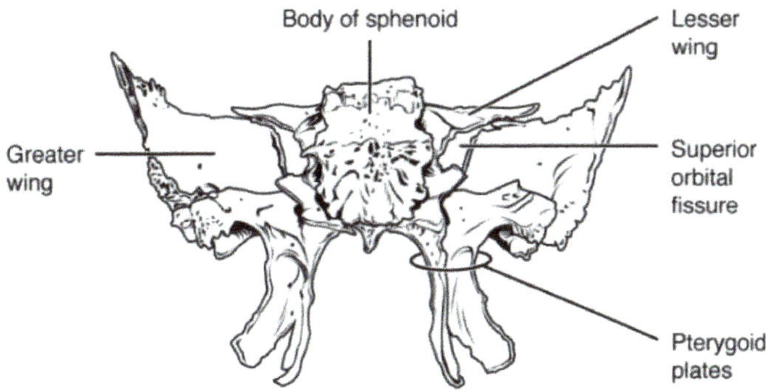

(b) Posterior view

Ethmoid bone

posterior view

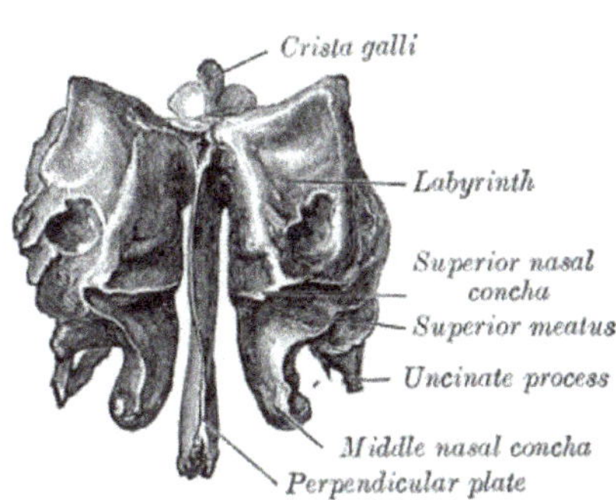

Maxilla

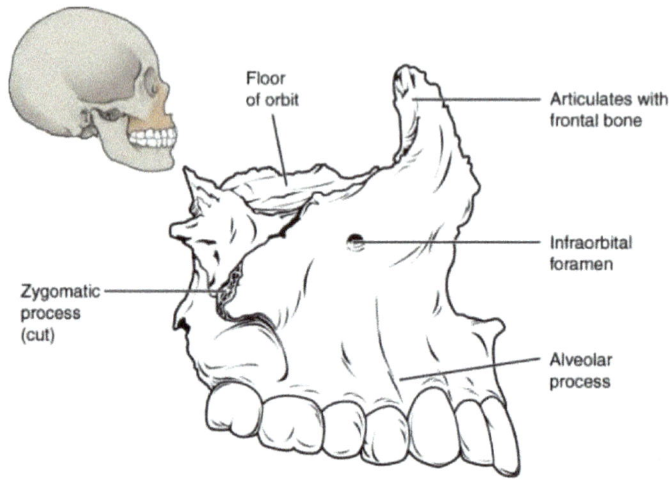

Anterior view of the ethmoid with the crista galli serving as an anchor site for the meninges of the brain.

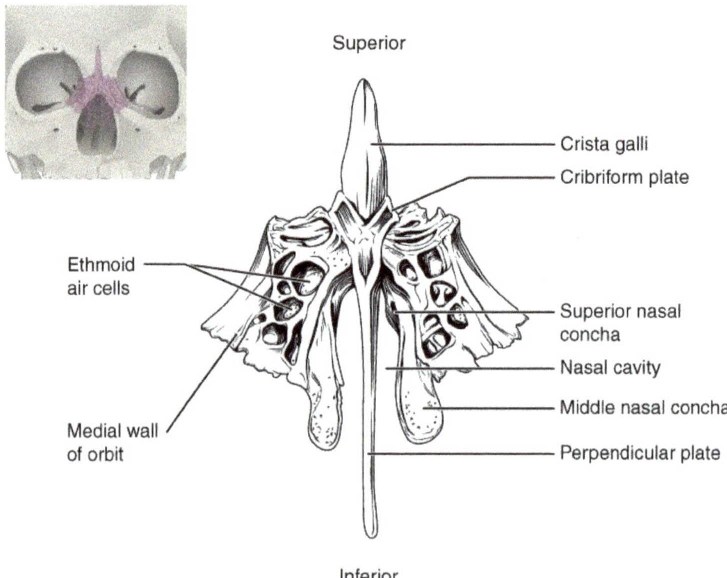

Bony nasal septum with the Vomer articulating superiorly with the perpendicular plate of the ethmoid to form the bony nasal septum posterior to the septal cartilage. Note the other structures as well.

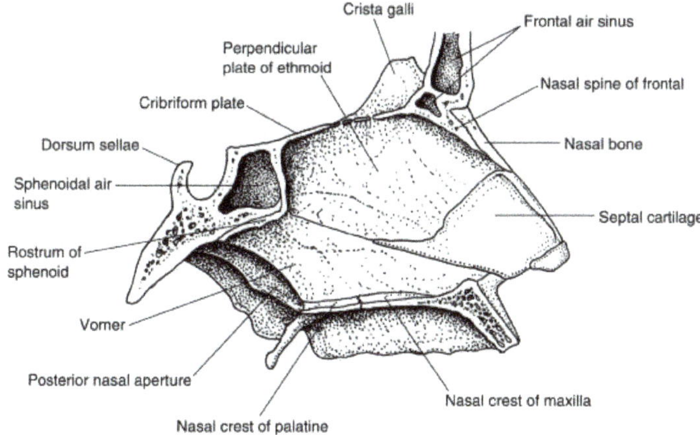

Vibration of the tympanic membrane (eardrum) from soundwaves is picked up by the malleus and transferred to the incus and finally to the stapes which then sends the signal to the vestibulocochlear nerve that carries the sound signal the thalamus and eventually to the auditory cortex in the temporal lobe for sound interpretation.

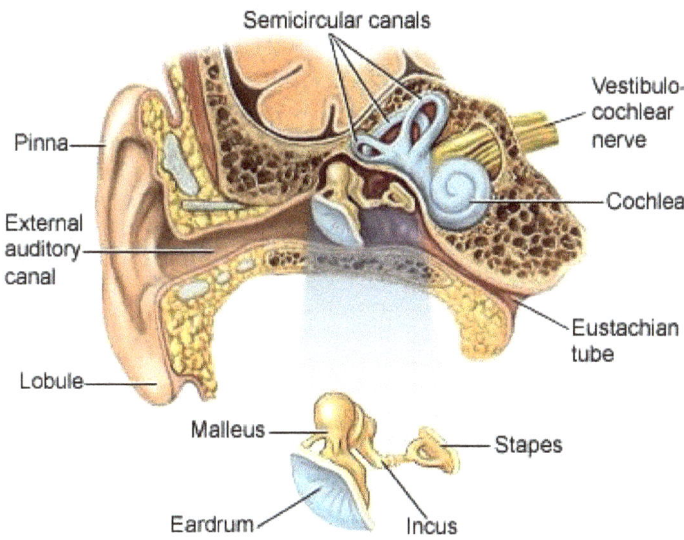

Mandible (lower jaw)

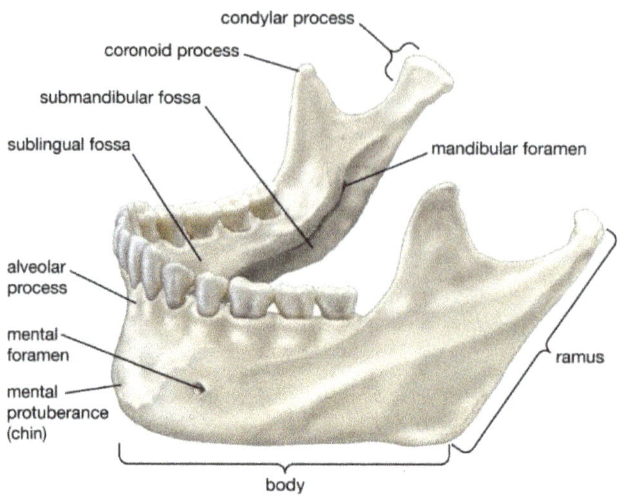

Ribcage (anterior and posterior views). Note ribs 1-7 are true ribs because each has its own costal cartilage whereas ribs 8 -10 share costal cartilage and thus, are false ribs. Ribs 11 and 12 are floating ribs because they are blind-ended and are additionally considered false since they lack costal cartilage altogether.

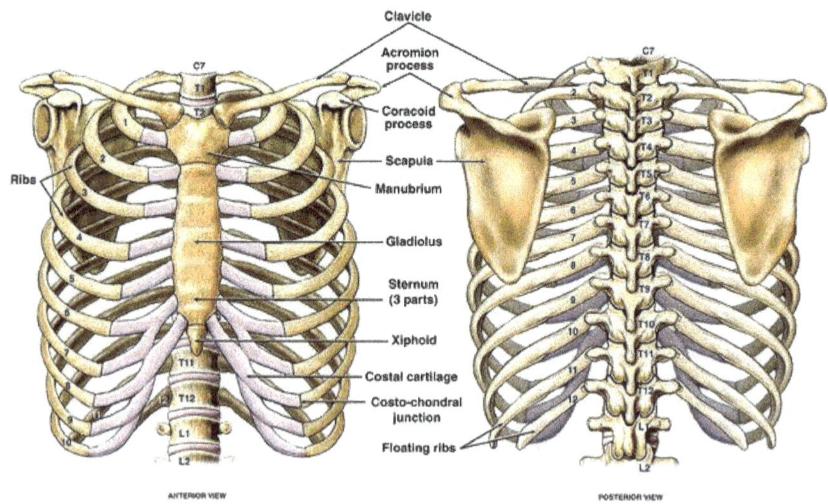

Anatomy of a typical rib below

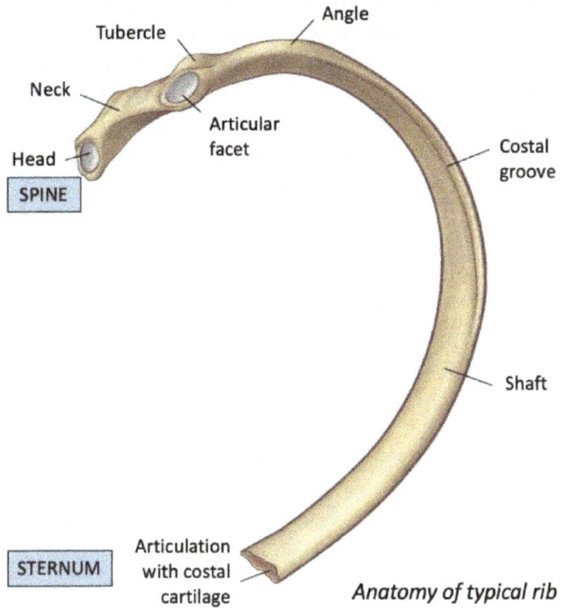

Anatomy of typical rib

Spine anatomy overview

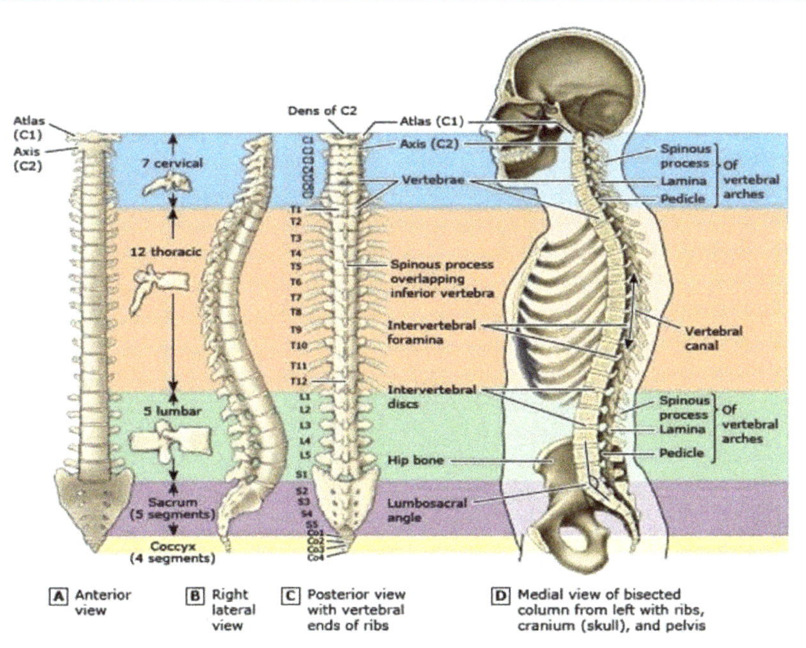

Atlas (C1) and Axis (C2). Atlas articulates with occipital condyles of occipital bone superiorly, and articulates with axis inferiorly.

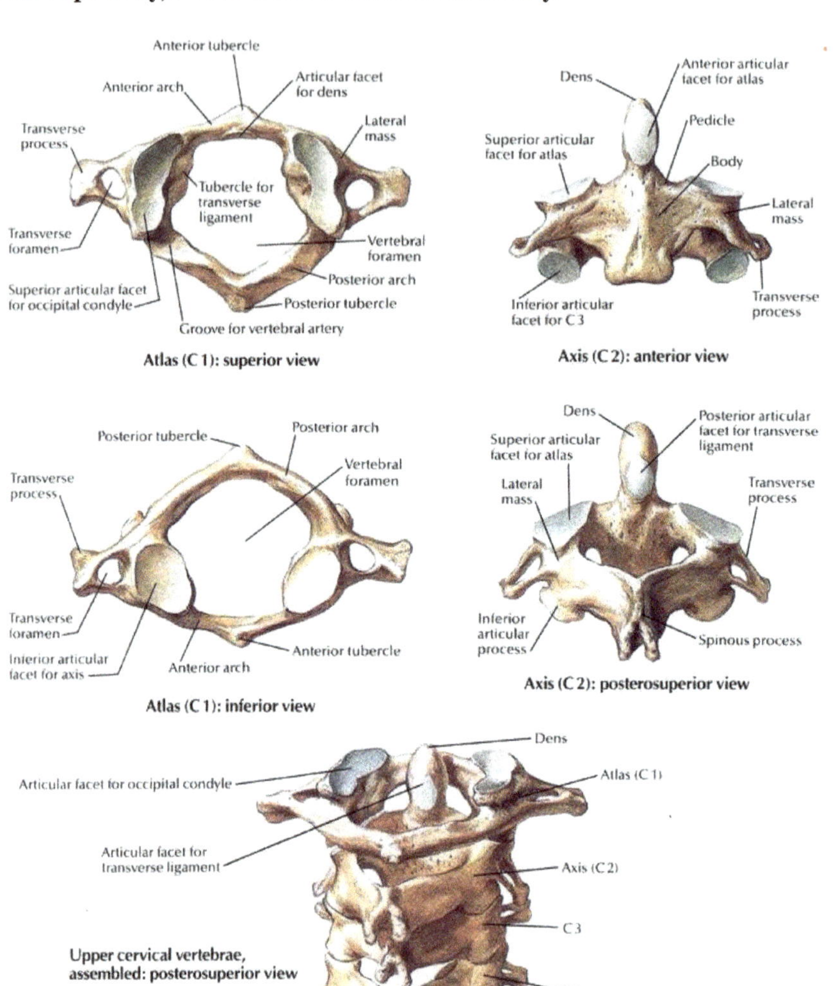

Thoracic vertebra (superior and lateral views)

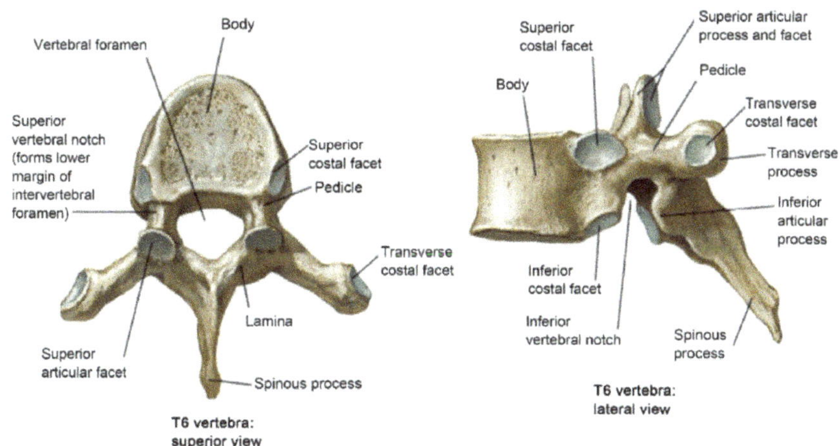

Lumbar Vertebrae

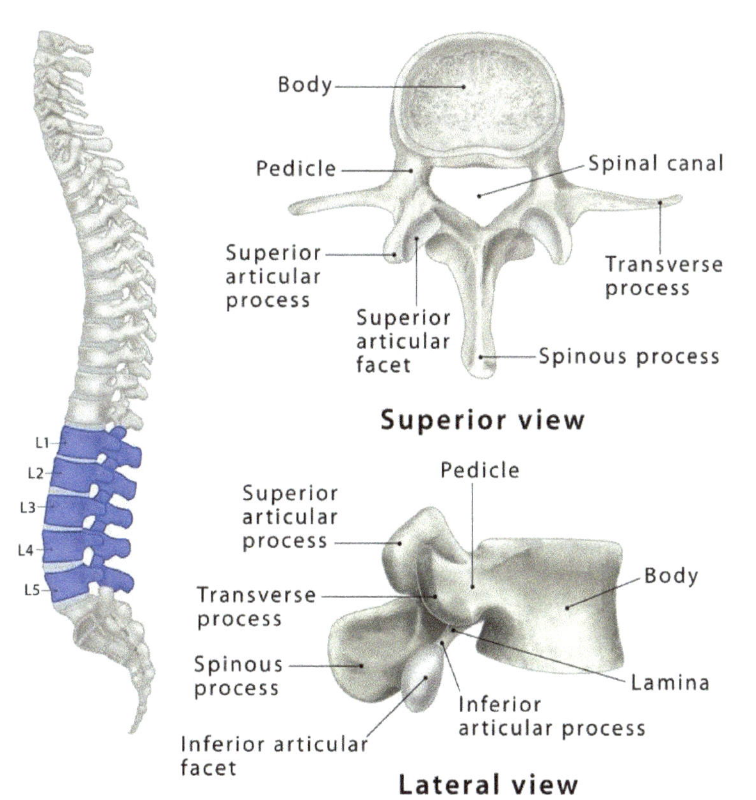

Sacrum and Coccyx (anterior and posterior views). Note the coccyx may move backwards during childbirth and defecation.

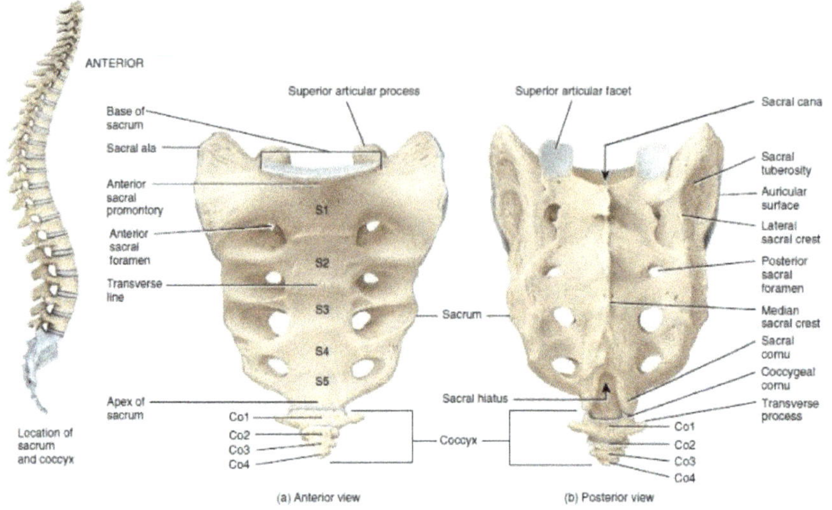

Practice by naming the structures below

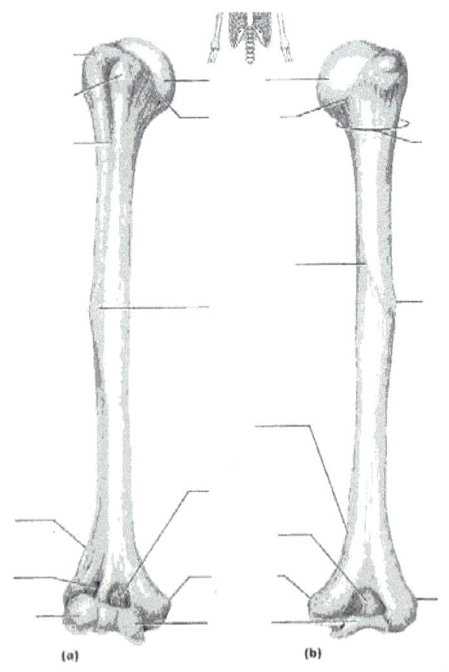

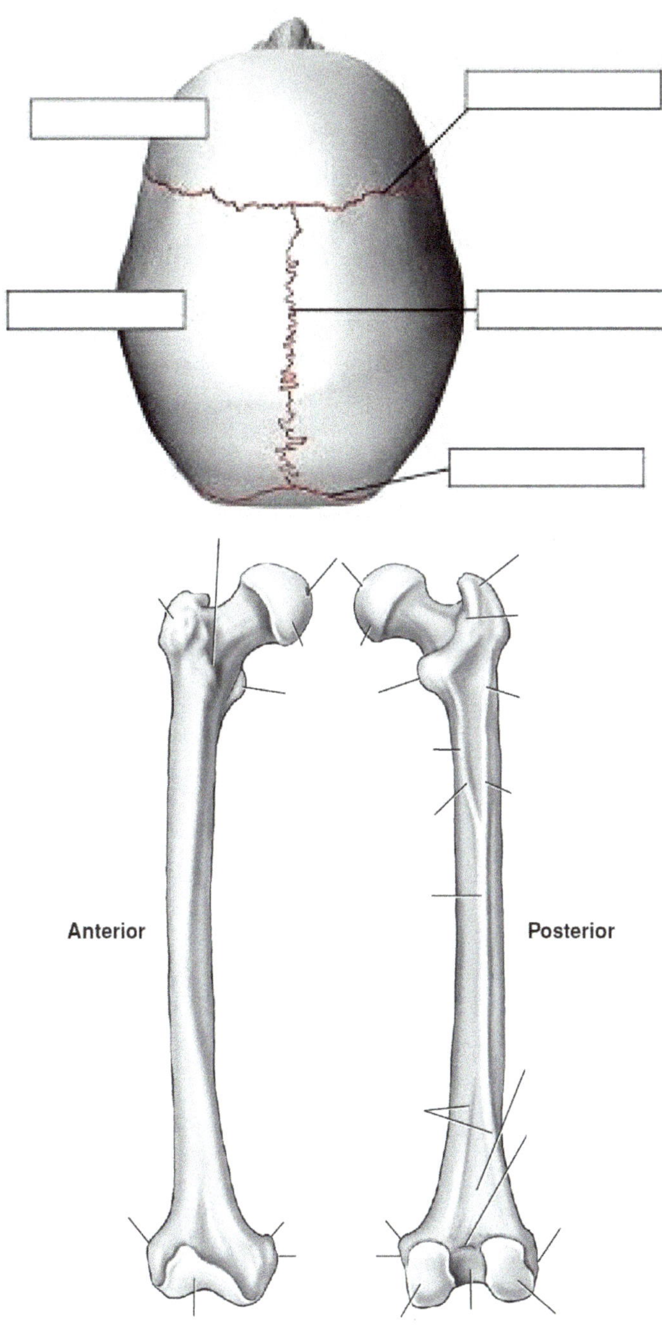
Anterior
Posterior

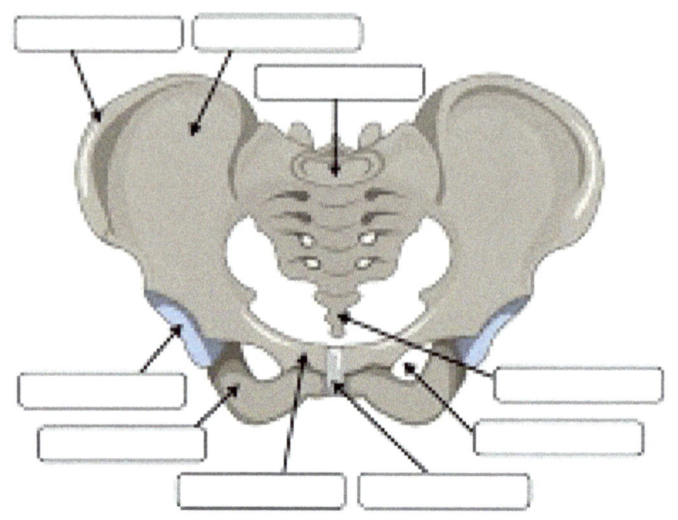

Chapter 7: Nervous System: Nervous Tissue

I. Introduction to the Nervous System

General functions of the nervous system

The nervous system has three primary functions:

1. Gather *sensory input* from the environment and the body itself.
2. Process and evaluate information and decide what to do, or, in a word, *integration* (process information).
3. Initiate responses via *motor output.*

Another function that underlies the functions described above is *communication*. The nervous system enables different parts of the body to communicate with each other. What other organ system in the body has a similar job of communication within the body? How does this system compare with and differ from the nervous system?

Organization of the nervous system

The nervous system can be broadly divided into two major areas:

A. The **central nervous system (CNS)** includes the brain and spinal cord.

B. The **peripheral nervous system (PNS)** is all of the nervous system outside the CNS.

Divisions Nervous System

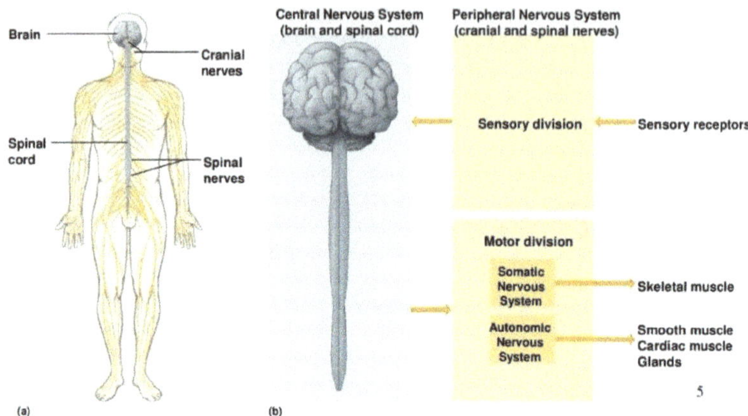

The PNS can be further divided into two components:

A. The **afferent division** carries sensory information from **receptors** to the CNS.

B. The **efferent division** carries commands from the CNS to **effectors**, which are either muscles or glands.

The afferent division can be further divided into two components:

A. **Somatic sensory** information, such as vision and taste, is perceived consciously.

B. **Visceral sensory** information is not perceived consciously, such as your body monitoring blood pressure or levels of carbon dioxide.

The efferent division can be further divided into two components:

A. The **somatic nervous system** innervates skeletal muscles and is associated with voluntary control.
B. The **autonomic nervous system** innervates the viscera (intestine, heart, vessels) and is associated with involuntary activities.

The autonomic system has two divisions:

A. The **sympathetic nervous system**, which is best known for the "fight or flight response."

B. The **parasympathetic nervous system**, which restores the body to normal function.

II. Nervous Tissue: Neurons

Neuron structure

The major functions of the nervous system are accomplished by cells called **neurons**. A neuron is composed of a **cell body (or soma)** and various projections from the body. The nucleus and other materials responsible for normal cell function are within the cell body. This includes an extensive rough endoplasmic reticulum (also known as **chromatophilic substance** or **Nissl bodies**), which is the site of protein and membrane production. Bundles of microscopic filaments inside the cell body, called **neurofibrils**, help maintain the neuron's shape. **Dendrites** are projections from the cell body responsible for receiving information from other neurons.

Basic diagram of a multipolar neuron. The soma or body contains among others the organelles such as: mitochondria, Nissl bodies (clusters of rough endoplasmic reticulum [RER] and ribosomes) for protein synthesis.

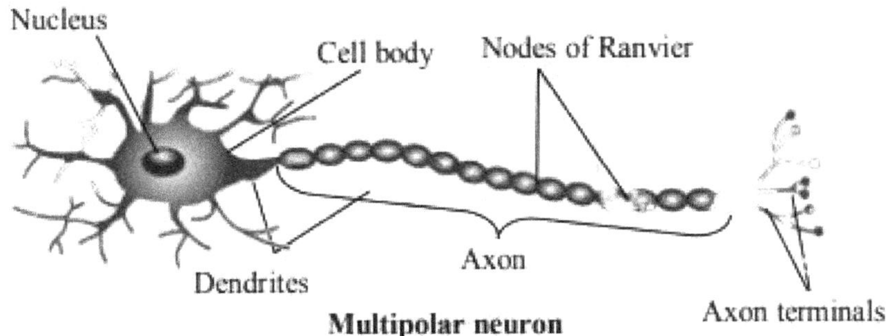

Diagram of the multipolar neuron below showing direction of the signal transmission at the depolarization region (node of Ranvier). Also note that saltatory (jumping) conduction of the myelinated neuron helps speed up signal transmission (meter/second).

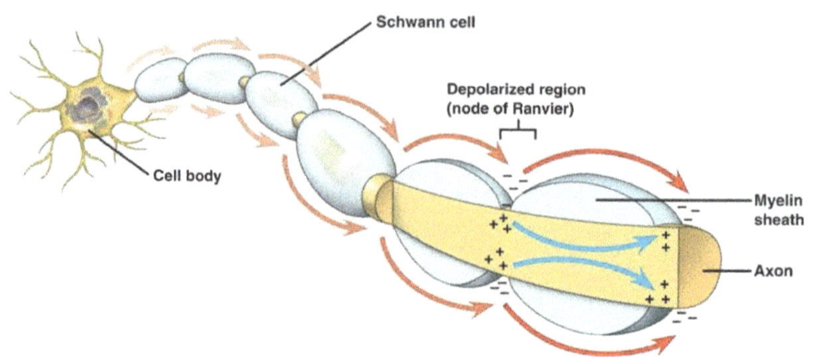

The **axon** is a projection from the cell body that sends information to other neurons (or effectors). Since axons are typically long and thin, they are often referred to as **nerve fibers** or simply as "fibers." Where the axon joins the cell body is the **axon hillock**. The plasma membrane of the axon is referred to as the **axolemma**. While each neuron has only one axon, the axon may split into branches called **axon collaterals**. At the very end of each axon (or axon collateral), there are many smaller branches called **telodendria**. At the end of each telodendrion is a knob-like swelling called a **synaptic knob** (or *synaptic terminal,* or *synaptic end bulb,* or *bouton*). A synaptic knob is where an axon connects to either another neuron or to an effector.

Portions of an axon may be wrapped by glial cells (*e.g.,* **Schwann cells**) that form a **myelin sheath** around the axon. The exposed (visible) portion of the Schwann cell is called the **neurilemma**, and gaps between the Schwann cells are the **neurofibril nodes** (also known as nodes of Ranvier). The functions of these structures will be discussed during the section on membrane potentials.

Practice drawing and labeling the diagram of a neuron.

Classification of neurons

Structurally, neurons can be classified into four basic types. Be familiar with the appearance of each.

A. **Multipolar neurons**, the most common type of neuron in the human body, have many dendrites and one axon extending from the cell body.

B. **Bipolar neurons** have two processes extending from the body: one axon and one dendrite, both of which typically have branches. These are rare and found only in special sense organs such as the olfactory sense (in nasal cavity for odor detection), and the retina of the eyes for vision.

C. **Unipolar neurons** have a single process that extends from the cell body. They are usually sensory neurons. One end (the **peripheral process**) has dendrites to receive information; the other end (the **central process**) functions as an axon and sends information to the CNS.

D. **Anaxonic neurons** have dendrites but no axons. They do not conduct action potentials and will not be discussed any further.

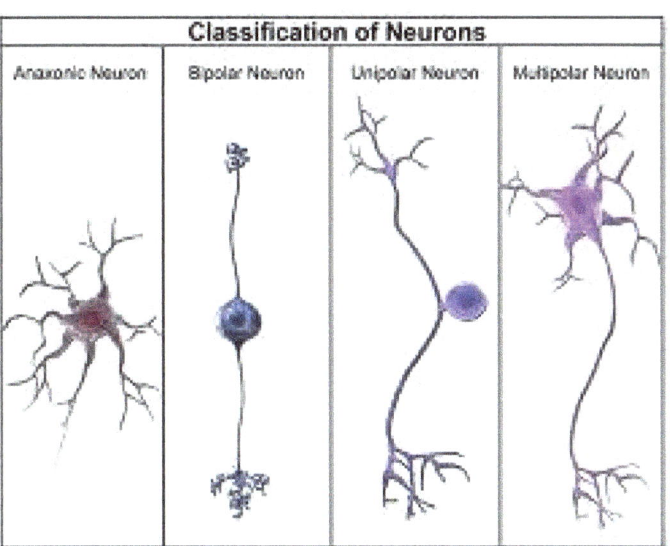

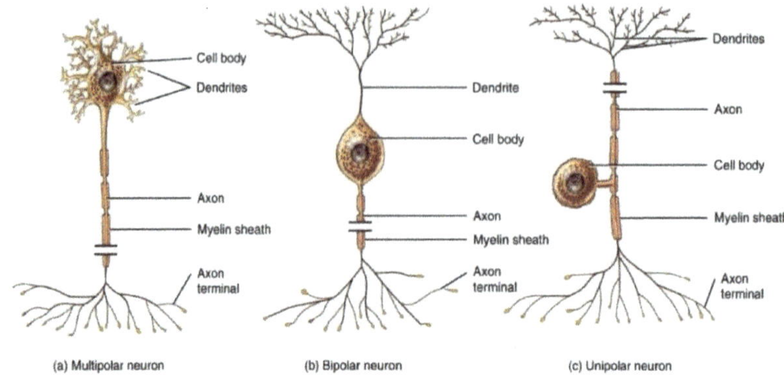

(a) Multipolar neuron (b) Bipolar neuron (c) Unipolar neuron

Functionally, neurons fall into three general categories:

A. **Sensory neurons** transmit information collected from external or internal stimuli to the CNS. Most sensory neurons are unipolar, and some are bipolar. Axons of sensory neurons (called **afferent fibers**) travel toward the brain or spinal cord.

B. **Motor neurons** carry signals to effectors. Axons of motor neurons (called **efferent fibers**) travel from the brain or spinal cord toward the effectors. **Somatic motor neurons** innervate the skeletal muscles, and **visceral motor neurons** innervate all other effectors. Motor neurons are multipolar.

C. **Interneurons** (or "*association neurons*") make connections between other neurons. Interneurons are found within the CNS and some ganglia. Nearly all Interneurons are multipolar.

Simple diagram of transverse section of the spinal cord below shows interrelationship of the three categories of neurons. Please note that signal transmission goes from sensory neuron (also known as pseudounipolar or unipolar) to interneuron in gray matter for integration (i.e., information processing) and finally to cell body of motor neuron (multipolar) that passes the signal to the effector organ (in this case, the effector is the skeletal muscle but in other cases it could be glands, or any other organs).

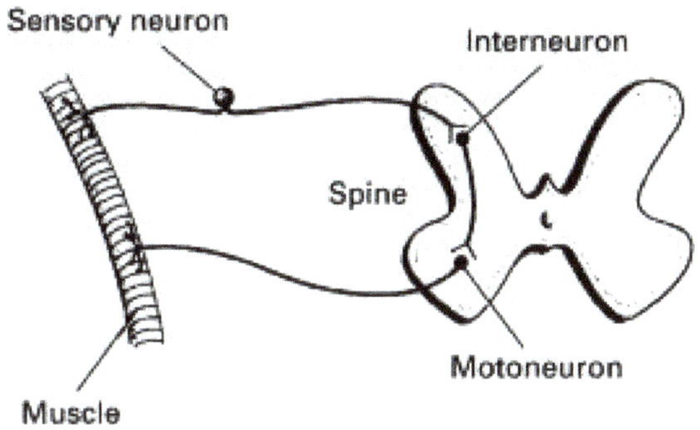

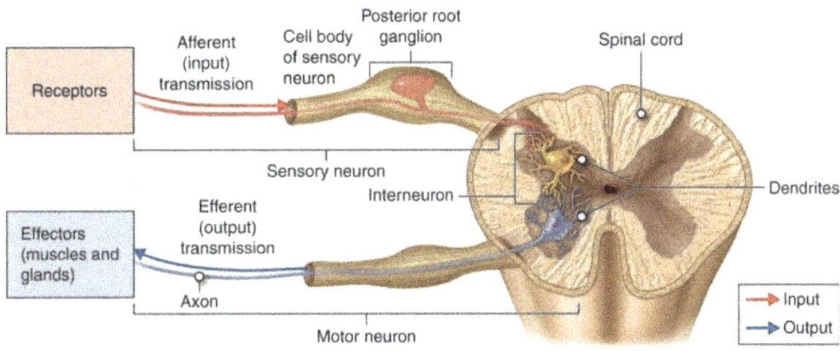

Relationship of neurons and nerves

A **nerve** is a cordlike bundle of axons traveling through the body and connecting to the CNS. By definition, a nerve is part of the PNS. Within a nerve, each axon is surrounded by a layer of connective tissue called the **endoneurium**. Groups of axons are bundled together into collections called **fascicles**. Each fascicle is wrapped in a layer of connective tissue called the **perineurium**. Finally, the entire nerve is wrapped in a layer of connective tissue called the **epineurium**.

Schematic illustration of a typical peripheral nerve

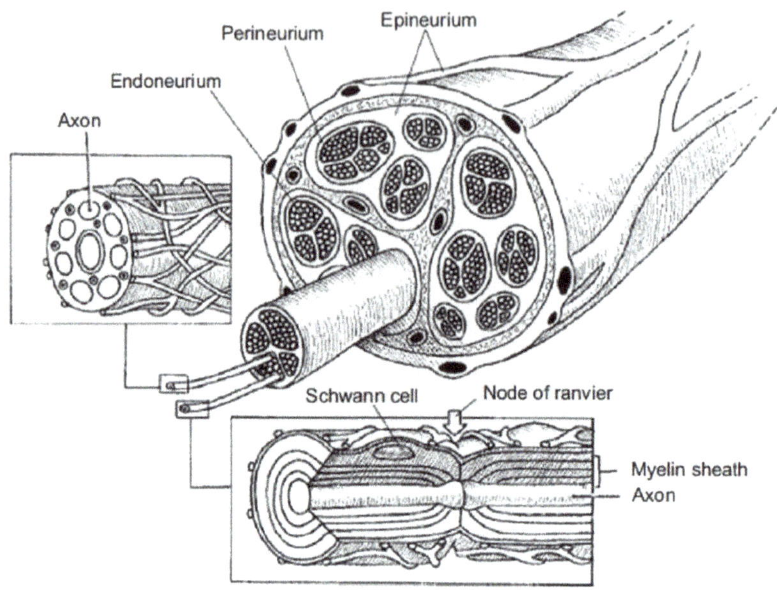

A **ganglion** is a collection of neuron bodies located in the peripheral nervous system.

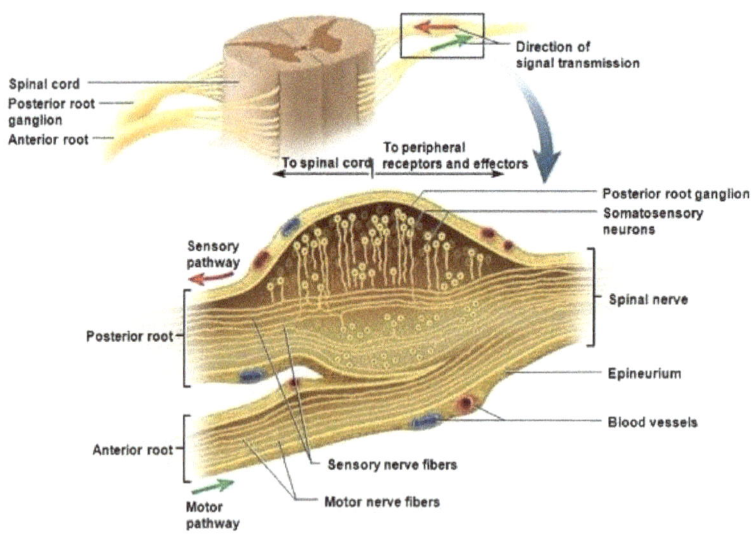

III. Synapses

The **synapse** is where parts of two neurons meet or where a neuron connects to an effector. The **presynaptic cell** at a synapse is the neuron that is sending a message down its axon to the synapse. The **postsynaptic cell** at a synapse is the neuron (or effector) receiving the message. If the postsynaptic cell is a neuron, it typically receives the message on its body or at a dendrite. As you will see later, transmission of information across the synapse is generally accomplished by chemicals called **neurotransmitters** (or *neurotransmitter substances*). The narrow gap between the presynaptic and postsynaptic cells is the **synaptic cleft**.

Nerve terminal below transmitting message to postsynaptic neuron by exocytosis of neurotransmitters with the assistance of calcium ions.

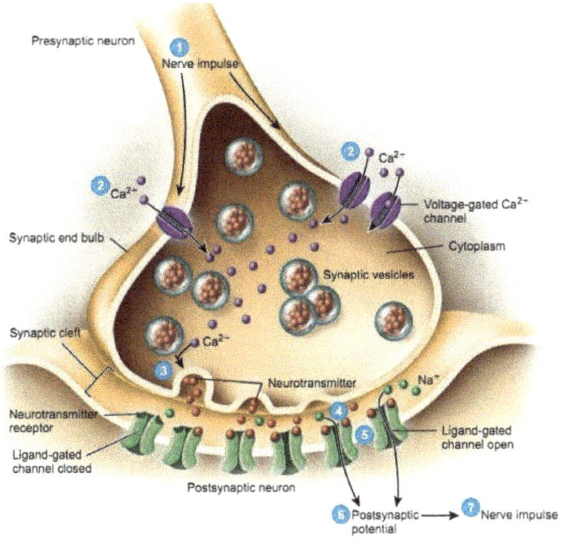

IV. Nervous Tissue: Glial Cells

Glial cells (also called “neuroglia”) are involved with supporting neurons. If you consider the neurons to be the "actors" of the nervous system, then neuroglia are the "stagehands." Various functions of neuroglia are as follows:

Glial cells (see chart below)

1. They provide physical, structural support for the neurons.
2. They provide nourishment for neurons.
3. They may guide the growth of young neurons.
4. Some neuroglia provide electrical insulation by wrapping around axons to create the **myelin sheath**. These glial cells and the function of myelin will be discussed in detail later.

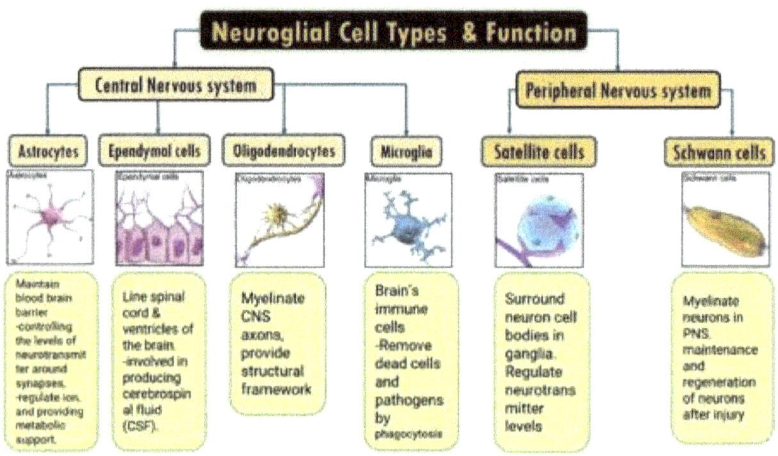

V. Introduction to Neuron Physiology

Neurons and Ohm's law

The word "electricity" has been defined in many ways. For our purposes, it will mean the movement of charged particles. **Current** is a quantitative measure of the number of charges moving from one point to another per unit of time.

Typically, the net movement of charged particles from one point to another (a current) is caused by a difference in "electric potential" between the two points. For example, a charged battery has a positive and negative electrode. When a circuit connects the two electrodes, current flows through the circuit. Another term for electric potential is **voltage**. The battery in your car has a potential difference between the positive and negative electrodes of approximately 12 Volts. The terms "potential" and "potential difference" are often used in physiology with essentially the same meaning as voltage.

A third term important for a basic understanding of electricity is resistance. **Resistance** is a measure of opposition to the flow of electrical charges. You should soon learn that a plasma membrane provides considerable resistance to the flow of electricity (current), but the resistance can be reduced by opening channels.

Ohm's law shows that there is a mathematical relationship among current, voltage, and resistance:

Current = voltage/resistance

According to the above equation, it then becomes quite clear what happens to the current if voltage or resistance increases or decreases.

Neurons at rest

The job of a neuron is to send signals to other neurons and/or effectors. These signals depend upon the creation of a voltage across the plasma membrane, often called the **membrane potential**. There are various important principles to understanding membrane potentials; here are two of them:

1. Intracellular fluids and extracellular fluids have different ionic compositions. Extracellular fluids contain high concentrations of sodium and chloride, whereas intracellular fluids contain high concentrations of potassium and negatively charged proteins.
2. Charged particles cannot move freely across cell membranes. How might a charged particle like Na^+ get across a membrane?

Pumps in the plasma membrane maintain the differences in sodium and potassium concentrations, called the **Na^+/K^+-ATPase**. Each pump can move three sodium ions out of the cell and two potassium ions into the cell as it breaks down one molecule of ATP. What process from Chapter 4 is this?

At the same time sodium and potassium are being pumped, they also leak back across the membrane. The plasma membrane has channels for both ions that are always open. These are sometimes called "**leak channels**." The membrane has considerably more leak channels for potassium than sodium. Potassium leakage to the outside of the cell leads to excess positive charge on the outside and, thus, excess negative charge on the inside. The amount of potassium that leaks out of the cell is limited because as positively charged potassium leaks out, the inside of the cell becomes negatively charged. Since negative charges attract positive charges, eventually, this attraction offsets the tendency of potassium to diffuse down its concentration gradient. The equilibrium point for potassium is about -90 mV (measured inside the cell).

Sodium also leaks across the plasma membrane, but there are much fewer leak channels for sodium than potassium. The movement of sodium through these channels makes the voltage inside the cell somewhat more positive (or you could say less negative) than it would be if only potassium crossed the membrane. In the typical human neuron, the resulting voltage inside the cell is about -70 mV, referred to as the **resting membrane potential**.

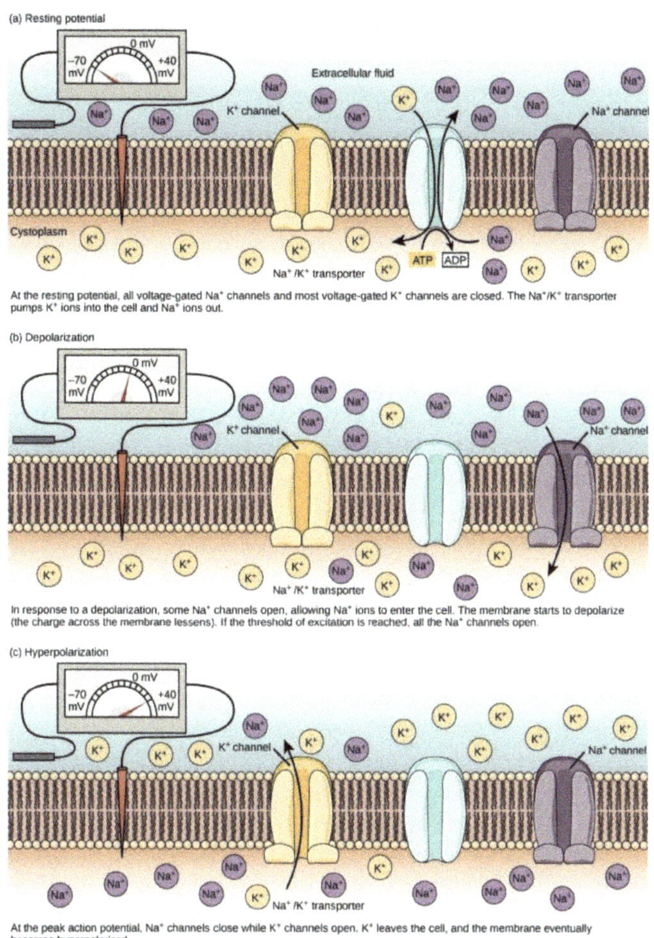

At the resting potential, all voltage-gated Na^+ channels and most voltage-gated K^+ channels are closed. The Na^+/K^+ transporter pumps K^+ ions into the cell and Na^+ ions out.

In response to a depolarization, some Na^+ channels open, allowing Na^+ ions to enter the cell. The membrane starts to depolarize (the charge across the membrane lessens). If the threshold of excitation is reached, all the Na^+ channels open.

At the peak action potential, Na^+ channels close while K^+ channels open. K^+ leaves the cell, and the membrane eventually becomes hyperpolarized.

VI. Physiologic Events in the Neuron Segments

At rest, a neuron tends to sit at its resting potential of -70 mV. However, the purpose of a neuron is not to sit at rest forever but to send a signal. The signal is sent when the voltage changes to what is called the **threshold**. To reach the threshold, the voltage must change. The voltage can be changed by opening more channels.

Two types of channels will be important in changing the voltage: **voltage-gated channels** and **chemical-gated channels**. A voltage-gated channel opens in response to a certain change in the potential across the membrane, while a chemical-gated channel opens in response to the binding of a certain chemical to the channel.

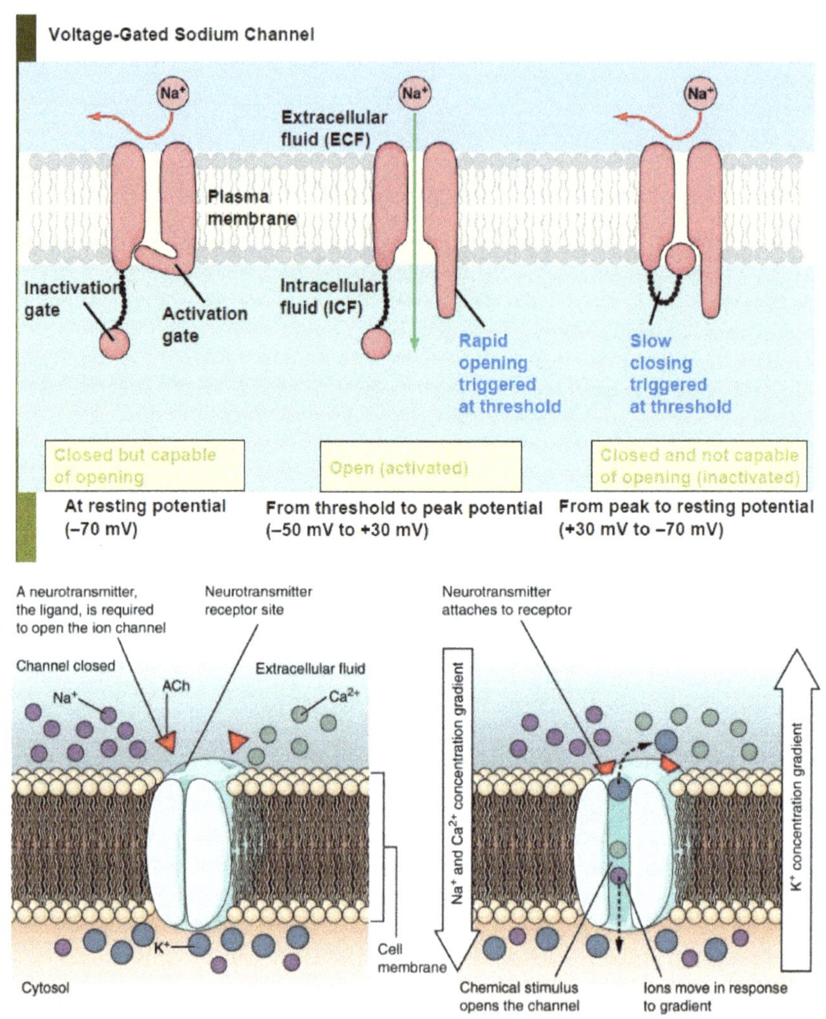

There are various chemical- and voltage-gated channels in the plasma membrane of a neuron, but we will continue to focus on channels for sodium and potassium. *Under the conditions we have described so far, the net movement of sodium through ion channels will always be into the cell, and the net movement of potassium through ion channels will always be out of the cell.* Thus, the opening of gated sodium channels allows more positive charge to enter the cell, reducing the voltage across the plasma membrane. A reduction in voltage (bringing the voltage toward zero) is called **depolarization** of the membrane. Opening gated potassium channels allows more positive charge to exit the cell, increasing the voltage across the membrane. An increase in voltage beyond the resting potential (more negative than -70 mV) is called **hyperpolarization**.

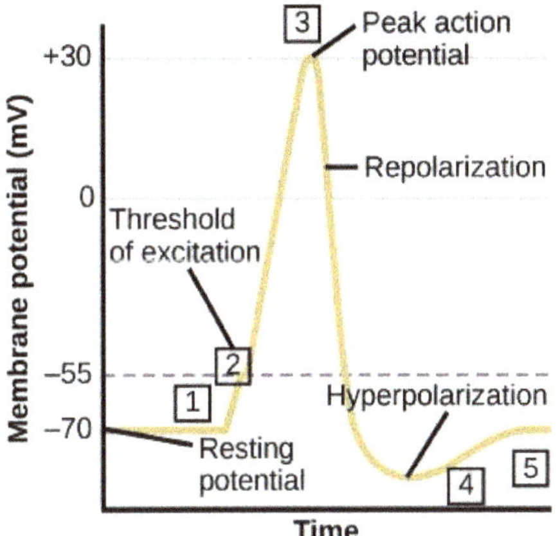

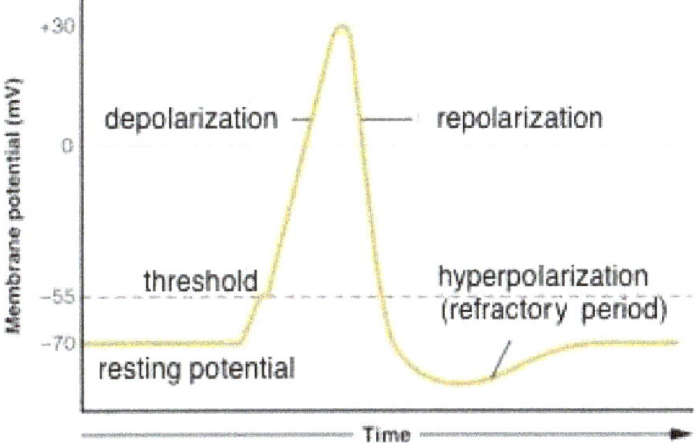

Receptive segment

Graded potentials are local changes in the membrane potential. They are called "graded" potentials because the magnitude of the change varies with the intensity of the stimulus.

The opening of chemical-gated channels (both sodium and potassium channels) in the cell membrane produces graded potentials. For example, opening gated sodium channels will allow sodium ions to diffuse into the cell . This influx of positive charge will cause a local change in the membrane potential, making it more positive. This shift in the membrane potential toward 0 mV is a depolarization. The greater the number of sodium channels that open, the greater the magnitude of the depolarization.

These channels allow the passage of both sodium and potassium. However, when a neuron is below the threshold, K^+ is much closer to its equilibrium point than Na^+. Thus, the driving force to move sodium through the channel is much greater than the driving force for potassium. More positive charge enters the cell than leaves, and the cell is depolarized.

The opening of chemical-gated potassium channels will allow more potassium to move out of the cell. This makes the membrane potential more negative, which is hyperpolarization. (This can also be accomplished by opening gated chloride channels, which allow a net movement of Cl into the neuron).

Graded potentials are generally caused by chemical signals and involve either gated sodium channels or gated potassium channels. Once the chemical stimulus is removed, the membrane quickly returns to its resting membrane potential. Restoration of the normal potential of –70 mV is called **repolarization**. Can you think of a specific mechanism that restores the normal resting membrane potential?

Graded potentials occur on the **receptive segment** of a neuron, where chemical-gated channels are located (both sodium and potassium channels), including the dendrites and often the soma. Since they occur on the membrane of a postsynaptic cell in response to activity at the synapse, graded potentials may also be called **postsynaptic potentials**. For the rest of this chapter, the terms "graded potential" and "postsynaptic potential" are interchangeable.

A postsynaptic potential can be either excitatory or inhibitory depending upon the neurotransmitter that is released into the synapse. If the stimulus is excitatory in nature, then it acts by opening chemical-gated sodium channels in the postsynaptic membrane. Excitatory postsynaptic potentials (EPSPs) cause depolarization, which may lead to an action potential.

If the stimulus is inhibitory in nature, it opens chemical-gated potassium channels (or chloride channels) in the postsynaptic membrane. Inhibitory postsynaptic potentials (IPSPs) cause hyperpolarization.

Initial segment

The **initial segment** of a neuron is where there is a transition from chemical-gated channels to voltage-gated channels. This region is typically found at the axon hillock. If the membrane potential at the initial segment reaches the threshold value, then an action potential begins.

As mentioned previously, postsynaptic potentials may be of varying magnitude. Another feature of the postsynaptic potential is **summation**. The effects of PSPs are cumulative. For example, if the postsynaptic cell simultaneously receives an IPSP and an EPSP of equal magnitude, they will cancel. Likewise, if the cell simultaneously receives multiple EPSPs, they will result in a larger magnitude of depolarization. If EPSPs reach the threshold, then an action potential begins.

Summation must be considered in both spatial and temporal terms. **Spatial summation** occurs when a large number of presynaptic neurons fire action potentials at the same time, resulting in summation as a large number of graded potentials occur simultaneously. **Temporal summation** occurs when one or more presynaptic neurons fire multiple action potentials in rapid succession, resulting in summation as PSPs occur repeatedly. Summation allows neurons to act in concert to provide an almost limitless number of patterns of neural activity.

As long as the voltage at the initial segment stays below the threshold, this neuron does not fire an action potential. If the voltage at the initial segment reaches the threshold, then an action potential begins. Action potentials are always of the same intensity; there are no "partial" action potentials. Thus, we can say that an action potential is an "all or none" event. (see illustration below).

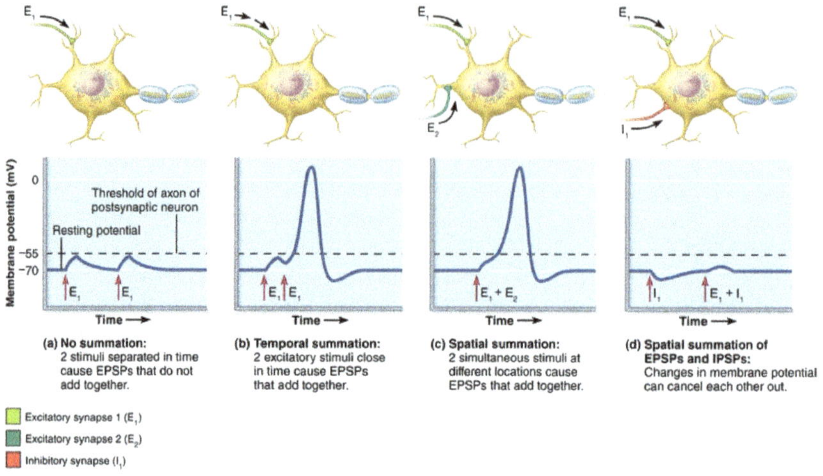

(a) No summation: 2 stimuli separated in time cause EPSPs that do not add together.

(b) Temporal summation: 2 excitatory stimuli close in time cause EPSPs that add together.

(c) Spatial summation: 2 simultaneous stimuli at different locations cause EPSPs that add together.

(d) Spatial summation of EPSPs and IPSPs: Changes in membrane potential can cancel each other out.

Conductive segment

The **conductive segment** is the part of the neuron with voltage-gated channels, generally the axon. The axolemma is sometimes referred to as the "excitable" portion of the neuron. An **action potential** is a change in the membrane potential that, once initiated, spreads across the entire conductive segment. An action potential is initiated when graded potentials cause a change in the voltage of the plasma membrane that is sufficient to open voltage-gated channels in the axonal membrane. Once initiated, the action potential spreads across the entire excitable membrane of the cell.

The membrane voltage required to trigger an action potential is called the **threshold** level. Threshold varies for different neurons, but it is typically around –55 mV. Once a cell reaches the threshold, it will fire an action potential. Notice that the threshold level is always *less negative* than the resting potential. Thus, *depolarization* is always required to trigger an action potential. On the other hand, hyperpolarization of a neuron takes it away from its threshold and makes it less likely to fire an action potential.

Segments of a neuron

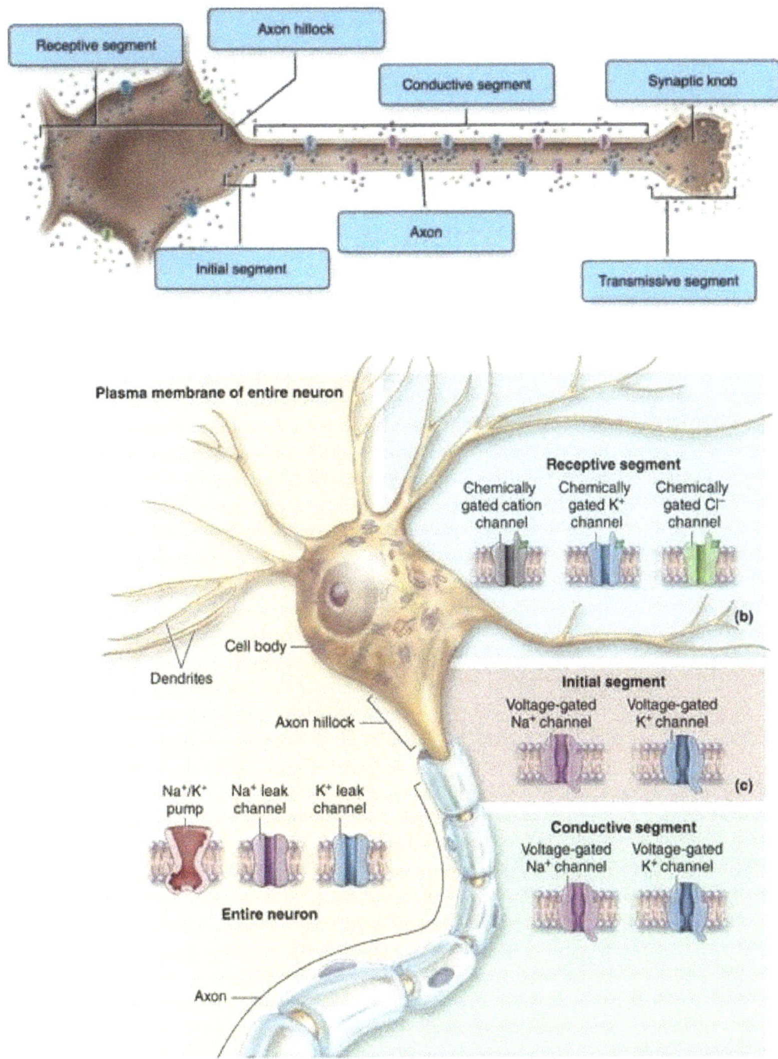

Here are the steps in the generation of an action potential:

A. **Depolarization**. If the initial segment reaches the threshold, **voltage-gated Na^+ channels** will begin to open. When the Na^+ channels open, the positive charge flows in rapidly, causing rapid depolarization of the cell. Depolarization generally results in a change of potential inside the neuron from –70 mV to approximately +30 mV.

Depolarization is limited by two factors: Influx of Na^+ is eventually resisted by the positive potential that builds up in the cell. Also, voltage-gated sodium channels rapidly close.

B. **Repolarization**. At about the same time when the sodium channels are closing, **voltage-gated potassium channels** open. These channels allow potassium to flow out of the cell, leading to the repolarization of the membrane. The membrane's electrical potential returns to a value near the resting potential.

C. **Hyperpolarization**. Voltage-gated potassium channels remain open longer than necessary to restore the potential to –70 mV. Therefore, enough potassium exits the cell to cause slight hyperpolarization. This is sometimes called the "undershoot."

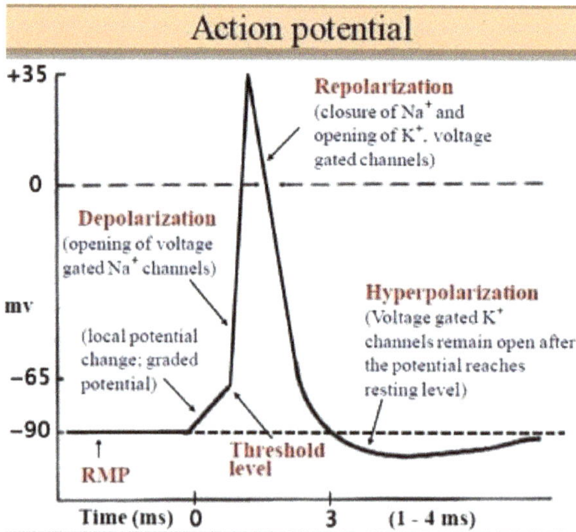

In the graph below, the 'Failed initiations' (i.e., graded potentials below threshold) would indicate that graded potential did reach the threshold, thus action potential (depolarization) would not occur along the axon.

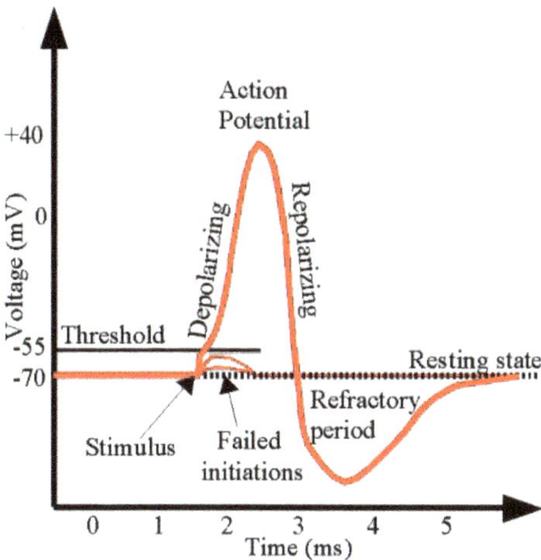

Eventually, the resting potential is restored. What components of a neuron are responsible for restoring the resting membrane potential?

You should become familiar with what an electrical recording of an action potential looks like, where depolarization and repolarization are occurring, and at which points the sodium and potassium channels are opening and closing.

Refractory period

Once a neuron has begun an action potential, a short period must pass before it can generate another action potential. This period of time is known as the **refractory period** (refer to graph above). More specifically, during the time that the voltage-gated sodium channels are open and then briefly closed in an inactive state, the neuron simply cannot respond to another stimulus. This specific period of time is the **absolute refractory period**. After the voltage-gated sodium channels have reset, the potassium channels remain open for a while. During this period (v-g Na^+ channels closed, v-g K^+ channels open), the neuron can be stimulated to fire another action potential. Still, it takes a stronger stimulus than normal because the neuron is hyperpolarized. This specific period of time is the **relative refractory period**.

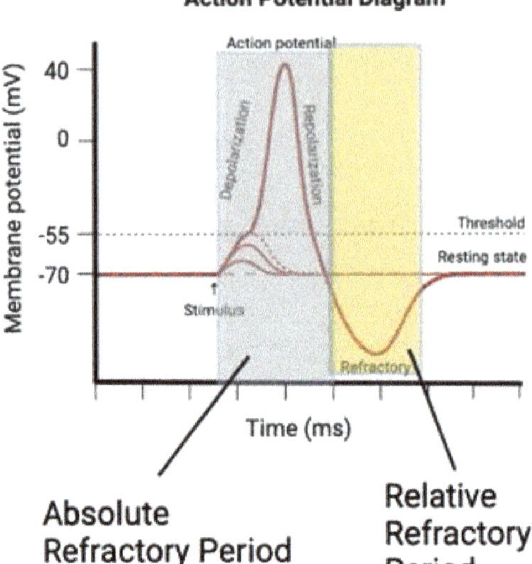

Continuous conduction and saltatory conduction

The movement of the action potential along the axon is called **propagation**. As positively charged sodium ions rush into the cell during depolarization, some will spread down the axon. This easily depolarizes more of the axonal membrane to the threshold, causing more voltage-gated sodium channels to open. This allows more sodium ions to rush into the axon and spread further down the axon.

The paragraph above is an example of a positive feedback loop. This loop allows the action potential to spread (propagate) along the axon until it reaches the axon terminals.

This conduction process in which the action potential travels down the axon in a continuous wave is called **continuous conduction**.

The **conduction velocity** (CV) of an axon measures how fast the action potential travels along the axon. The CV is a function of the diameter of the axon; the larger the diameter of the axon, the faster the conduction. Among the largest known axons are those of the squid, and squid axons may conduct action potentials at velocities of about 35 meters per second. In smaller, unmyelinated axons of a human, CV is on the order of 1 m/s. Although vertebrates have axons with smaller diameters than the squid, many vertebrate axons have much faster conduction velocities. This is because of myelin. Glial cells (**oligodendrocytes** in the CNS and **Schwann cells** in the PNS) wrap around axons to form the myelin sheath. Each wrapping consists of multiple lipid bilayers. Gaps between regions of the myelin sheath are called neurofibril nodes (nodes of Ranvier), and voltage-gated channels are found at the nodes.

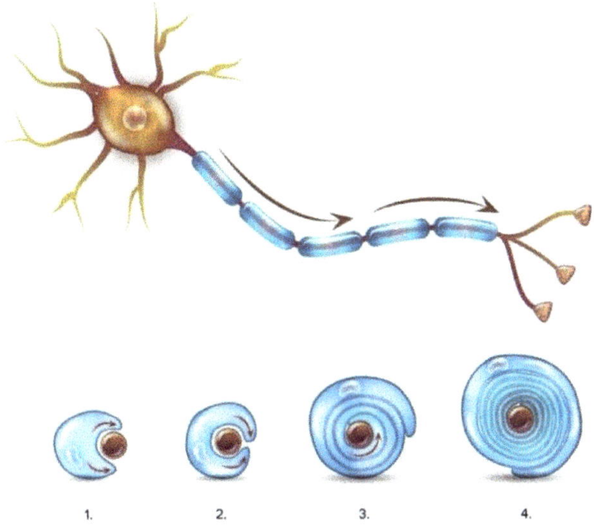

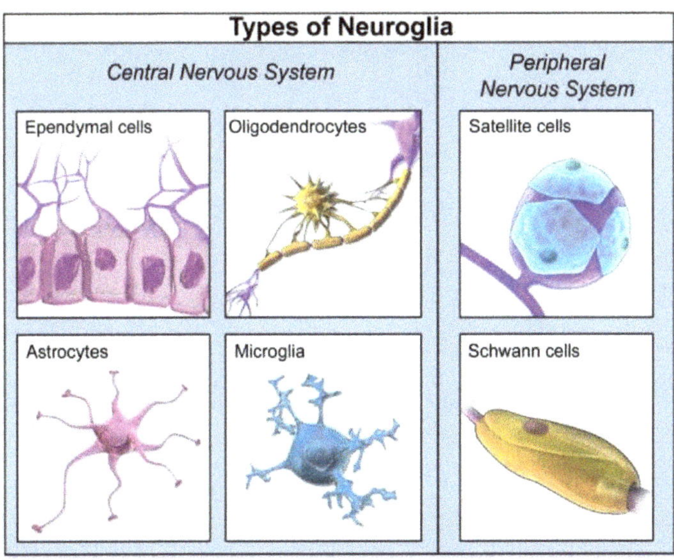

To keep things reasonably simple, just be aware that the presence of myelin allows positive current (Na^+) to travel down the axon at a faster velocity. However, regardless of how fast it travels, the current that enters the axon at the axon hillock cannot make it all the way down the axon. Thus, nodes are required to periodically allow more sodium to enter the axon through voltage-gated channels, ensuring that the signal gets all the way to the axon terminal. Since the action potential only appears at the nodes, it is as if the action potential is jumping from node to node. This process is called **saltatory conduction**; "saltere" means "jump." The largest myelinated axons in a human have conduction velocities on the order of 150 m/s (meters per second).

Transmissive segment

Neurons make connections with other neurons or their effectors at junctions called **synapses**. Most synapses are **chemical synapses** in which the signal is transferred from the presynaptic cell to the postsynaptic cell as a chemical signal. These chemical signals are called **neurotransmitters**.

Recall that telodendria of an axon typically end in knob-shaped structures called synaptic knobs. Each axon terminal contains vesicles called **synaptic vesicles**, which are filled with neurotransmitters. On the outside of the knob, there is a high concentration of Ca^{++}.

When an action potential reaches the axon terminal, the following processes occur to send a signal across the synapse:

1. The change in voltage to threshold opens **voltage-gated Ca^{++} channels** in the membrane of the synaptic knob. This allows calcium into the knob.

2. The vesicles inside the knob fuse to the membrane, and the neurotransmitters are dumped into the **synaptic cleft** by the process of exocytosis.

3. Neurotransmitters bind to receptors on the receptive segment of the postsynaptic cell.

4. Chemical-gated channels on the postsynaptic cell open in response to the neurotransmitter binding. This causes a PSP on the postsynaptic cell, and we're back where we started!

Chemical-gated channels tend to remain open as long as the chemical is bound to the receptor. The neurotransmitters must be removed from the synaptic cleft for the signal at the synapse to end. There are three basic processes by which this can happen:

1. Neurotransmitters can be broken down by enzymes. For example, acetylcholine in the synaptic cleft is broken down by the enzyme acetylcholinesterase.
2. Neurotransmitters can be removed from the synaptic cleft by the presynaptic cell or by astrocytes.
3. Neurotransmitters can diffuse out of the synaptic cleft into the interstitial space.

The reflex arc with interneuron depicted in the diagram below

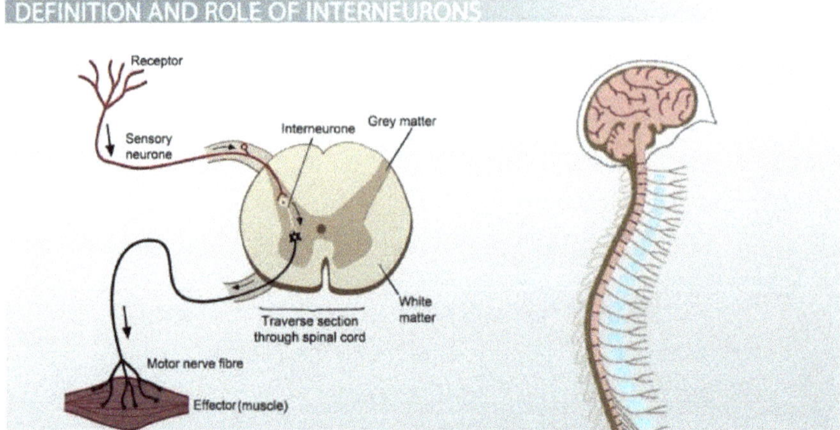

Position of spinal cord in the veterbral foramen

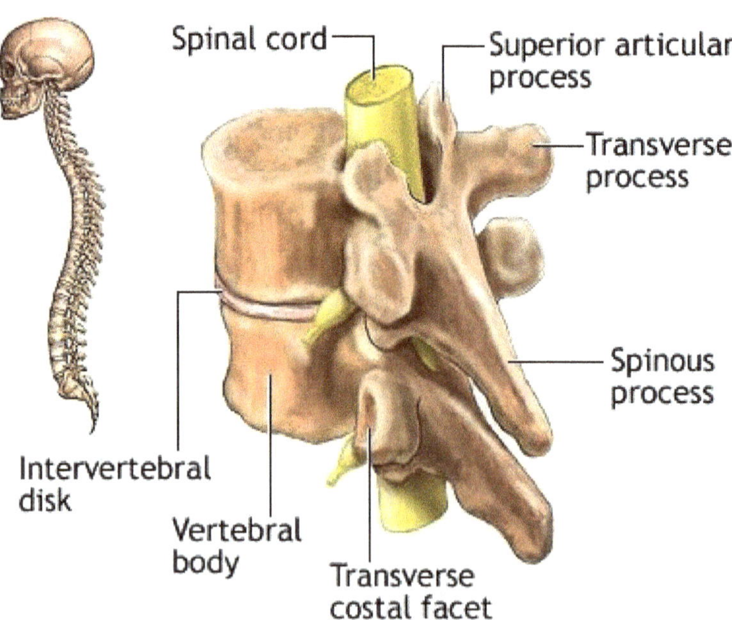

Diagram of transected spinal cord. Practice drawing this diagram.

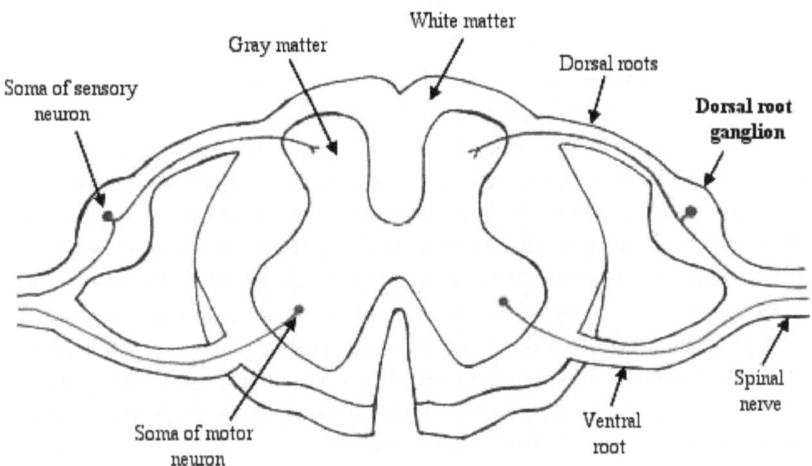

Why does the dorsal root ganglion look different from the ventral root? (Hint: this area contains a large number of sensory neurons and specific structure labeled in the diagram)

Chapter 8: The Brain and Cranial Nerves

I. The Brain

Regions and organization

The brain can be divided into four major regions:

A. The **cerebral hemispheres** (which make up the **cerebrum**) form the largest and most readily visible part of the brain. Conscious thoughts, sensations, intellect, memory, and complex movements originate in the cerebrum.

B. The **diencephalon** lies below the cerebrum and forms critical connections between the cerebrum and the brain stem. It contains the **thalamus** and the **hypothalamus**.

C. The **brain stem** is the "lower" portion of the brain and connects the "upper" portions to the spinal cord. It consists of the **midbrain**, the **pons**, and the **medulla oblongata**.

D. The **cerebellum** lies inferior to the posterior portion of the cerebrum. It coordinates the body's movements based on sensory information it receives from the body.

Note the components of the brainstem below.

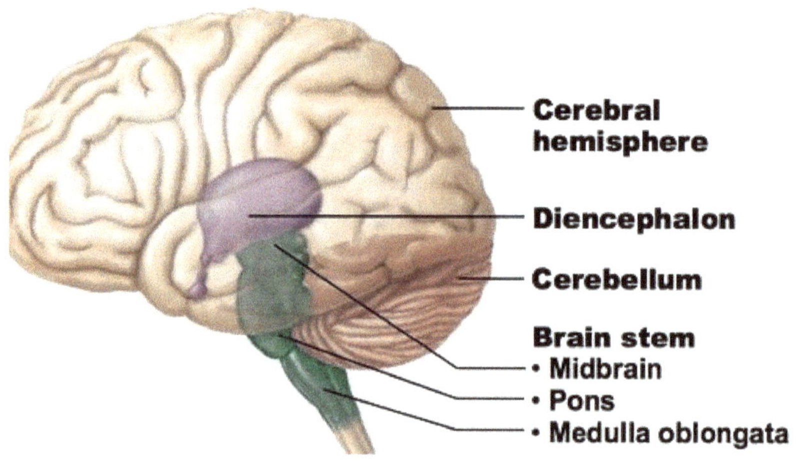

Ventricles

The brain contains cavities called **ventricles**. Each cerebral hemisphere contains a large **lateral ventricle**. The **third ventricle** lies within the diencephalon, and the **fourth ventricle** lies between the pons and the cerebellum.

The ventricles are connected to each other, and the fourth ventricle is continuous with the central canal of the spinal cord. Like the spinal cord's central canal, the brain's ventricles are filled with cerebrospinal fluid (CSF). The CSF circulates from the ventricles and central canal into the subarachnoid space of the surrounding meninges via openings at the fourth ventricle.

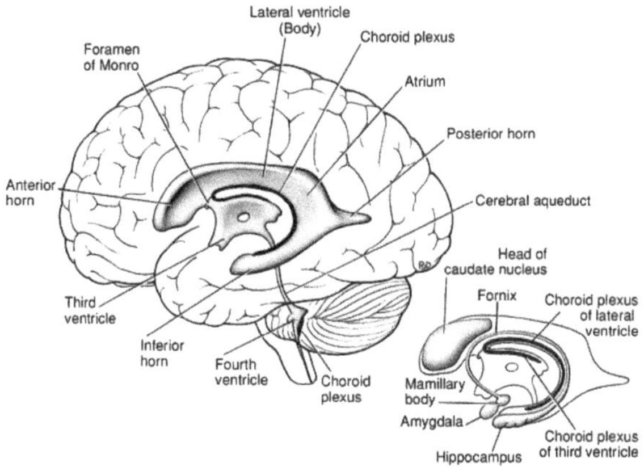

Midsagittal section shows ventricles and other important structures in the diagram above. Frontal view below.

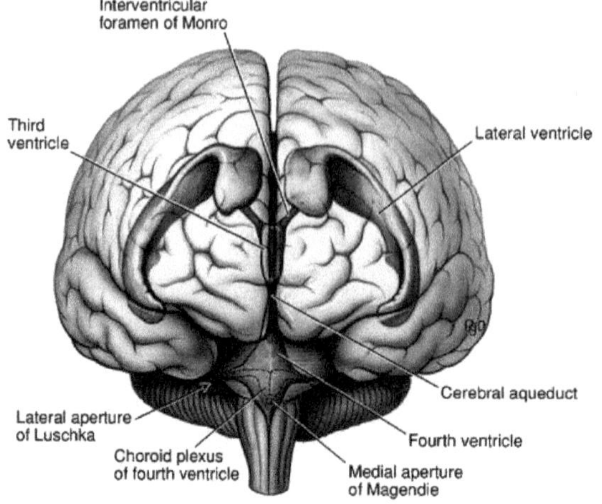

Frontolateral view of ventricles in diagram below

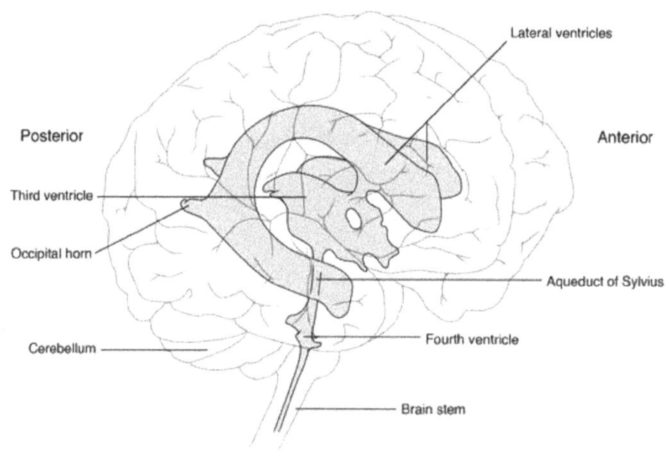

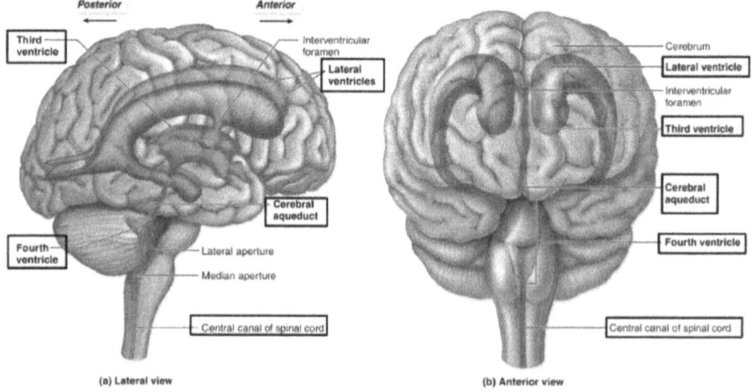

(a) Lateral view

(b) Anterior view

Cerebral hemispheres

The surface of the cerebrum is called the **cerebral cortex**, and it is composed of 2-4 mm of gray matter with billions of neuron bodies. Underneath the cortex lies the **cerebral white matter**. The cerebral cortex has many folds, called **gyri**, which increase its surface area. The grooves between the gyri are called **sulci**. Particularly deep grooves are called **fissures**. One major brain landmark is the **longitudinal fissure**, which divides the cortex into left and right hemispheres.

Superior view of brain

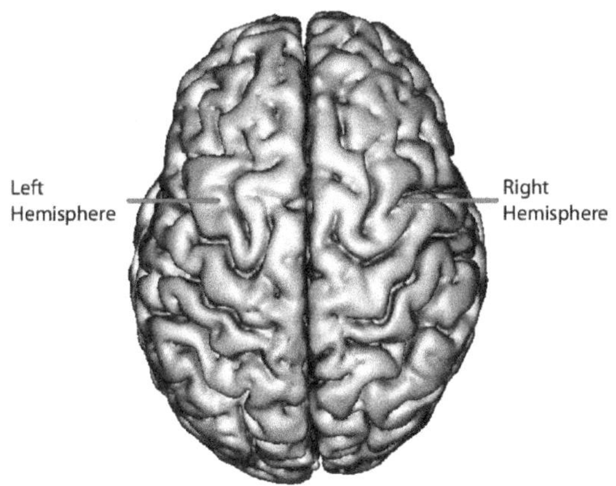

Midsagittal section of brain below displaying the inner structures below

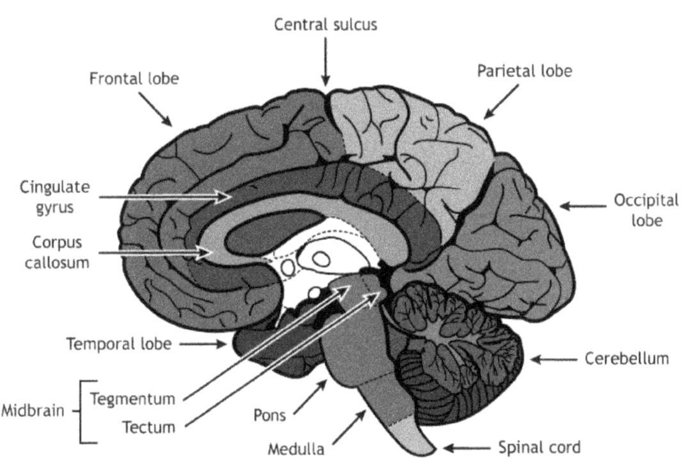

Each hemisphere can be divided into five major regions:

A. The **frontal lobe** is anterior.

B. The **parietal lobe** is dorsal and in the middle.

C. The **temporal lobe** is lateral and in the middle.

D. The **occipital lobe** is posterior.

E. The **insula** lies deep near the intersection of the frontal, temporal, and parietal lobes.

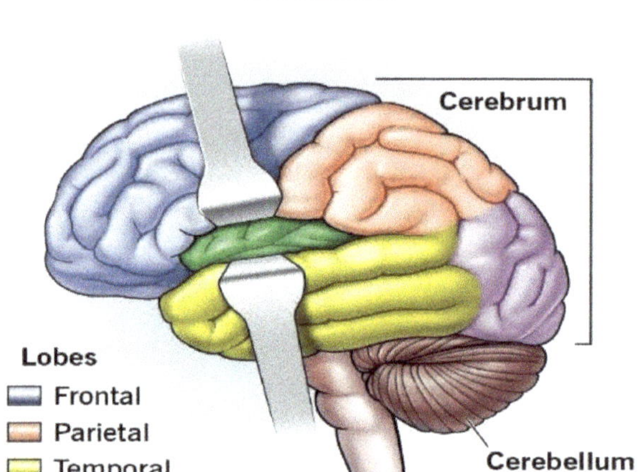

The **central sulcus** forms a border between the frontal and parietal lobes, the **parieto-occipital sulcus** forms a border between the parietal and occipital lobes, and the **lateral sulcus** forms a border between the frontal and temporal lobes.

Sulcus(*i*) and Gyrus(*i*) of the brain [sulcus=singular; sulci=plural, likewise for gyrus/gyri]

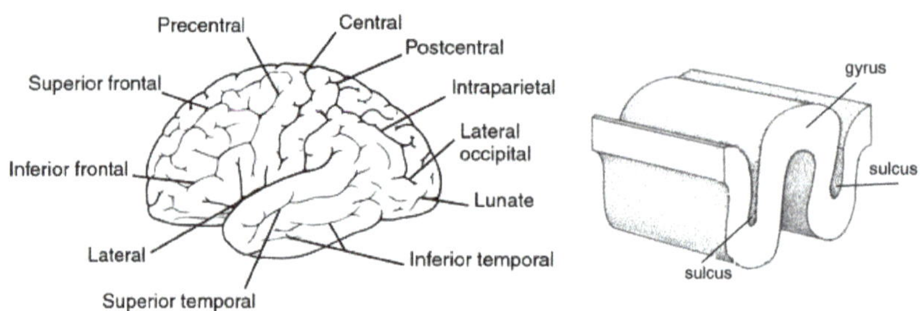

The central sulcus separates the motor and sensory areas of the cortex. The **precentral gyrus** is located immediately anterior to the central sulcus. This is the primary motor-generating area of the cortex. The movements of muscles associated with speech are generated in **Broca's area**. A portion of the brain, known as **Wernicke's area**, is also involved in speaking and understanding language. (See images below).

The **postcentral gyrus** is located immediately posterior to the central sulcus. This is the primary somatosensory processing area of the cortex. It receives information from receptors for touch, pressure, pain, vibration, and temperature. The sensations of sight, sound, smell, and taste are processed in other parts of the cortex, specifically, the **visual cortex** (occipital lobe), **primary auditory cortex** (temporal lobe), **olfactory cortex** (temporal lobe), and the **gustatory cortex** (insula), respectively.

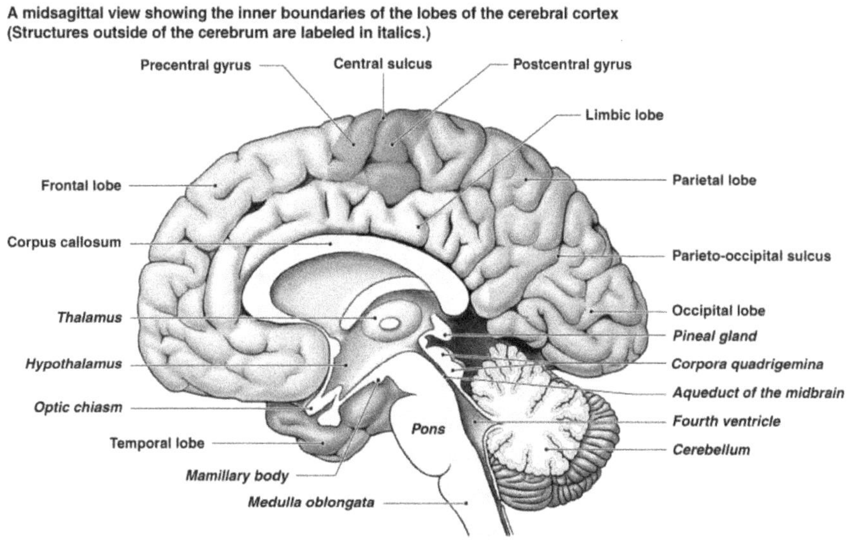

Functional Areas of the Cerebral Cortex

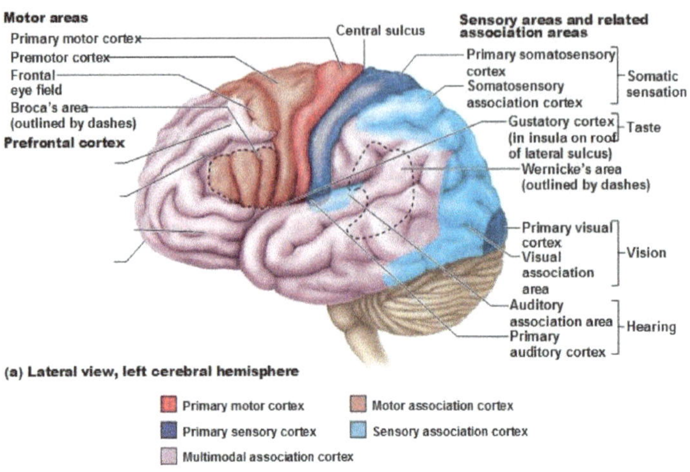

(a) Lateral view, left cerebral hemisphere

The **prefrontal cortex** (frontal lobe) is a very complicated region of the cerebrum. It is involved with intellect, cognition, and personality.

Cerebral white matter consists primarily of myelinated axons, which generally fall into three categories (see illustration below):

A. **Association fibers** connect areas of the cortex within a single hemisphere

B. **Commissural fibers** connect the two hemispheres. The **corpus callosum** is the largest structure containing commissural fibers.

C. **Projection fibers** link the cerebrum to other parts of the brain.

Deep to the cerebral white matter on each side of the brain are areas of the cerebrum called the **basal nuclei**. Although the functions of the basal nuclei are very complicated and poorly understood, they are clearly important in starting, stopping, and monitoring the intensity of voluntary movements. Parkinson's disease appears to affect the basal nuclei. Can you explain how symptoms of Parkinson's are related to the functions of the basal nuclei?

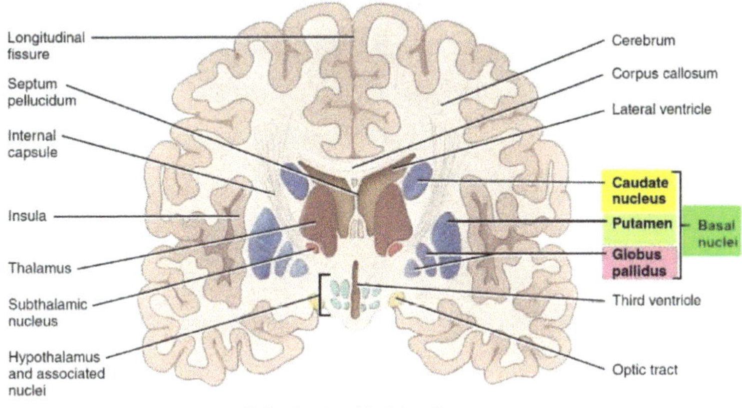

(b) Anterior view of frontal section

Diagram below showing prefrontal section of cerebrum composed of gray matter (unmyelinated neurons) for integration and white matter made up of myelinated neurons for signal propagation to different areas of brain.

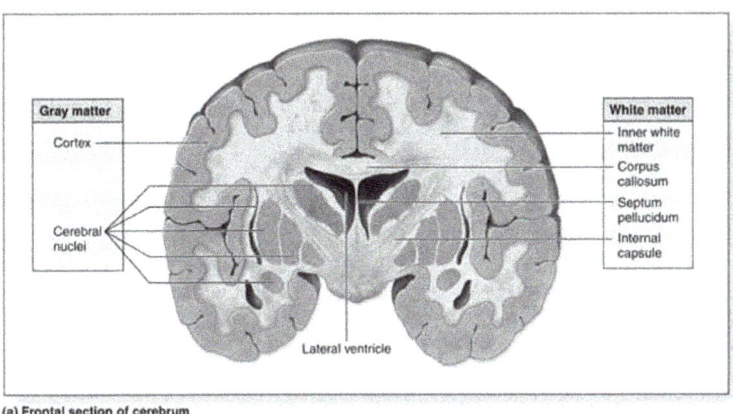

(a) Frontal section of cerebrum

Diencephalon

The diencephalon is important for integrating conscious and unconscious sensory information and motor commands.

The **thalamus** consists of two egg-shaped regions of the brain. It is the principal relay station for directing sensory information from the spinal cord, medulla, and cerebellum to the cortex. The thalamus processes and relays auditory, visual, taste, and somatic sensory information.

The **hypothalamus** is a small region situated below the thalamus. It has many functions, including control and integration of various autonomic functions. Some specific functions of the hypothalamus are as follows:

Process and relay olfactory information

Release various hormones, including oxytocin and antidiuretic hormone (ADH)

Integration of the autonomic nervous system

Control over heart rate

Control of digestive tract activity

Rage and aggression

Regulation of body temperature

Hunger and satiety centers

Thirst

Sleep patterns

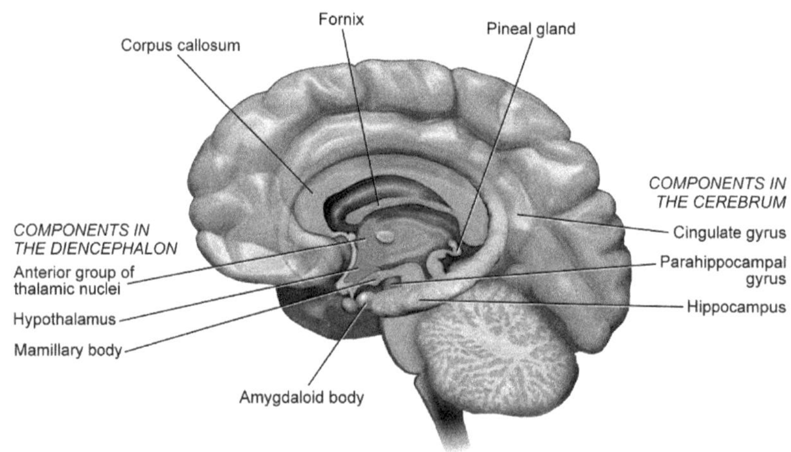

Brain stem

The brain stem consists of three specific regions: the midbrain, the pons, and the medulla oblongata. All regions of the brain stem contain fibers that connect the higher portions of the brain (cerebrum and diencephalon) to the cerebellum and spinal cord. The **cerebral aqueduct**, which connects the third ventricle to the fourth ventricle, passes through the brain stem.

The **midbrain** is located inferior to the diencephalon. It has reflex centers that control eye, head, and neck movements in response to visual and auditory stimuli. On the ventral side of the midbrain are the paired cerebral peduncles. They contain motor fibers that extend from the cortex to the lower parts of the CNS. The midbrain also contains the spinothalamic sensory tracts, which come from the spine and lead to the thalamus. The superior cerebellar peduncles connect the midbrain to the cerebellum.

The **pons** is the swelling of the brain stem between the medulla and the midbrain. It contains longitudinal fibers that connect the medulla with the higher parts of the brain. Transverse fibers link the cerebellar hemispheres with the midbrain. Respiratory centers that help control breathing are located in the pons.

The **medulla oblongata** is the part of the brain that links the brain to the spinal cord. There is not a clear anatomical distinction between the medulla and the spinal cord. Still, the foramen magnum marks the division between what is considered the brain and what is considered the cord. The medulla contains ascending and descending tracts that run between the brain and the cord. On the ventral side of the medulla are two prominent bulges called the **pyramids**. They contain the largest motor nerve tracts. Just superior to the spinal cord, these tracts cross from right to left and vice versa. This crossing over of the motor tracts is known as **the decussation of the pyramids**. This is why motor signals that control the left side of the body originate from the right side of the cortex. Likewise, sensory information crosses from one side of the body to the other on the dorsal side of the medulla oblongata. The medulla has a cardiovascular center, which regulates cardiac output (heart rate and stroke volume) and the diameter of blood vessels. There is also a respiratory center, which regulates breathing rate.

The diencephalon and brainstem (anterior view)

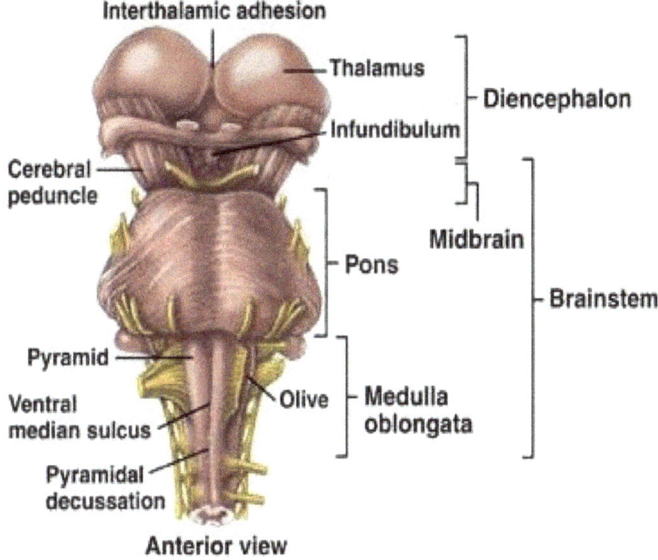

Posterior view of diencephalon and brainstem

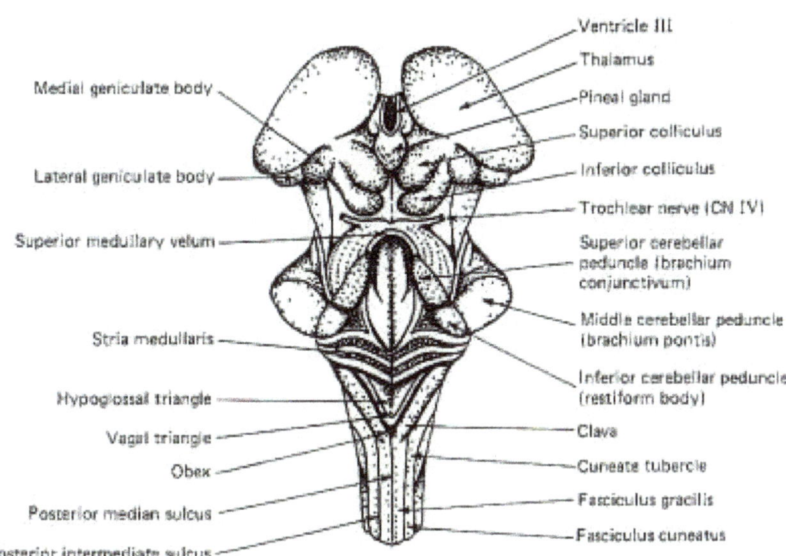

Lateral view of diencephalon and brainstem below

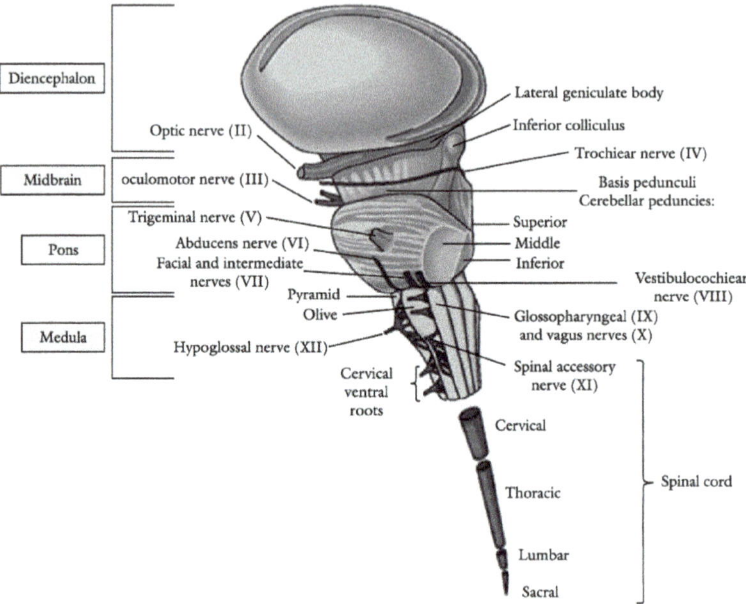

Cerebellum

The cerebellum has two primary functions: (1) adjusting body posture to maintain balance and (2) fine-tuning movement. When viewed dorsally, the cerebellum is shaped somewhat like a butterfly (see illustration below). The portion along the midline is called the **vermis**, and the wings are the **cerebellar hemispheres**.

Cerebellum

- two **cerebellar hemispheres**
- anterior lobe and posterior lobe separated by **primary fissure**
- **vermis** is narrow band of cortex
  - separates hemispheres
  - helps maintain balance

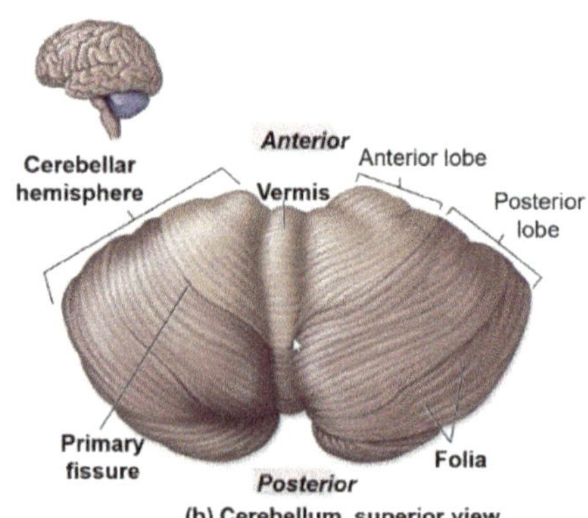

(b) Cerebellum, superior view

Anatomy of spinal cord: a) cross-section of pinal cord in vertebral foramen; b) anterior view of spinal cord

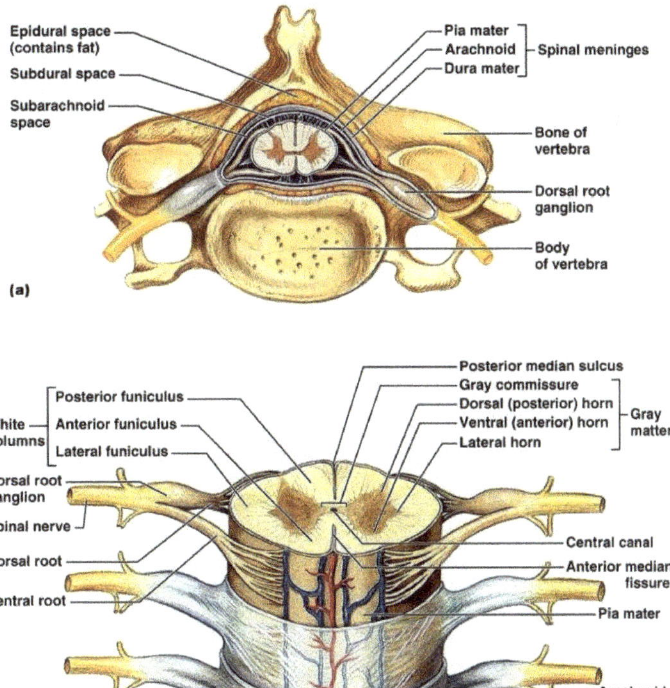

II. Protection and support of the Brain

Meninges

The **cranial meninges** are similar to the spinal meninges. In fact, they are contiguous with the spinal meninges.

A. The **dura** (tough) **mater** (mother) is the outermost and toughest layer. There are two major differences between the dura of the brain and the dura of the cord: (1) The dura of the brain consists of two layers; the **periosteal layer** (also known as *endosteal*) is attached directly to the inner surface of the skull, and the **meningeal layer** is deep to the periosteal layer. In places, the two layers are fused together; in other places, dural sinuses lie between the two layers. (2) There is no epidural space associated with the brain.

B. The **arachnoid** is the middle layer (so named because of its spider web-like appearance). A subarachnoid space filled with CSF lies between the arachnoid and the pia mater. Note that the arachnoid is smooth, and it does not follow the folds of the cerebrum.

C. The **pia** (tender) **mater** is the innermost layer, and it is tightly associated with the surface of the brain.

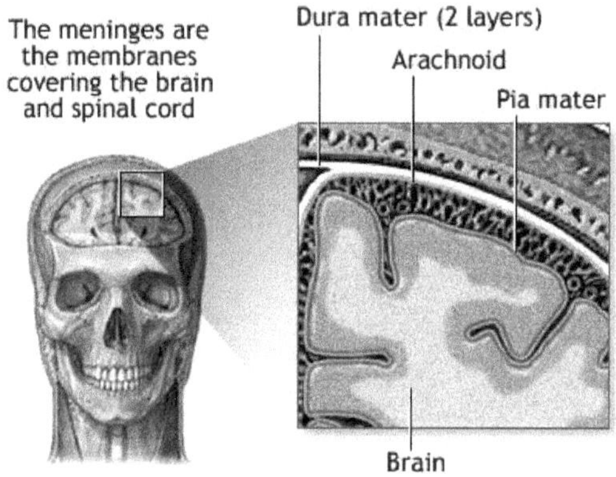

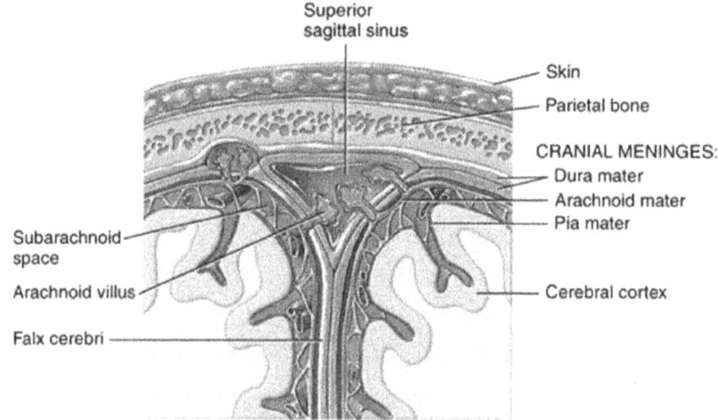

Cerebrospinal fluid

Cerebrospinal fluid (CSF) forms a liquid cushion around the brain and spinal cord. It is also found in the ventricles of the brain and central canal of the spinal cord. CSF is similar in composition to blood plasma but with less protein and different concentrations of various ions.

CSF is produced from blood in structures called the **choroid plexuses** (network of capillaries – see illustration below on left hand side), which are located in each of the ventricles. The choroid plexuses contain specialized **ependymal cells**, which secrete the CSF, remove waste products from the CSF, and regulate its composition. Structures called the **arachnoid villi** absorb CSF back into the bloodstream.

CSF is produced at a rate of about 500 ml per day, although the total volume at any one time is just 150 ml. Thus, there is a rapid turnover of CSF. **Hydrocephalus** (see CT scan of brain below on right hand side) may result in infants with problems with reabsorption of CSF.

Cerebrospinal Fluid (CSF) – Choroid plexus

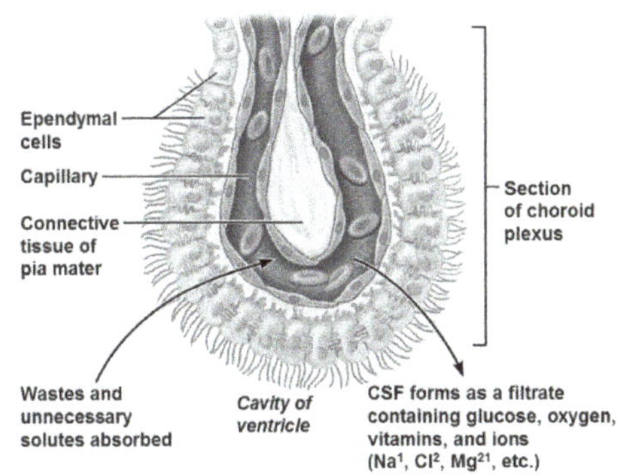

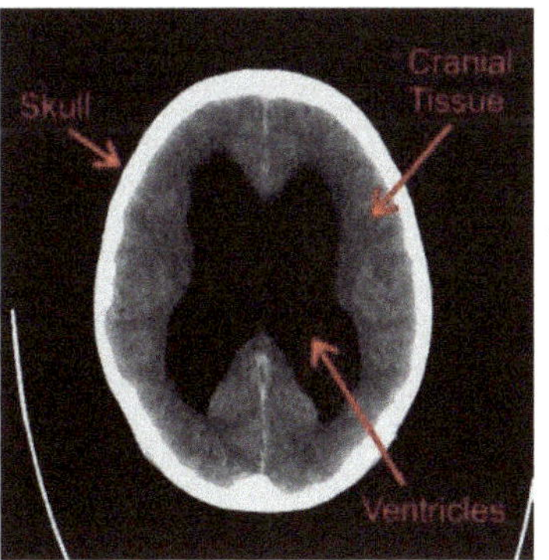

Blood-brain barrier

Unlike capillaries in most other parts of the body, capillaries within the CNS are lined with epithelial cells that are tightly connected to prevent movement of materials between the blood and the interstitial spaces. These epithelial cells form what is called the **blood-brain barrier** (see the illustrations below). Although lipid-soluble molecules (*e.g.,* oxygen, CO_2, lipids, and small alcohols) are able to diffuse freely between the CNS and the blood, the movement of other molecules is restricted.

Structures of the blood-brain barrier

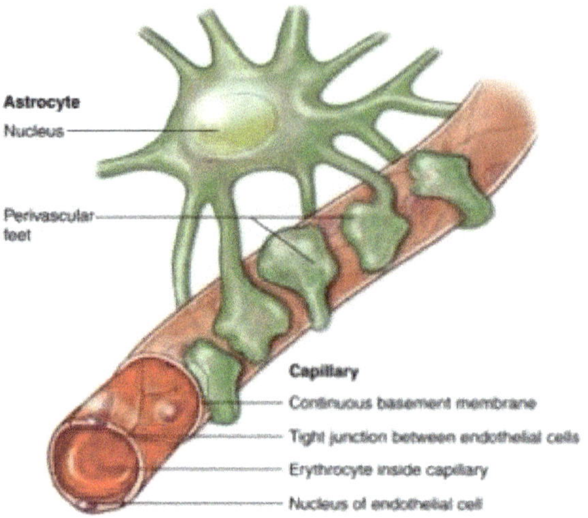

Use the illustration below to identify the numbered structures of the brain tissue and the blood-brain-barrier.

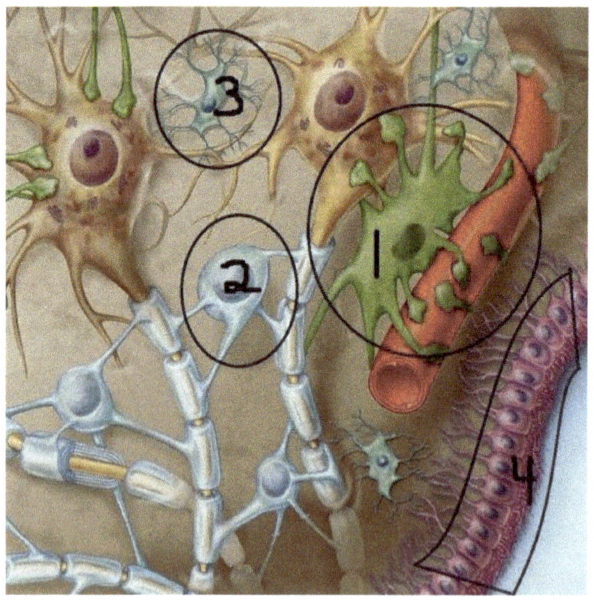

Diagram of the blood supply to the brain below (inferior view of the brain)

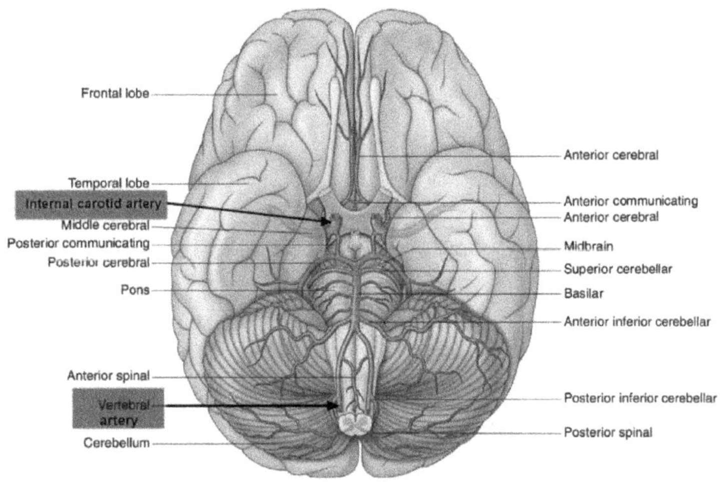

III. Cranial nerves

The cranial nerves originate from the brain, and they exit the skull through various foramina. Each cranial nerve may carry sensory information, motor information, or a mix of the two. Individual cranial nerves are numbered, beginning with the most anterior nerve.

Learn the names (in order) of the twelve cranial nerves found in humans, learn at least one function of each, and know whether it is sensory, motor, or mixed.

I. Olfactory—smell (sensory)
II. Optic—vision (sensory)
III. Oculomotor—movement of the eye (motor)
IV. Trochlear—movement of the eye (motor)
V. Trigeminal—sensory and motor to the face (sensory and motor)
VI. Abducens—movement of the eye (motor)
VII. Facial—sensory and motor to the face (sensory and motor)
VIII. Vestibulocochlear—balance and hearing (sensory)
IX. Glossopharyngeal—sensory and motor to head and neck (sensory and motor)
X. Vagus—sensory information from the esophagus, respiratory tract, visceral organs, motor commands to heart, glands, stomach, and more (sensory and motor)

XI. Accessory—movement of neck and upper back (motor)
XII. Hypoglossal—the movement of the tongue (motor)

Use the following mnemonic to remember the twelve cranial nerves and their associated function.

- CN I........**O**......Olfactory.................Sensory - **Some**
- CN II........**O**......Optic.....................Sensory - **Say**
- CN III......**O**......Oculomotor............Motor - **Marry**
- CN IV......**To**......Trochlear........,,.....Motor = **Money**
- CN V.......**Touch**...Trigeminal...Both...(mixed) - **But**
- CN VI ..... **And**......Abducens............Motor - **My**
- CN VII.....**Feel**......Facial...................Both - **Brother**
- CN VIII....**Very**....Vestibulocochlear....Sensory - **Says**
- CN IX......**Green**...Glossopharyngeal.......Both- **Big**
- CN X.......**Vegetables**...Vagus...............Both - **Brains**
- CN XI......**A**,,,,,,,,,,,,,,,,,,,,,Accessory........Motor - **Matter**
- CN XII.....**H**............Hypoglossal........Motor - **Most**

Inferior view of brain and cranial nerves

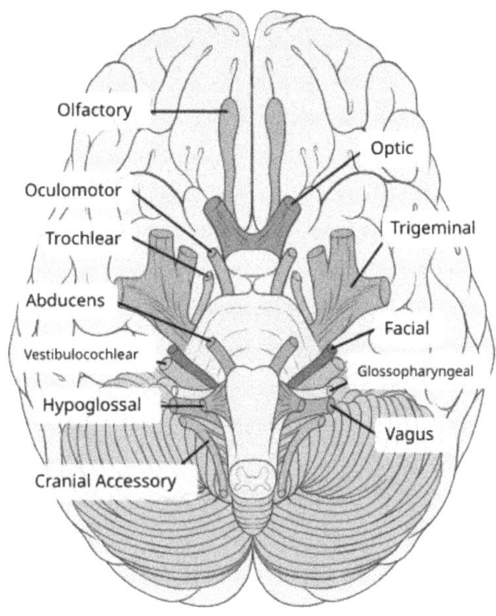

Inferior view of the brain with focus on the optic nerve and its structures

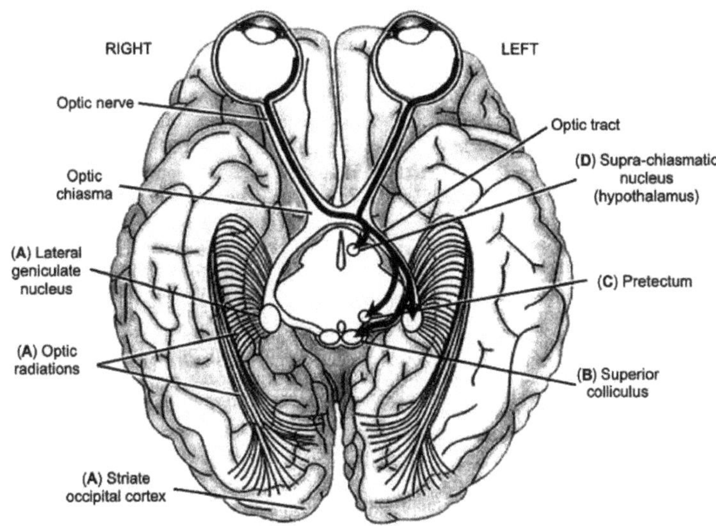

Defects in visual fields: the diagram below shows the visual defects that would occur if the optic chiasm (1) and the right optic tract (2) were severed. Damaged optic chiasm would result in loss of peripheral vision; and the damaged right optic tract would cause left visual field impairment.

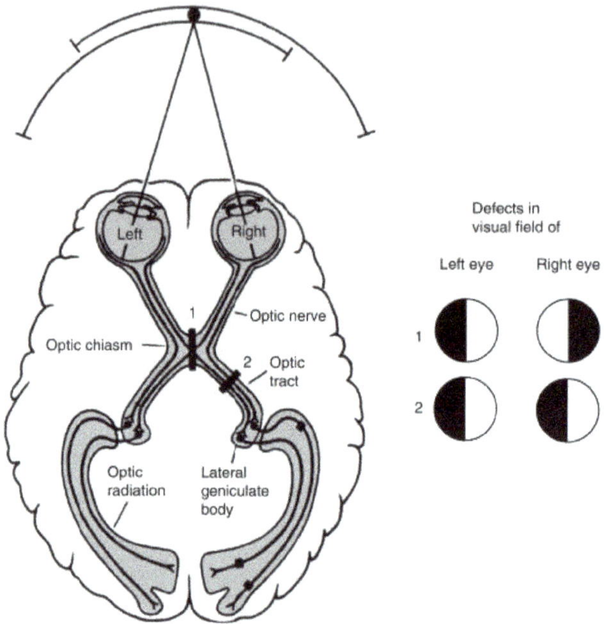

Diagram below showing the 3 branches (V1, V2, and V3) of the trigeminal nerve (CN V) and the corresponding area each branch innervates.

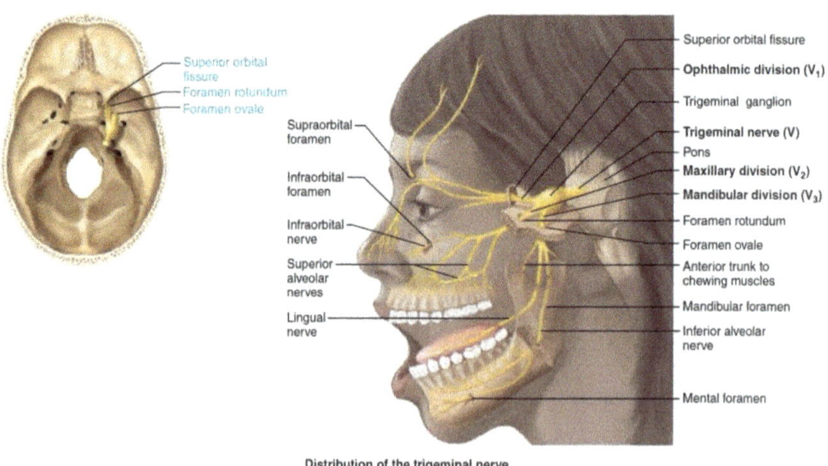

Distribution of the trigeminal nerve

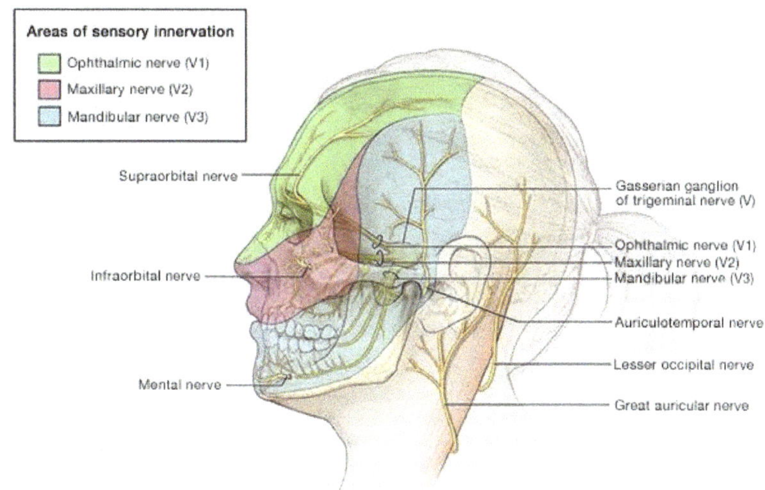

Diagram summarizing areas of control by the 12 cranial nerves

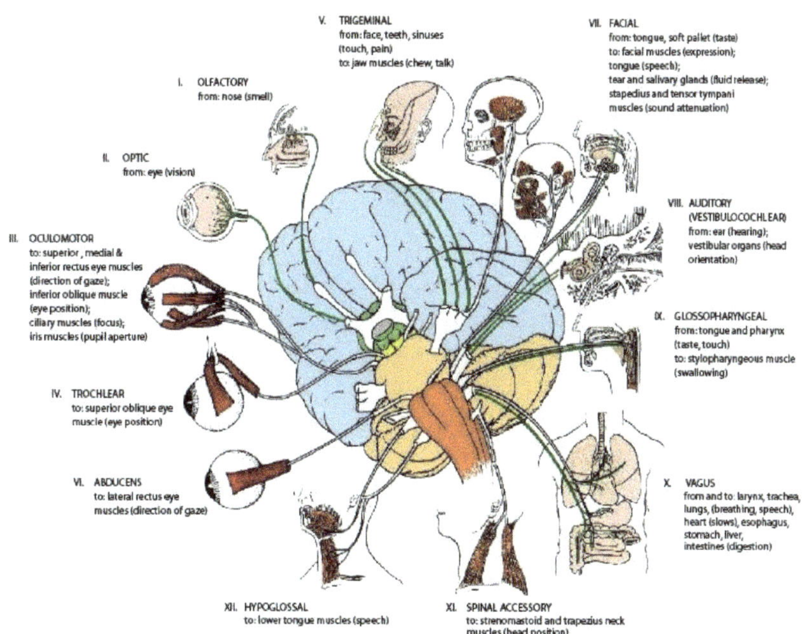

Chapter 9: Nervous System: Autonomic Nervous System

I. Comparison of the Somatic and Autonomic Nervous Systems

The somatic nervous system is fairly straightforward. It consists of **somatic motor neurons** with bodies located in the brain or spinal cord and axons that extend through cranial or spinal nerves. The axons of somatic motor neurons synapse with skeletal muscles. All somatic motor neurons release the neurotransmitter **acetylcholine (ACh)** from their synaptic knobs. ACh is always excitatory at synapses with skeletal muscle fibers; a synapse between a motor neuron and a muscle fiber is called a neuromuscular junction. Thus, whenever the CNS decides to activate a skeletal muscle, it simply sends commands through somatic motor neurons, which stimulate the contraction of the particular skeletal muscle.

Somatic motor neurons can be activated consciously. However, they can also be activated unconsciously to maintain posture, breathe, carry out a reflex, etc.

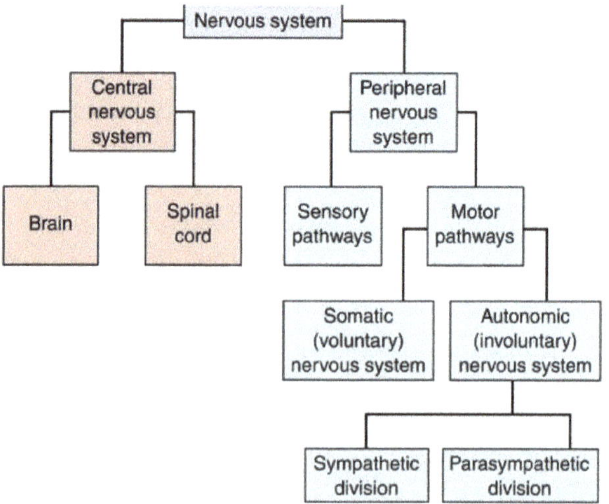

The **autonomic nervous system (ANS)** consists of **visceral motor neurons**, which carry motor commands to glands, smooth muscle, and the heart. In so doing, the ANS is involved in maintaining blood pressure, water and salt balance, dissolved gas concentration, digestion, and other functions. However, the ANS does not *control* any of these functions; control is accomplished by various parts of the CNS, such as the hypothalamus.

In the somatic nervous system, motor neuron bodies originate in the CNS, and their axons extend all the way to their effectors. In contrast, the ANS uses pairs of visceral motor neurons (see illustration below). The body of the first visceral motor neuron lies in the CNS, and the body of the second visceral motor neuron lies in an **autonomic ganglion** (*e.g.,* the sympathetic trunk ganglia).

Visceral motor neurons that originate in the CNS are called **preganglionic neurons**. Axons of preganglionic neurons, called **preganglionic axons** (or fibers), synapse with **ganglionic neurons** in the autonomic ganglia. Axons of the ganglionic neurons, which are called **postganglionic axons** (or fibers), extend to effectors. Preganglionic axons are thin and lightly myelinated, and postganglionic axons are generally thinner and unmyelinated.

How do the conduction velocities of these axons compare with the conduction velocities of somatic motor neuron axons?

All preganglionic axons of the ANS release ACh. The effect of ACh on the ganglionic neuron is always excitatory.

Postganglionic axons of the parasympathetic division of the ANS also release ACh, which may be either stimulatory or inhibitory, depending upon the type of ACh receptor present on the effector.

Postganglionic axons of the sympathetic division of the ANS generally release **norepinephrine (NE)**, which may also be either stimulatory or inhibitory depending upon the type of receptor present on the effector. Be aware that some sympathetic postganglionic axons release ACh.

The ANS has pre- and postganglionic neurons

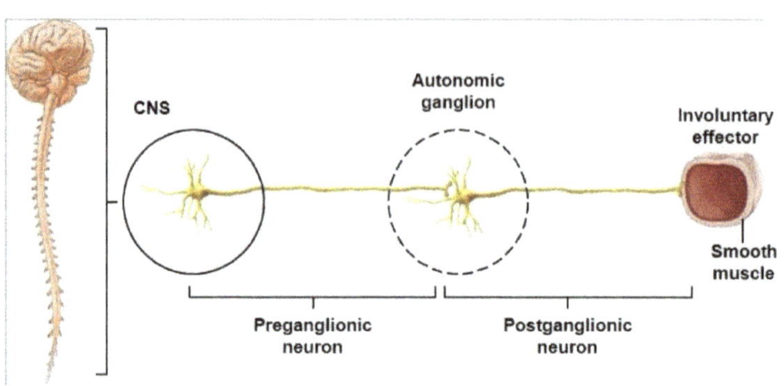

Neurotransmitters and receptors of the preganglionic and postganglionic neurons below

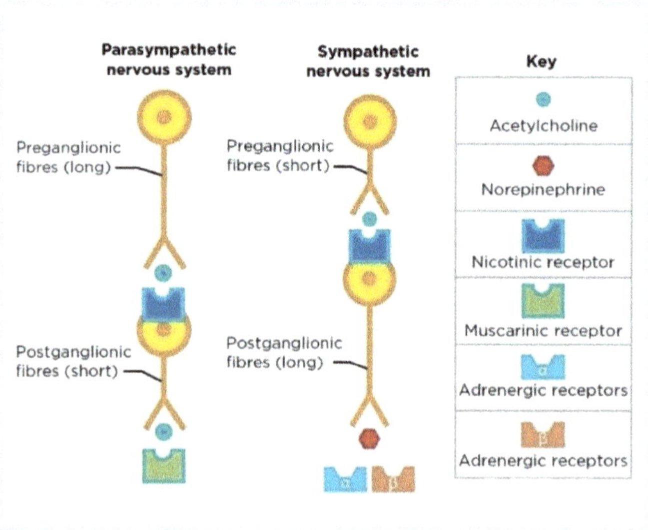

II. Divisions of the Autonomic Nervous System

Functional differences

The autonomic nervous system is divided into the **sympathetic** and **parasympathetic** divisions. The two divisions often innervate the same effectors but generally have opposing effects.

The parasympathetic division maintains the body during "normal," non-stressful situations. Some texts may refer to it as the "resting and digesting system." It reduces the activities of some organs (*e.g.*, the heart) and generally elevates the activity of the digestive system.

The sympathetic division generally prepares the body for action in what is often called the "fight-or-flight" response. The activities of various effectors (*e.g.*, sweat glands and the heart) increase, and the digestive system's activity is reduced. Although specific sympathetic effectors can be activated independently, in a crisis, the entire division is generally activated.

Anatomic differences

Although the sympathetic and parasympathetic divisions share many effectors and work together to maintain homeostasis, their anatomical pathways are separate: Their motor pathways originate in different regions of the CNS and use different ganglia.

Parasympathetic axons originate from cranial nerves III, VII, IX, X, and sacral spinal nerves. The term **craniosacral control** is often used to refer to the parasympathetic division. Ganglia used by the parasympathetic division are located in or near the target organs, and preganglionic axons travel more or less directly to the ganglia. Therefore, the preganglionic axons are long, and the postganglionic axons are short.

Sympathetic axons originate from the thoracic and lumbar regions. The term **thoracolumbar control** is often used in reference to the sympathetic division. The bodies of these neurons are located in the lateral horns of the spinal cord. Ganglia used by the sympathetic division are located near the spinal cord. Therefore, the preganglionic axons are generally short, and the postganglionic axons are usually long (see above diagram).

Parasympathetic (Craniosacral region) vs. Sympathetic (Thoracolumbar region). Note the mechanism of action of these regions on the organs.

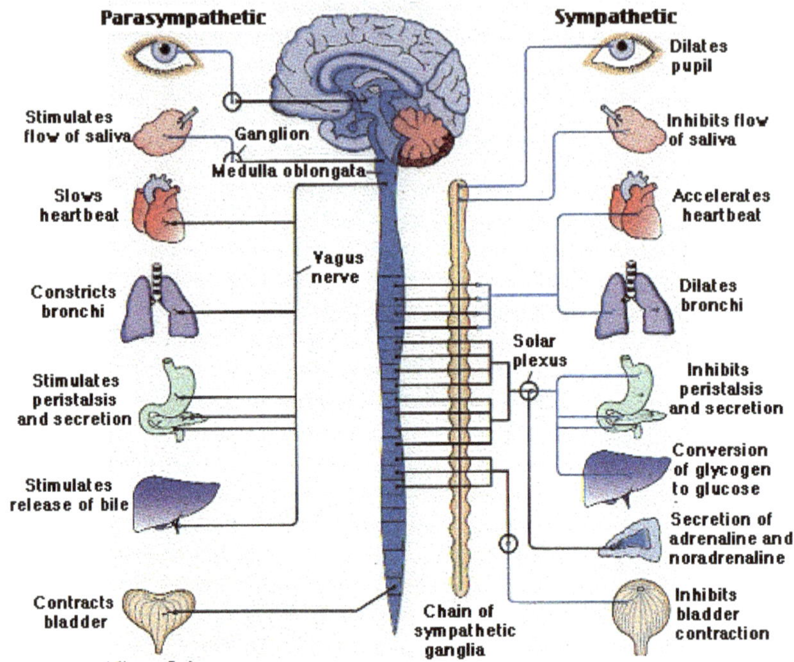

III. Parasympathetic Division

This section reinforces the concept of craniosacral control. Although you do not need to memorize specific pathways, you should understand the general point illustrated above. Notice that the vagus nerve carries parasympathetic motor commands to most organs in the ventral body cavity.

IV. Sympathetic Division

This section reinforces the concept of thoracolumbar control. Although you do not need to memorize specific pathways, you should understand the general point illustrated above. Notice that all preganglionic axons of the sympathetic enter the **sympathetic trunk**.

Many axons form synapses with ganglionic neurons in the sympathetic trunk ganglia. Some preganglionic axons pass through the sympathetic trunk via nerves called **splanchnic nerves** to make synapses with ganglionic neurons in other ganglia, such as the **prevertebral ganglia**. Preganglionic axons that send sympathetic commands to the **adrenal medulla** travel directly to the adrenal medulla. These preganglionic axons do not synapse with ganglionic neurons. Instead, they stimulate the adrenal gland directly, releasing both norepinephrine and epinephrine into the blood. What is the significance of releasing these chemicals into the blood?

V. Comparison of Neurotransmitters and Receptors of the Two Divisions

Refer to above illustrations.

Neurons that release acetylcholine are called **cholinergic neurons**. Somatic motor neurons, all preganglionic visceral motor neurons, and parasympathetic postganglionic neurons are all cholinergic. The sympathetic postganglionic neurons that innervate sweat glands are also cholinergic. Cholinergic neurons can affect their targets differently because there are different receptors for acetylcholine. **Nicotinic receptors** are found on the motor end plates of skeletal muscle fibers, all ganglionic neurons, and hormone-producing cells of the adrenal medulla. Nicotinic receptors are always excitatory, depolarizing the postsynaptic cell. **Muscarinic receptors** are found on all effectors targeted by parasympathetic postganglionic neurons and on sweat glands. Muscarinic receptors are generally excitatory, but there are exceptions; muscarinic receptors on cardiac muscle are inhibitory.

Neurons that release norepinephrine are called **adrenergic neurons**. Most sympathetic postganglionic neurons are adrenergic. Adrenergic neurons can affect their targets differently because there are different receptors for norepinephrine. **Alpha receptors** are found in blood vessels that serve the skin, abdominal organs, kidneys, and salivary glands. The binding of NE to these alpha receptors causes vasoconstriction. **Beta receptors** are found in the bronchioles of the lungs and blood vessels that serve the heart and skeletal muscles. The binding of NE to these beta receptors causes dilation. Beta receptors on cardiac muscle cause increases in heart rate and strength of contraction. Beta receptors associated with smooth muscle of the digestive tract are inhibitory.

The sympathetic division of the ANS tends to have widespread effects throughout the body. One reason for this is the stimulation of the adrenal medulla by the sympathetic. When stimulated, the adrenal medulla releases epinephrine and norepinephrine into the bloodstream. Thus, these two chemicals travel throughout the body. Why does a widespread sympathetic response throughout the body make sense?

VI. Interactions Between the Parasympathetic and Sympathetic Divisions

As mentioned previously, the parasympathetic and sympathetic divisions often innervate the same effectors, but they generally have opposing effects. This system by which the two divisions work together by counteracting each other is referred to as **dual innervation**. For example, the sympathetic division dilates bronchioles in the lungs, and the parasympathetic division constricts bronchioles. However, some effectors of the ANS are innervated only by the sympathetic system (*e.g.,* the arrector pili muscles, sweat glands, liver, and kidneys).

In addition to its effects on specific organs (*e.g.,* heart, lungs, sweat glands), the sympathetic division affects metabolism throughout the body. For example, sympathetic stimulation causes (1) increased metabolic activity of cells, (2) increased blood glucose levels, and (3) mobilization of fats from adipose tissue into the bloodstream (see chart below).

System/function	Parasympathetic	Sympathetic
Cardiovascular	Decreased cardiac output and heart rate	Increased contraction and heart rate; increased cardiac output
Pulmonary	Bronchial constriction	Bronchial dilatation
Musculoskeletal	Muscular relaxation	Muscular contraction
Pupillary	Constriction	Dilatation
Urinary	Decreased urinary output; sphincter contraction	Increased urinary output; sphincter relaxation
Gastrointestinal	Decreased motility of stomach and gastrointestinal tract; decreased secretions	Increased motility of stomach and gastrointestinal tract; increased secretions
Glycogen to glucose conversion	Increased	No involvement
Adrenal gland	Release epinephrine and norepinephrine	No involvement

SANS Effects	PANS Effects
"fight or flight"	"rest and digest"
↑ alertness/vigilance	↓ alertness/vigilance
↑ heart rate and contractility	↓ heart rate and contractility
↑ breathing rate with bronchodilitation	↓ breathing rate with bronchoconstriction
↑ cardiac and skeletal muscle blood flow	↓ cardiac and skeletal muscle blood flow
↓ gut blood flow	↑ gut blood flow
↓ cutaneous blood flow	↑ cutaneous blood flow
↑ blood sugar	↓ blood sugar
↑ temperature	↓ temperature
↓ gut contractility	↑ gut contractility
↓ bladder contractility	↑ bladder contractility
↓salivation	↑ salivation
↓lacrimation	↑ lacrimation
↓digestion	↑ digestion

SANS= sympathetic autonomic nervous system
PANS= parasympathetic autonomic nervous system

Cortical homunculus showing areas of cerebral cortex having sensory and motor control of specific areas of the body

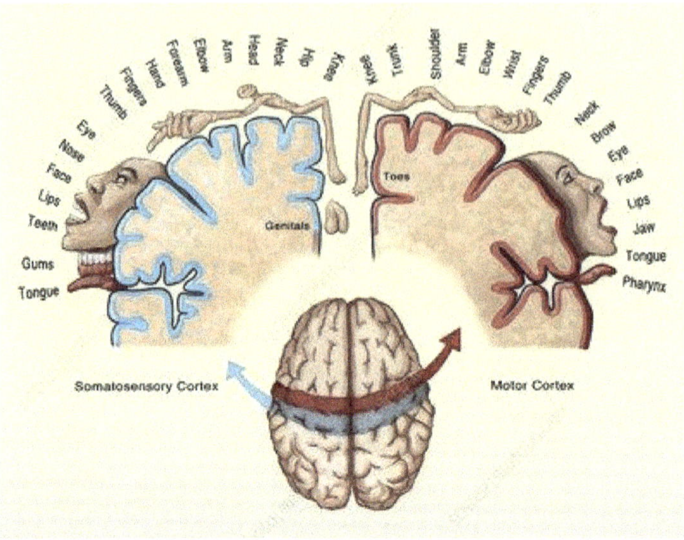

Let's try a few questions.

1. Where is the soma of a multipolar (motor) neuron located?
2. What are some of the functions of the medulla oblongata?
3. Can you list some of the functions of the hypothalamus?
4. Where can you find the cell body of a unipolar (sensory afferent) neuron?
5. What is the difference between the gray matter and white matter of the CNS?
6. What is the function of gray matter in the CNS?
7. What purpose does white matter serve in the CNS?

A Quick Exercise on Reflexes

A reflex is a rapid, automatic, predictable response to a stimulus. There are 5 components to a reflex: 1) receptor, 2) sensory neuron, 3) integration center, 4) motor neuron, and 5) effector. The diagram below is an example of a monosynaptic, ipsilateral, spinal, somatic reflex. Note the 5 components of this reflex arc, and also the stimulus which is the reflex hammer tapping on the patellar ligament and it eventually elicits a response (**stretch reflex** = contraction of the quadriceps femoris muscle that results in a kicking motion of the lower leg).

Diagram below (stretch reflex) is a monosynaptic, spinal, ipsilateral, and somatic reflex

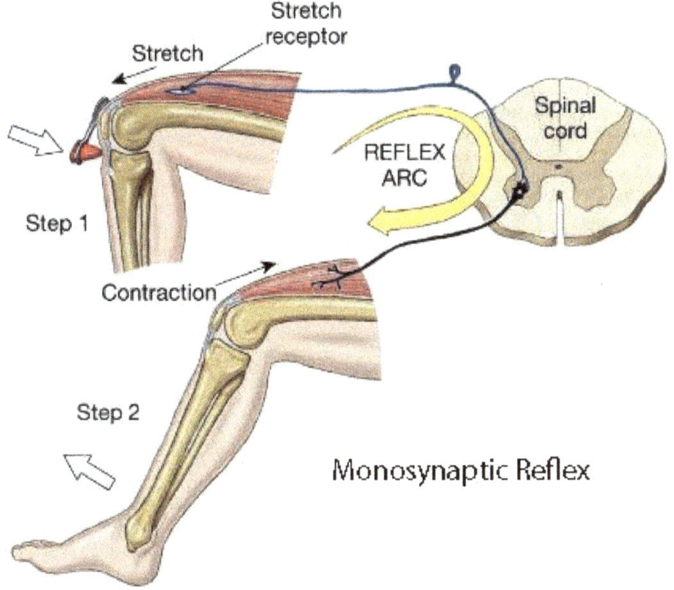

Generally, there are 4 classes of reflexes (not in any specific order):

1. **Monosynaptic / Polysynaptic**: Do sensory neurons synapse directly with motor neurons or are there interneurons in the reflex arc? Look at the above diagram, what makes it monosynaptic? [Remember mono = 1 and poly = greater than one]
2. **Spinal / cranial**: Is the spinal cord or brain the reflex integration center?
3. **Ispsilateral / contralateral**: Are receptors and effectors on the same side of the body, or on opposite side of the body?
4. **Somatic / autonomic**: Is the effector a skeletal muscle, or is it cardiac muscle, smooth muscle or a gland?

According to the exercise below. What is the classification of the '**withdrawal reflex**' below? Note that touching the flame will result in contraction of the biceps brachii and eventually the 'withdrawal' of the hand from the flame. And here again, all 5 components of the reflex arc are present.

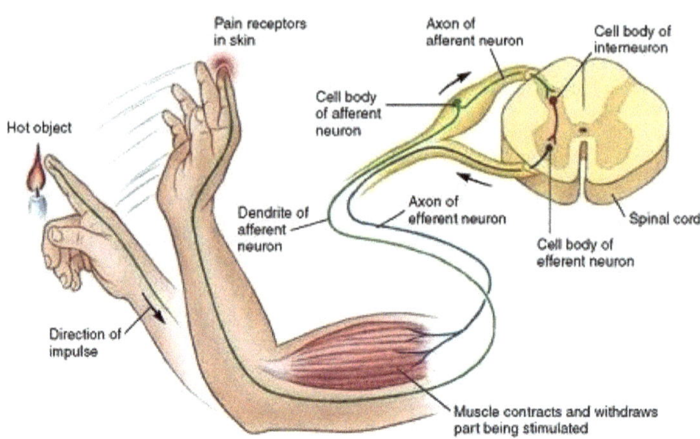

This diagram below is a ' **withdrawal and cross-extensor reflex**,' note this person has stepped onto a piece of broken glass, and has withdrawn their foot off the adverse stimulus while supporting their body weight on the other foot to avoid a fall. So, what classification will you assign to this reflex?

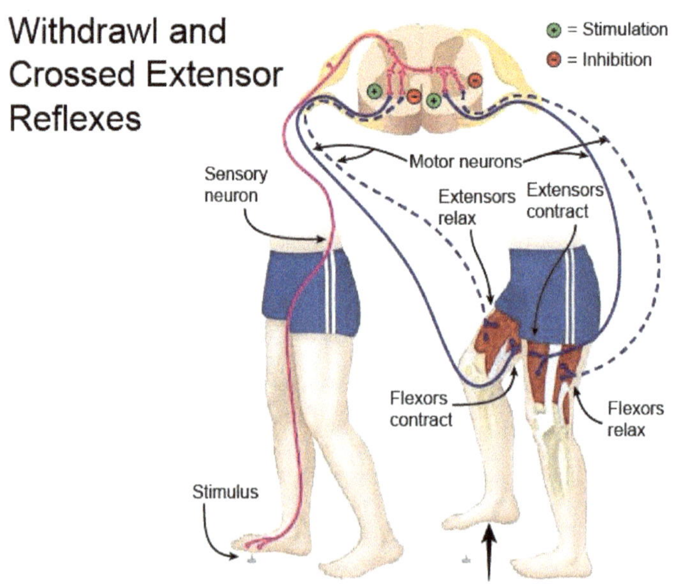

And finally, the pupillary reflex below belongs to which classification?

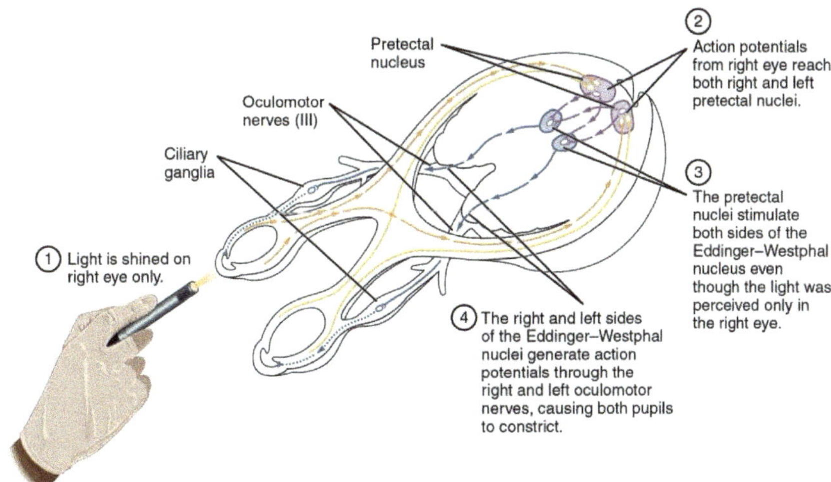

Chapter 10: Muscle Tissue

There are three types of muscle tissue found in animals:

A. **Skeletal muscle** generally causes movement of elements of the skeleton. It is" voluntary" because it can be controlled consciously. When viewed under a microscope, the cells have a striped pattern, and skeletal muscle is often referred to as **striated** muscle.

B. **Cardiac muscle** makes up the heart and is not under voluntary control. Unlike skeletal muscle, cardiac muscle cells do not need input from the nervous system to contract. They will contract on their own. Cardiac muscle cells are connected by gap junctions, and depolarization spreads from cell to cell. This leads to the contraction of the entire heart as a single unit. Similar to skeletal muscle, cardiac muscle is striated.

C. **Smooth muscle** is not under voluntary control and does not have striations (hence the name Smooth). Smooth muscle tissue is found throughout the body, including the digestive tract, uterus, urinary bladder, blood vessels, and respiratory passages.

In this section, the focus will be on skeletal muscle, the others will be discussed more fully in part 2.

	Main features	Location	Type of cells	Histology
Skeletal muscle	- Fibers : striated, tubular and multi nucleated - Voluntary - Usually attached to skeleton			
Smooth muscle	- Fibers : non-striated, spindle-shaped, and uninucleated. - Involuntary - Usually covering wall of internal organs.			
Cardiac muscle	- Fibers : striated, branched and uninucleated. - Involuntary - Only covering walls of the heart.			

I. Introduction to Skeletal Muscle

Functions of skeletal muscle

Skeletal muscles perform the following functions:

A. *Body movement.* Many skeletal muscles have the function of moving elements of the skeleton, such as the bones in the arms and legs. However, some skeletal muscles act on soft tissue structures; for example, the orbicularis oris purses the lips, the diaphragm inflates the lungs, and the epicranius raises the eyebrows.

B. *Maintenance of posture.* Although skeletal muscles can be activated consciously, your nervous system continually activates them unconsciously to maintain posture and body position.

C. *Protection and support.* Layers of skeletal muscle form the walls and floor of the abdominopelvic cavity, protecting and supporting visceral organs in this cavity.

D. *Storage and movement of materials.* Circular bands of skeletal muscle at the ends of the digestive and urinary tracts help regulate the release of feces and urine from the body.

E. *Heat production.* The bulk of heat generated by regular cellular metabolism comes from skeletal muscle. This is particularly noticeable when you exercise or shiver in an effort to keep warm.

Special characteristics of muscle tissue

The functions of muscle depend upon some specific characteristics of muscle tissue:

A. *Excitability.* A skeletal muscle cell can respond to stimulation by a motor neuron with a change in electrical potential. This typically results in the muscle cell generating an action potential.

B. *Conductivity.* Action potentials can propagate along the plasma membrane of a skeletal muscle cell, similar to an axon.

C. *Contractility.* Upon stimulation, a skeletal muscle cell will pull (generate tension) on its attachments and attempt to shorten. Note that the word "contract" in this chapter does not necessarily mean "shorten." When a muscle contracts, it generates tension; whether or not the muscle is able to shorten depends on the magnitude of the load against which the muscle is contracting (imagine trying to lift a brick and trying to lift a car).

D. *Extensibility.* A skeletal muscle cell can be stretched well beyond its resting length without damage.

E. *Elasticity.* After contracting or being stretched, a skeletal muscle cell tends to return to its resting length.

II. Anatomy of a Skeletal Muscle

<u>Gross anatomy of skeletal muscle</u>

The basic structure of a skeletal muscle is shown in the image below. A "muscle" is a bundle of muscle cells, which are often referred to as **muscle fibers** because they are long and thin. In order to give the muscle both strength and flexibility, the muscle fibers are wrapped in a series of layers of connective tissue. The muscle itself is surrounded by the **epimysium**. Inside are smaller bundles of muscle fibers called **fascicles**. Each fascicle is surrounded by a membrane called the **perimysium**. Individual muscle fibers are surrounded by the **endomysium**. The cell membrane of a muscle fiber is called the **sarcolemma**. At each end of a muscle, fibers of the epimysium, perimysium, and endomysium join together to form a **tendon** or a sheet of connective tissue called an **aponeurosis**. The tendon or aponeurosis attaches the muscle to a bone or sometimes to another muscle or a soft tissue structure, such as the eyeball.

Anatomy of a skeletal muscle

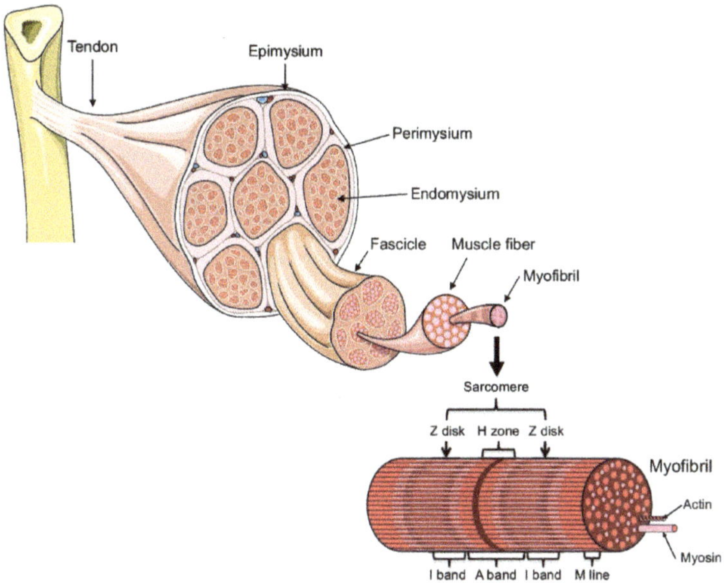

Microscopic anatomy of skeletal muscle

Skeletal muscle fibers are elongated and cylindrical in shape. Each fiber is multinucleate, forming from the fusion of multiple cells, called myoblasts, during development. At regular intervals, there are invaginations of the sarcolemma called **transverse tubules** (or **T-tubules)**. These invaginations extend to about the middle of the cell and are important in communicating with the **sarcoplasmic reticulum (SR)**, which is the endoplasmic reticulum of the muscle cell. Each T-tubule joins with two sections of the SR, called the **terminal cisternae**, to form a structure called a **triad** (see images below). The special function of the SR is to store and release calcium. In the SR membrane are **SR-Ca^{++}-ATPase pumps**, which pump calcium from the cytoplasm into the lumen of the SR. Voltage-gated Ca^{++} channels in the membrane allow calcium to diffuse from the lumen of the SR into the cytoplasm.

Microscopic anatomy of skeletal muscle cell

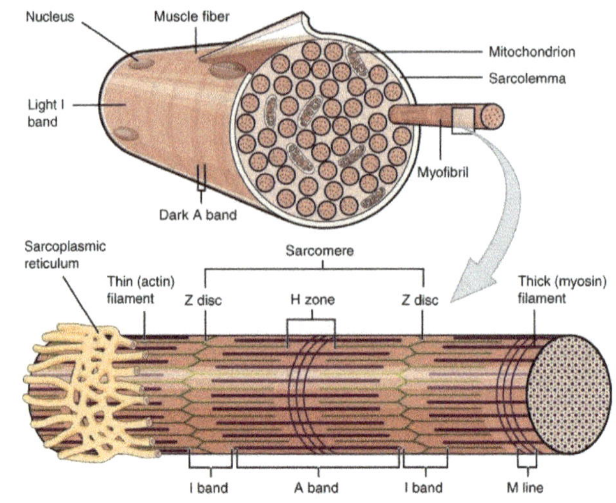

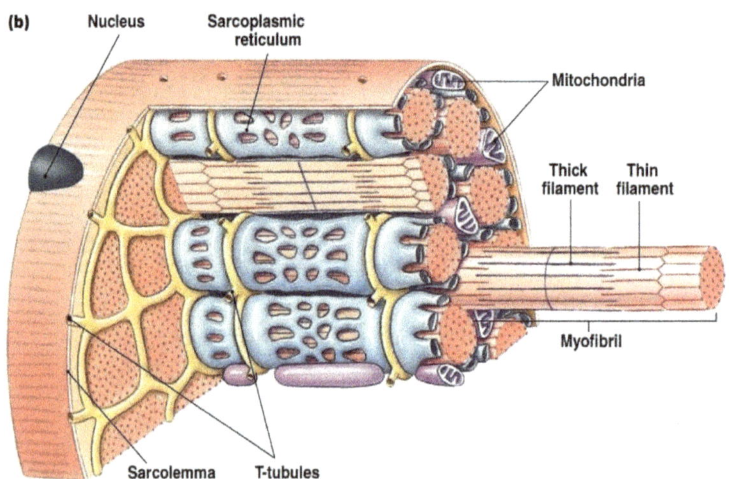

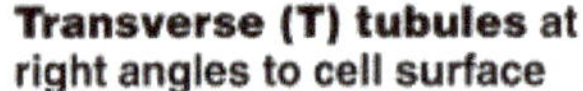

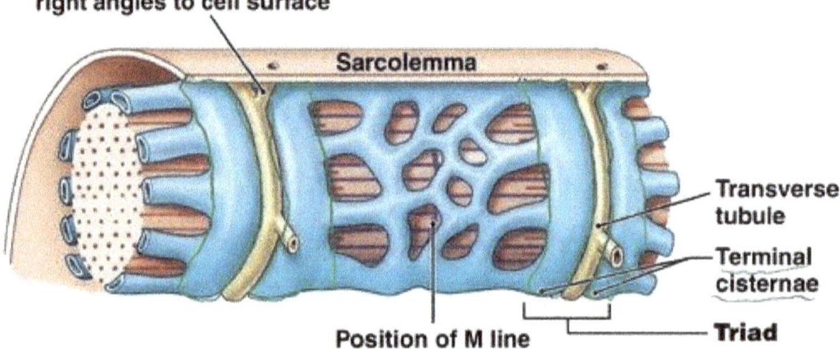

Within the muscle fiber are bundles of protein called **myofibrils**. The myofibrils can be divided into functional units called

sarcomeres. A sarcomere consists of thick filaments composed of **myosin** and thin filaments composed of **actin**. Thin filaments are held in place in the myofibril by **Z discs**. Z discs occur at regular intervals down the length of a muscle cell. **M lines** hold the thick filaments in place. Thick filaments have structures (cross-bridges) that grab onto the thin filaments and pull the Z discs toward each other, causing the muscles to shorten. These cross-bridges are the "heads" of the myosin molccules.

The region of a sarcomere containing thick filaments is called an **A band**; the regions containing only thin filaments are called the **I bands**. When muscle tissue is viewed through a microscope, the thick filaments block much of the light, and the A bands appear dark. Light passes more easily through the thin filaments, and the I bands appear light. This creates the striated appearance of the muscle. Look at the pictures below and you can see that when a muscle contracts, the I bands get narrow, and the A bands stay the same length. Can you out why?

Sarcomere structures (in relaxation and contraction phases). H zone (not labeled) is the gap between the thin filaments and contains the M line/band.

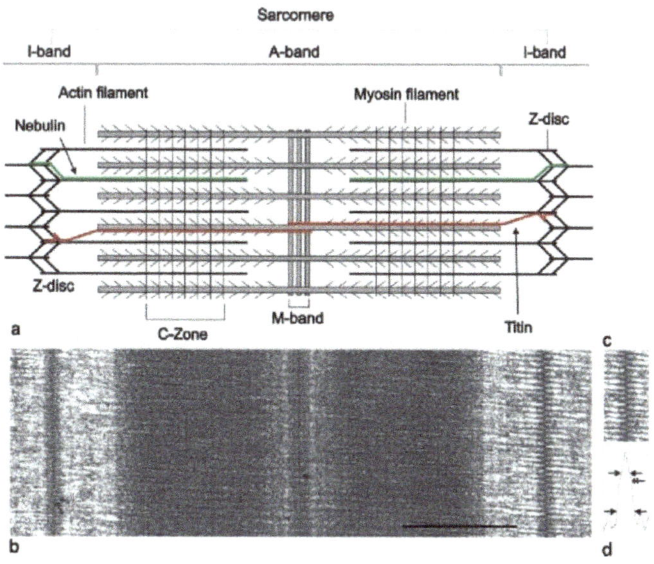

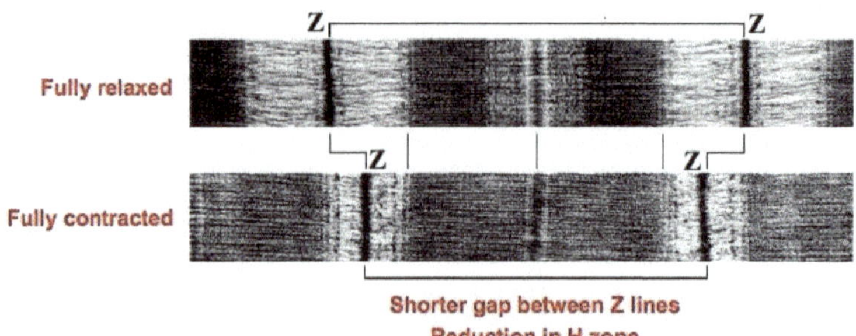

Innervation of skeletal muscle fibers

Skeletal muscle fibers are connected to somatic motor neurons. A somatic motor neuron typically innervates multiple skeletal muscle fibers within a muscle. The neuron and the fibers it innervates comprise a motor unit. When the neuron releases neurotransmitters, all of the fibers in the motor unit will contract. A given skeletal muscle will contain many motor units. Activating different numbers of motor units allows a muscle to contract with varying amounts of force.

Single motor unit

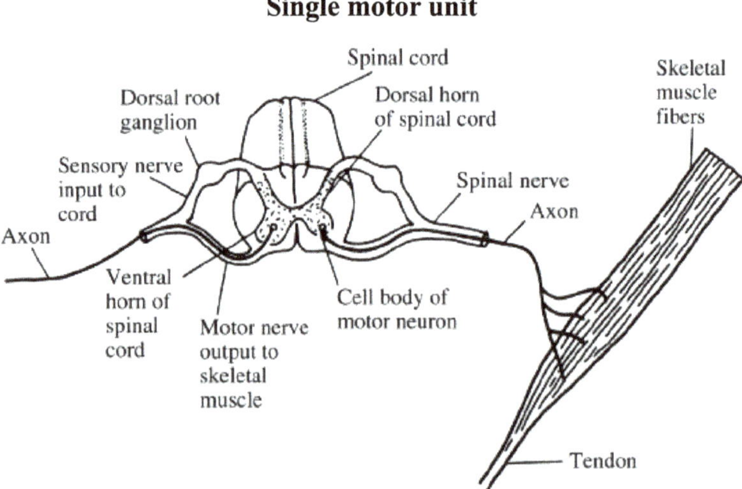

The number of muscle fibers per motor unit varies from muscle to muscle; muscles used for fine movement (*e.g.,* the muscles that move your eyeballs) tend to have fewer fibers per unit. Muscles used for gross movements, like your quadriceps muscles, tend to have many fibers per motor unit.

The synapse formed where a motor neuron joins to a skeletal muscle fiber is called a **neuromuscular junction** (NMJ). The NMJ consists of three structures: (1) the synaptic knob of the motor neuron, which releases the neurotransmitter **acetylcholine** (ACh), (2) a special region of the muscle fiber's sarcolemma called the motor end plate, which has receptors for ACh, and (3) a synaptic cleft between the synaptic knob and the motor end plate. The receptors on the motor end plate are chemical-gated channels that allow passage of Na^+ and K^+. Each muscle fiber has one neuromuscular junction located about midway along the length of the cell.

III. Physiology of Skeletal Muscle Contraction

It was thought for quite some time that muscle acts like a rubber band, but this was not an accurate view. With the use of electron microscopes confirming the "sliding filament theory" of muscle contraction, the steps are the following: (1) the cross-bridges within the myofibrils change orientation when the muscle contracts and (2) the thick and thin filaments do not change in length during contraction. An overview of the process of contraction is shown in the illustration below.

Neuromuscular junction (below): excitation of a skeletal muscle fiber when a somatic motor neuron fires an action potential, it releases ACh into the NMJ.

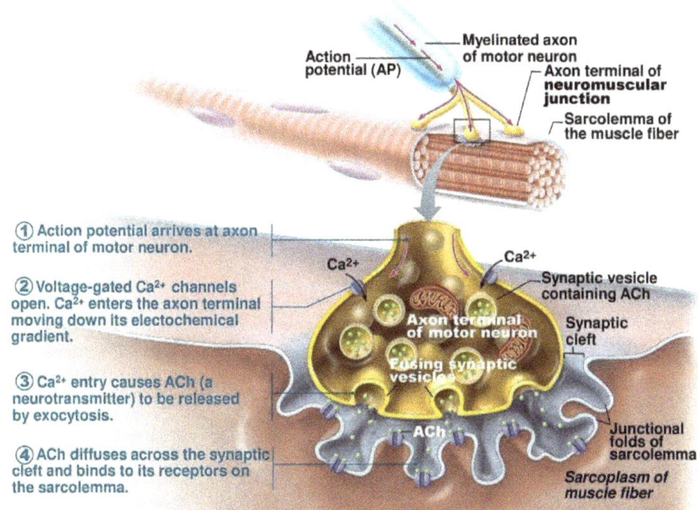

Physiology of muscle contraction diagram below

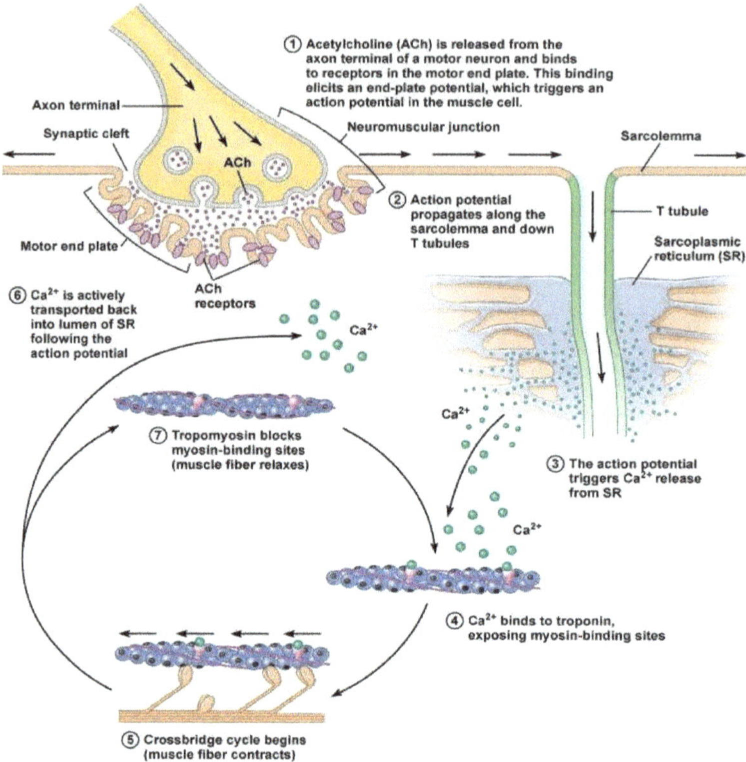

When acetylcholine binds to its receptors on the motor end plate, chemical-gated channels open, allowing Na^+ to enter the muscle fiber and K^+ to exit the fiber. However, the resting potential of a skeletal muscle fiber is typically -90 mV, which is near the equilibrium point for potassium, and there is a much greater diffusion of sodium into the muscle fiber than diffusion of potassium out. Thus, ACh depolarizes the muscle fiber.

Sarcolemma, T-tubules, and sarcoplasmic reticulum

The depolarization that occurs at the NMJ is called an **end plate potential**. This effect is similar to the production of a graded potential or postsynaptic potential on the body/dendrites of a neuron. However, there are some key differences. The end plate potential generally does not vary in magnitude; it is always depolarizing, and it should hit the threshold without the need for temporal or spatial summation. How does this compare to graded potentials on a neuron?

The end plate potential should always hit the threshold and generate an action potential. The sarcolemma contains voltage-gated Na^+ and K^+ channels that enable depolarization and repolarization similar to the action potential of an axon. The action potential propagates along the sarcolemma in a manner similar to continuous conduction along an unmyelinated axon. One difference is that since the NMJ is in the middle of the muscle fiber, the action potential propagates toward both ends of the fiber.

As the action potential spreads across the sarcolemma, it moves into the T-tubules. The action potential then causes the opening of the voltage-gated calcium channels located in the membrane of the SR. This allows calcium to flood the myofibrils of the muscle fiber and initiate muscle contraction (see illustration above).

These processes that lead from stimulation by the motor neuron to the initiation of muscle contraction are collectively referred to as **excitation-contraction coupling**.

Illustration below showing exposure of myosin-binding sites by calcium ions, a catalyst to cross-bridge formation.

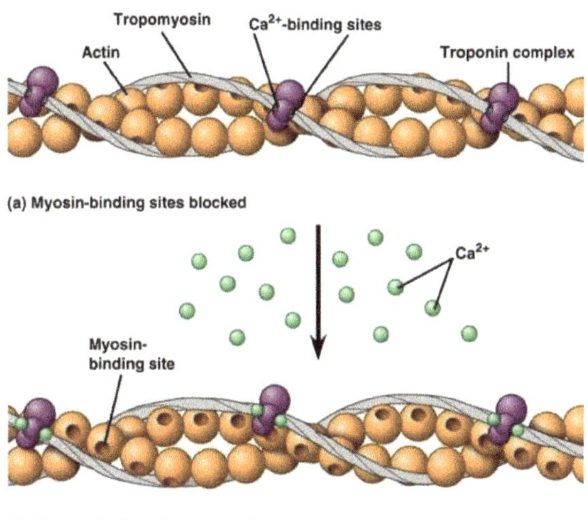

Sarcomere: cross-bridge cycling

In the absence of calcium, the myosin binding sites on the actin molecules of a thin filament are blocked by the **troponin-tropomyosin complex** (see illustration above). Calcium released from the SR binds to the troponin, and this causes a shift in the tropomyosin that exposes the myosin binding sites. Once the myosin binding sites are exposed, the muscle cell will contract according to a process referred to as the **cross-bridge cycle**.

Diagram showing involvement of calcium in the cross-bridge formation

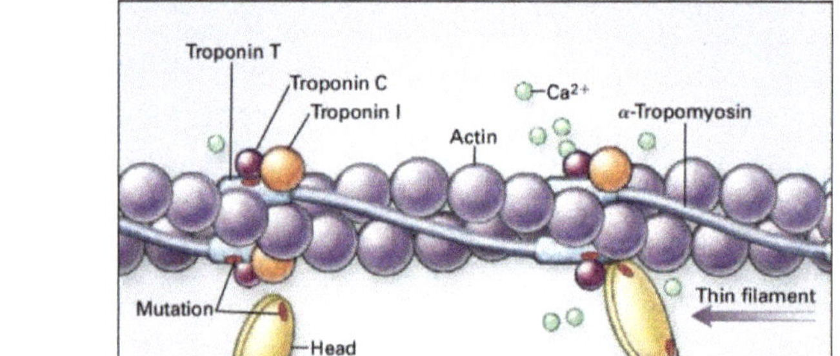

Here is a summary of the cross-bridge cycle, broken down into four steps:

Step 1: **Crossbridge formation**. At this point in the cycle, the myosin head has ADP and P_i bound to it. With the binding sites exposed, the myosin head (*i.e.,* the cross-bridge) attaches to a thin filament.

Step 2: **The power stroke**. The myosin head releases ADP and P_i and generates the power stroke. The cross-bridge swivels, pulling the thin filament.

Step 3: **Release of the myosin head**. The myosin head binds a molecule of ATP, causing it to detach from the thin filament.

Step 4: **Reset of the myosin head**. The myosin head breaks ATP down to ADP and P_i, the myosin returns to its previous orientation, and it is ready to return to Step 1.

The cross-bridge cycle is repeated many times per second during a muscle contraction by many myosin heads. It continues as long as calcium is bound to troponin and the muscle cell has an adequate supply of ATP.

Cross-bridge cycle diagram below

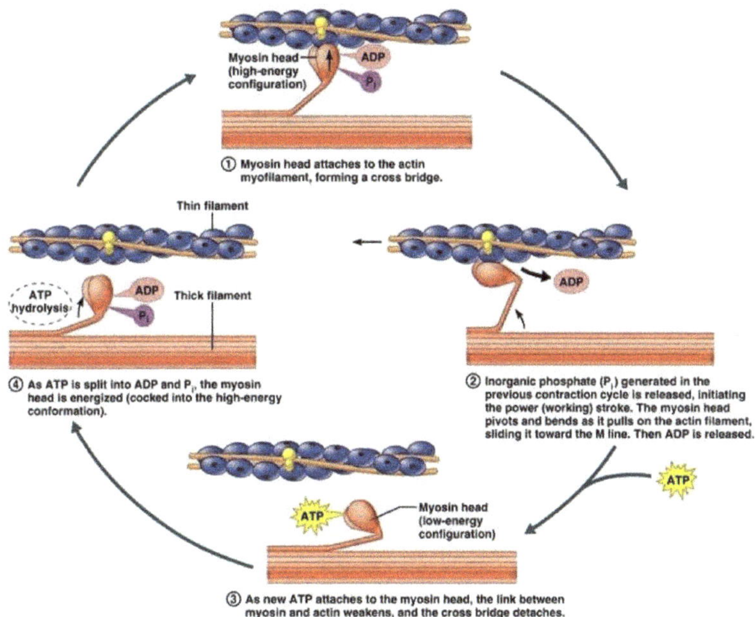

Skeletal muscle relaxation

After muscle stimulation ends, the SR-Ca++-ATPase pumps calcium back into the sarcoplasmic reticulum. Removal of calcium from the myofibrils to the SR results in relaxation. This process is one of active transport; calcium is pumped into the SR against its concentration gradient. Thus, relaxation requires an input of energy in the form of ATP. What are the two molecules using most of the ATP in a working skeletal muscle fiber?

IV. Skeletal Muscle Metabolism

Supplying Energy for Skeletal Muscle Contraction

In order for a muscle fiber to contract, it needs energy. The two main consumers of energy in a working muscle are myosin and the SR-CA^{++}-ATPase. Both of these molecules get their energy by breaking molecules of ATP into ADP and inorganic phosphate (P_i), according to the following reaction:

$$ATP \rightarrow ADP + P_i$$

Activation of a muscle cell from rest can increase its energy demand 1000-fold within a fraction of a second. However, the amount of ATP in a muscle cell does not change much, even during a maximal contraction. How does the cell match the supply of ATP to the demand for ATP?

Imagine you are planning dinner for yourself and three other people tonight. How much food do you plan to prepare? What if fifty people show up at your house for dinner?

The amount of ATP in a resting muscle fiber is relatively small. When the fiber is stimulated to contract, this ATP would last only a few seconds. If the muscle fiber is going to contract for longer than a few seconds, it must be able to generate more ATP, and lots of it. There are three basic mechanisms by which a muscle fiber can generate more ATP:

A. ***The phosphagen system***. A muscle cell can very rapidly generate ATP via the chemical reaction catalyzed by the enzyme **creatine kinase**:

$$PCr + ADP \rightarrow ATP + Cr$$

where PCr is **creatine phosphate** (also called *phosphocreatine*), and Cr is **creatine**. (The molecule PCr is one of a group of molecules referred to as "phosphagens.") As the muscle fiber initially breaks down ATP to ADP and P_i, the ADP can be rapidly turned back into

ATP by transferring the phosphate group from PCr. This source of ATP is particularly important during the initial stages of activity.

In a resting muscle cell, the amount of PCr is about three to five times the amount of ATP. This extends the energy supply of a working muscle for another ten to fifteen seconds. However, the supply of PCr is limited, and a muscle that contracts for a longer period needs yet another source of energy.

B. ***Anaerobic metabolism***. Anaerobic metabolism involves the breakdown of **glucose** (a six-carbon molecule) into two molecules of **pyruvate** (a three-carbon molecule). When glucose is broken down, energy is released. This energy can be used to convert ADP and P_i back into ATP. Notice here that ATP, ADP, and P_i molecules are constantly recycled in a cell. When one molecule of glucose is broken down to pyruvate, enough energy is harnessed to produce two molecules of ATP. Glucose is stored in muscle cells and the liver as a molecule called **glycogen**. When muscle cells are active, glucose molecules can be released from the glycogen.

Under some conditions (you'll see in the next section), pyruvate can enter the mitochondria and be broken down into carbon dioxide. This process requires oxygen and releases much more energy that can be used to make more ATP. However, when a muscle cell needs energy quickly and oxygen cannot be supplied rapidly enough, pyruvate is converted to **lactic acid**.

Anaerobic metabolism is typically used to supply energy for short-term, intense exercise. Think of running a 400-meter sprint, returning a kickoff 90 yards in a football game, or running at full speed for four blocks to get away from the neighbor's pit bull. On the upside, anaerobic metabolism can generate ATP rather rapidly. On the downside, muscles generating ATP in this manner quickly become fatigued. There is still another source of energy to get energy to fuel a muscle for longer durations.

C. ***Aerobic cellular respiration***. As mentioned in the previous section, pyruvate can be broken down to carbon dioxide. This process occurs in the mitochondria, has many steps, and requires oxygen. The overall chemical reaction is one you should be familiar with:

$$\text{glucose} + 6\ O_2 \rightarrow 6\ CO_2 + 6\ H_2O$$

Although this reaction is usually shown with glucose, other sugars, fats, and proteins can also be used as substrates for cellular respiration. When a molecule of glucose is broken down to carbon dioxide, enough energy is released to allow the production of a total of 36 molecules of ATP.

The process of aerobic cellular respiration can go on for long periods; think of someone participating in a triathlon. For most normal people, this is the process we use to supply energy for a long walk or a leisurely ride on a bicycle. The upsides of this process are that it is very fuel efficient (36 ATP per glucose), and the process can last for long periods of time without leading to fatigue. The downsides are that it takes a little time to get the process going, and the process is limited by the cell's ability to supply and use oxygen.

Oxygen debt

When someone has been working hard and producing lactic acid, the lactic acid is transported from the muscles via the bloodstream to the liver. At the liver, lactic acid can be converted back into glucose. However, the conversion of lactic acid to glucose requires more ATP than is gained by breaking glucose into lactic acid (no free lunch). The ATP required to regenerate glucose from lactic acid must be obtained aerobically (with oxygen). So, one can exercise anaerobically, but the removal of lactic acid will eventually require oxygen. Furthermore, exercise depletes the supply of PCr in the muscle fibers. Oxygen is required to generate the ATP needed to convert creatine back into PCr (the creatine kinase reaction shown earlier is run backward). The oxygen needed to turn lactic acid into glucose and to regenerate PCr from creatine is called the **oxygen debt**. This explains why a person can run a 100-meter sprint without using oxygen, but the runner will then spend the next several minutes breathing hard.

V. Skeletal Muscle Fiber Types

Criteria for classification of muscle fiber types

Although all skeletal muscle cells function by the same principles, there are some differences in the muscles that allow them to perform different jobs. For example, the muscles in your back that are used to maintain posture have somewhat different properties than your biceps muscles, which may be used to lift heavy objects.

Skeletal muscle fibers are generally grouped into categories according to two properties:

1. *The speed with which the myosin can break down ATP*. Not all myosin molecules are the same. Some muscle fibers have myosin that can break down ATP faster than others. Muscle fibers with the faster myosin can contract more rapidly and with more power than those with the slower myosin. Accordingly, some muscle fibers are referred to as **fast twitch fibers**, and others are referred to as **slow twitch fibers**.

2. *The preferred source of energy for the fiber.* **Glycolytic fibers** rely mostly on anaerobic metabolism to supply ATP. These fibers have few mitochondria but large stores of glycogen. They tend to fatigue quickly. **Oxidative fibers** rely mostly on aerobic cellular respiration to supply ATP. These fibers have many mitochondria and are supplied with an extensive network of capillaries to provide oxygen. They also contain a protein called **myoglobin**, which helps store and transport oxygen within the fibers. Myoglobin (like the related protein, hemoglobin) is a red pigment, and it gives oxidative fibers a reddish appearance. As glycolytic fibers have little or no myoglobin, they tend to be pale and white in color.

Classification of muscle fiber types

Based on the two criteria given above, skeletal muscle fibers tend to be grouped into three categories:

A. **Slow oxidative fibers** (also called *type I* fibers) contain slow myosin and rely primarily on aerobic cellular respiration for ATP production. These muscle fibers typically have many mitochondria for the efficient production of ATP and plenty of myoglobin. These fibers are fatigue-resistant and good for long-term use. They do not generate much power.

B. **Fast glycolytic fibers** (also called *type IIb* fibers) contain fast myosin and rely primarily on anaerobic metabolism for ATP production. They have few mitochondria, and they fatigue rapidly. They are large in diameter and can contract quickly and generate a lot of power.

C. **Fast oxidative fibers** (also called *type IIa* fibers) are somewhere in the middle. They contain fast myosin, yet are also capable of producing ATP by aerobic cellular respiration. As you might guess, the properties of fatigue resistance and power generation are intermediate between the slow oxidative and fast glycolytic fibers.

Cross-section of the 3 types of muscle fibers below

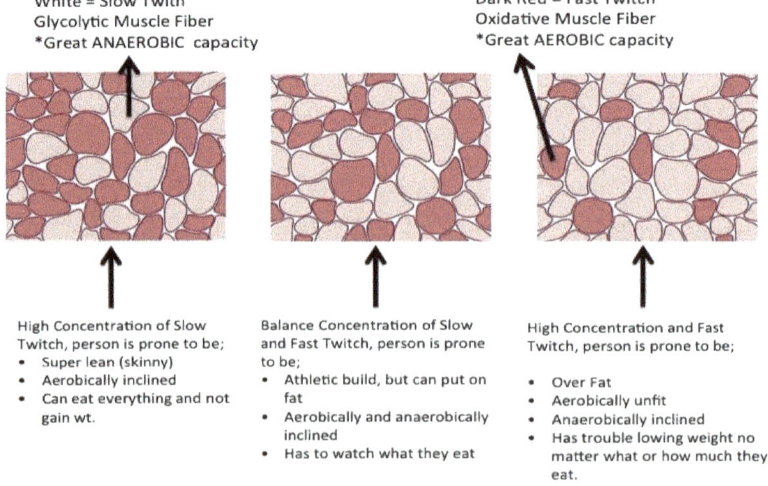

<u>Distribution of muscle fiber types</u>

In mammals, each skeletal muscle has a mix of the three fiber types. However, the proportions vary in different muscles. For example, the soleus muscle has a high proportion of slow oxidative fibers. This muscle is well suited to maintaining posture and taking long walks. On the other hand, the gastrocnemius has a high proportion of fast glycolytic fibers. This muscle is important in jumping and sprinting. Note that sprinters and weightlifters typically have large gastrocnemius muscles.

VI. Measurement of Skeletal Muscle Tension

Because a muscle cell attempts to pull during a contraction (it cannot push), the force generated by a muscle is often called "tension." Much of what we know about how muscles generate tension was learned using frog gastrocnemius muscles and machines known as "force transducers." The muscle is attached to the device and stimulated electrically; the force generated by the muscle is then measured by the device and recorded on a computer. Think of the virtual laboratory exercise we did earlier in the semester.

Muscle twitch

When a muscle is stimulated artificially with a brief electrical pulse, the muscle will generate a single, brief contraction called a **twitch**. The twitch may involve one or many motor units, depending upon the intensity of the stimulus. The record of a single twitch in a graph is called a **myogram** (see illustration below), and the twitch can be broken down into three phases:

A. The **latent period** begins at stimulation and ends as tension begins to develop. During this time, the action potential spreads across the muscle fiber, and calcium is released from the SR.

B. The **contraction phase** begins with tension development and ends with peak tension.

C. The **relaxation phase** begins as tension peaks and lasts until tension comes to an end.

Myogram of muscle fiber

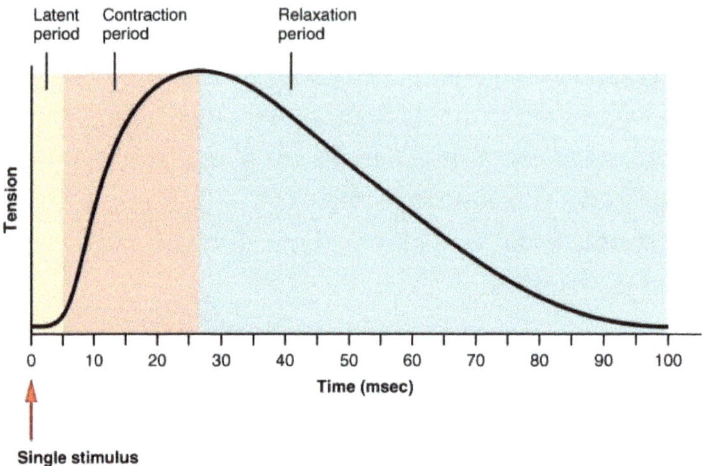

Changes in stimulus intensity: motor unit recruitment

Contraction of a single muscle fiber is (more-or-less) an all-or-none event. Either the fiber is relaxing, or it is contracting. A motor neuron cannot stimulate a muscle fiber halfway. However, the division of a muscle into motor units allows **graded responses** of the muscle (Fig. 10.19). The amount of tension produced by the entire muscle varies depending on the number of motor units that are **recruited** for the contraction. For example, if you decide to use your right arm to lift a bowling ball, then you will recruit more motor units than you would recruit if you were lifting a paper clip. When a muscle is acting normally in your body, your nervous system unconsciously controls the number of motor units recruited for a particular activity.

VII. Factors Affecting Skeletal Muscle Tension Within the Body

Isometric contractions and isotonic contractions

Different types of contractions can be described based on what happens to the muscle after contraction is initiated. An **isotonic contraction** occurs when the muscle exerts a constant force during the contraction, and the length of the muscle changes. An isotonic contraction is **concentric** if the muscle tension exceeds the resistance and the muscle shortens. An isotonic contraction is

eccentric if the resistance exceeds the tension and the muscle lengthens. (see chart below)

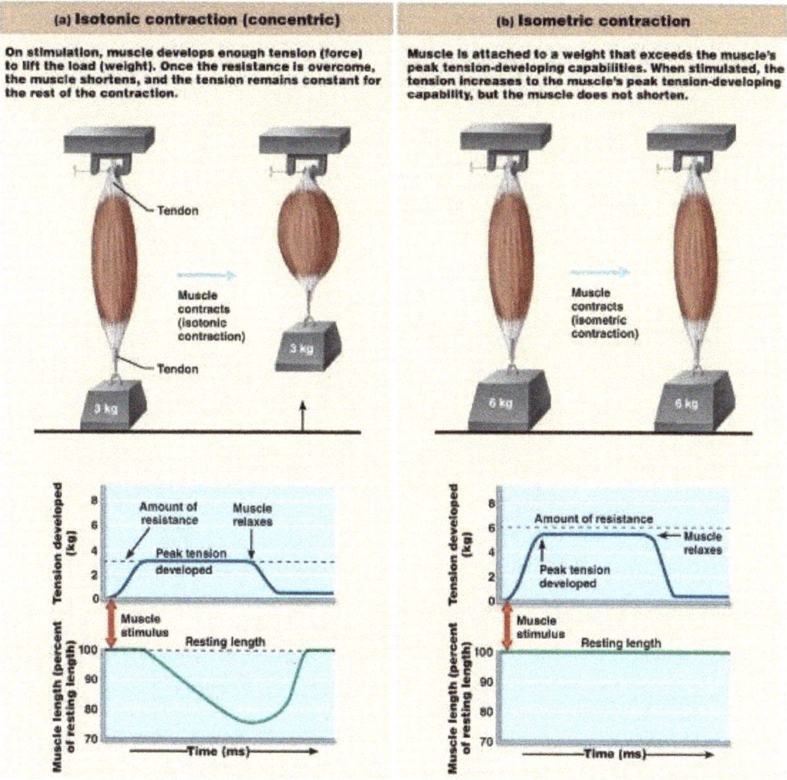

An **isometric contraction** occurs when the muscle remains at a fixed length during contraction. Imagine holding a bowling ball in front of you and keeping it in place; your muscles are contracting isometrically. I assume all of you would also be performing isometric contractions if you tried to lift a car. In these cases, either the muscle exerts a force equal to the resistance, or the resistance simply cannot be moved.*

Muscle fatigue

Muscle **fatigue** is the state that exists when stimulation of a muscle fails to produce the expected amount of tension. The exact causes of fatigue are hard to determine and may vary under different conditions. One factor that may contribute to fatigue is the build-up of lactic acid and the accompanying decline in pH. Another factor may be the inhibition of myosin cross-bridges by rising levels of P_i that accompany intense exercise. Fatigue may also result from ionic imbalances (*e.g.*, changes in Ca^{++} concentrations) that disrupt excitation-contraction coupling. Lack of ATP is sometimes cited as a cause of fatigue, but studies show that levels of ATP rarely drop that low during any normal activity.

VIII. Effects of Exercise and Aging on Skeletal Muscle

Effects of exercise

Although you may not have realized this before, one of the primary goals of exercise is to enhance certain skeletal muscle fibers or to change fibers from one type to another. The type of muscle that is recruited during exercise depends on the intensity and duration of the exercise.

Low-impact, long-duration exercise, known as **aerobic exercise**, tends to recruit primarily slow oxidative fibers. These are the smallest fibers, and they do not produce much force. However, they are very fatigue-resistant. Continued aerobic exercise (*e.g.*, jogging, cycling, swimming) enhances the oxidative capacity of these muscles and makes them even more resistant to fatigue. Endurance exercise also promotes a shift of fast glycolytic fibers to fast oxidative fibers.

As the demand for power increases, fast oxidative fibers are recruited. These fibers might be useful for a sprint or lifting moderately heavy weights for multiple repetitions. Activities such as basketball and soccer enhance the oxidative capacity of these fibers and slow oxidative fibers.

High-intensity, short-duration exercise, known as **resistance exercise**, recruits fast glycolytic fibers. Intense weight-lifting, for example, results in an increase in the number of myofibrils within fast glycolytic fibers with an accompanying increase in muscle mass. This increases power output from the muscles. There is also an increased glycolytic activity in the fibers, and there is a conversion of fast oxidative fibers to fast glycolytic fibers.

Exercise can alter the energy-producing capacity of muscle in humans (aerobic vs. anaerobic), and an increase in the size of fast glycolytic fibers can increase the proportion of fast myosin in a muscle. However, in humans, there is generally not a shift between fast and slow myosin as a result of exercise. Research indicates that each person is born with a certain amount of fast fibers and a certain amount of slow fibers, and exercise does little or nothing to change this.

IX. Smooth Muscle Tissue

Functional categories

Smooth muscle tissue is classified according to how it is stimulated to contract.

Multiunit smooth muscle is found in large airways of the respiratory system, large arteries, arrector pili muscles, and internal eye muscles. Muscle cells of this tissue are structurally independent, and individual cells are directly innervated by visceral motor neurons. These cells are arranged in motor units, similar to the arrangement of skeletal muscle.

Single-unit smooth muscle is the most common type of smooth muscle found in the organs of the digestive tract, the uterus, the bladder, and other hollow organs (except the heart). Visceral motor neurons may stimulate this muscle to contract, but it may also contract spontaneously. Cells of visceral smooth muscle are linked by gap junctions; thus, large areas of smooth muscle tissue tend to contract as single units.

Types of Smooth Muscle

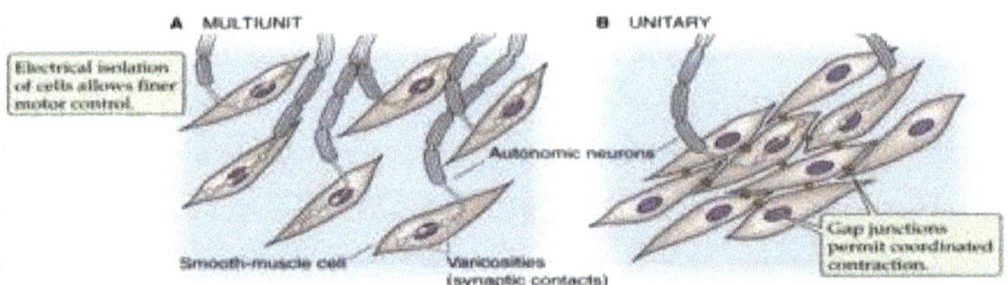

Multi-unit smooth muscle	Unitary (or single-unit) smooth muscle
• Discrete/separate fibres each with own nerve ending – independent contraction • Mainly innervated by nerve signals • E.gs: ciliary muscle of the eye; iris; piloerector muscles; vas deferens	• Sheets of electrically coupled cells – syncytium or visceral smooth muscle • Contract in unison • Connected by gap junctions • E.gs: GI tract, bile ducts, ureters, uterus & blood vessels

Skeletal Muscles of the Body

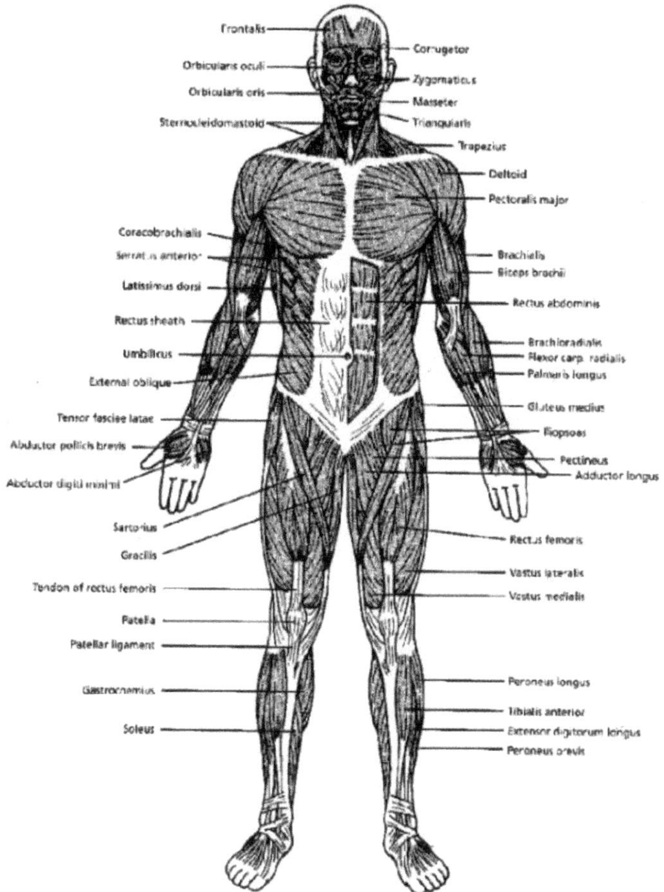

Overview of the Musculoskeletal System Anatomy and Physiology

Though most of the skeletal muscles are labeled in the following illustrations, it is not however, a requirement to know all of them. Therefore, the '**Must Knows**' for each body region to study will be emphasized here.

<u>Muscles of the head</u>

1. **Epicranius** or **Frontalis** (frontal belly of occipitofrontalis)- action is to raise the eyebrows or furrow the forehead (there is also an occipital belly not seen in the picture here, the two

bellies are connected by the aponeurosis (fibrous tissues seen in the picture).

2. **Temporalis** – this is a muscle of mastication (chewing muscle) – its origin is on the parietal bone and inserts on the coronoid process of the mandible. Its action is to elevate the mandible (during speaking, chewing, and closing the mouth). It is a synergist to the masseter (a synergist performs the same action as the other muscle it is synergist to)

3. **Orbicularis oculi** – the fibers run circumferentially around the eyes and its action is to close the eyes tightly

4. **Zygomaticus major** – originates from the zygomatic bone and inserts onto the superolateral fibrous tissue of the corners of the mouth. This muscle is also known as the ‘smiling muscle,’ its action is to draw the corners of the mouth superiorly and laterally as in a smile

5. **Orbicularis oris** – its fibers as seen encircles the mouth and its action is to close the mouth and puckers the lips.

6. **Masseter** is another muscle of mastication, like the temporalis it elevates the mandible. It originates on the zygomatic arch and inserts on the ramus of the mandible.

7. **Platysma** – superficial muscle and is antagonist to both the temporalis and masseter because it opposes their action. When contracted, the platysma depresses the mandible and hardens or tightens the cervical (neck) region, it may mimick a grimacing gesture.

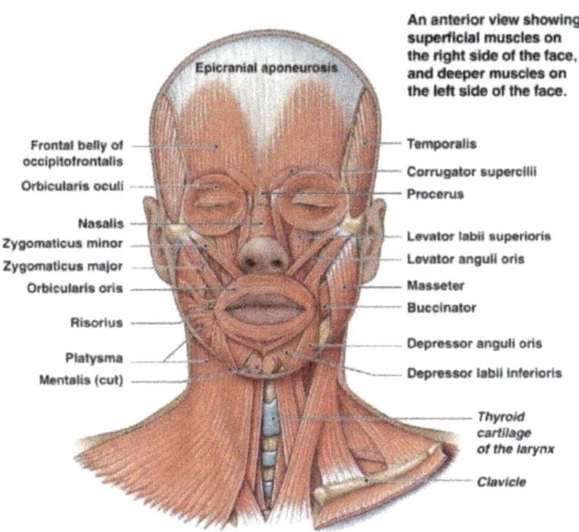

Muscles of the neck:

1. **Sternocleidomastoid** – its origin is from the mastoid process and insertion is on the medial clavicle (from mid clavicle to sternal end of clavicle) and the manubrium of the sternum. Contraction of a single sternocleidomastoid muscle causes contralateral rotation and ipsilateral flexion of the head. It is a synergist to the splenius capitis on the opposite side of the neck. In order words, when turning the head to the right, the left sternocleidomastoid muscle and the right splenius capitis which causes ipsilateral rotation of the head contribute to this movement.

When both the right and left sternocleidomastoid muscles contract (bilateral contraction) simultaneously, they cause flexion of the neck. (recall the body movements terminology from chapter 6 on the skeletal system).

2. **Splenius capitis** – located between the trapezius and sternocleidomastoid muscles as seen below. Its action is already mentioned above. Simultaneous bilateral contraction of the splenius capites results in neck extension.

3. **Trapezius** – although it is mainly a large, flat, diamond-shaped back muscle, its upper fibers run along the back of the neck as well. It originates from the occipital bone, nuchal

ligament, and spinous processes of C-1 to T12 vertebrae. It inserts into the lateral third of the clavicle, acromion, and spine of the scapula. It contains three groups of fibers:

- Upper fibers - upon contraction, they elevave the scapulae
- middle fibers – retract or adduct the scapulae
- lower fibers – depress the scapulae.

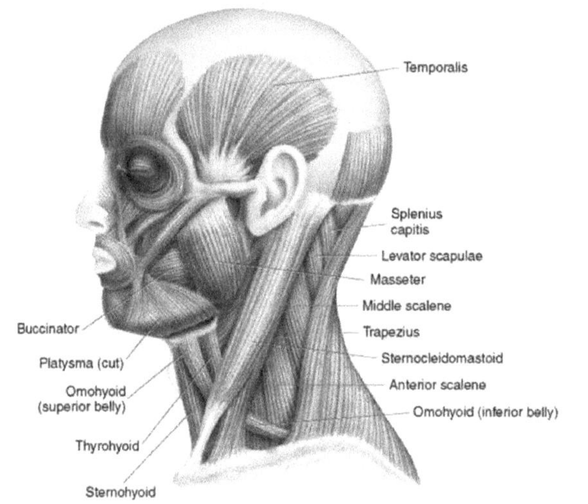

Lateral view of sternocleidomastoid muscle and neck muscles within the anterior and posterior triangles

Muscles of the upper body/torso (anterior view)

1. *Pectoralis minor* - deep to the pectoral major, it originates on ribs 3, 4, and 5 and inserts on the corocoid process of the scapula. It acts to protract the scapula.
2. *Pectoralis major* – superficial to the pectoralis minor. Its acts to flex and adducts the arm (flexion and adduction of the arm).
3. Internal intercostal muscles – lie deep to the external intercostals. They act to depress the ribcage which decreases the thoracic cavity volume and are used during active forceful expiration.

4. *Serratus anterior* – also a protractor / abductor of the scapula and therefore is a synergist to the pectoralis minor.

5. Rectus abdominis – (incorrectly spelled in diagram below) originates on the crest of the pubis and inserts on the xiphoid process and costal cartilage. It flexes the trunk.(as in abdominal crunches exercise).

6. *External oblique m.* – contralateral rotation when one side contracts and contraction of both sides cause flexion of the trunk.

7. *Internal oblique muscles* – one-sided contraction causes ipsilateral rotation of the abdomen. When both sides work together, their action is to compres the abdominal cavity.

8. *Transverse abdominis* – (incorrect spelling in the illustration) the fibers of this muscle are positioned transversely across the anterior and lateral abdomen. Upon contraction it compresses the abdominal cavity.

9. *External intercostal muscles* – are located between the ribs and act to elevate the ribs and expand the thoracic cavity during inspiration

On a side note, the upper fibers of the trapezius can be viewed anteriorly.

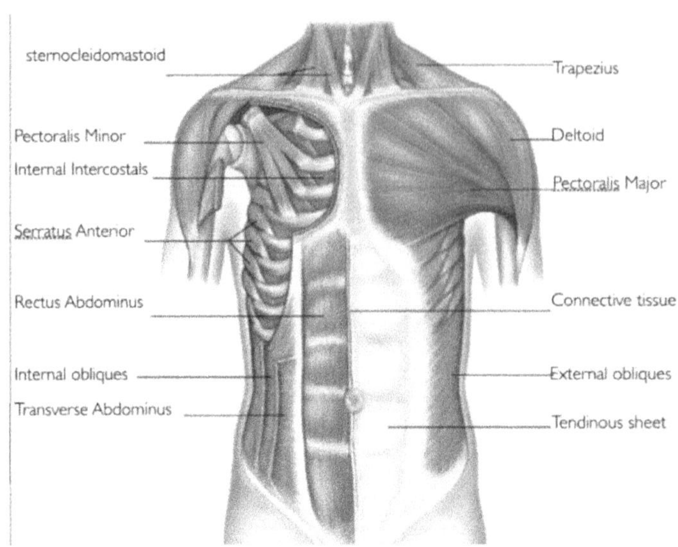

The upper extremity muscles

The Deltoid

Deltoid – the name of this muscle is derived from the Greek letter delta (Δ) as it resembles an upside down triangle. It is composed of three groups of fibers: anterior (seen from anterior view of the torso), middle (can be viewed laterally), and posterior (can be seen from the posterior view of the torso). The deltoid originates from the lateral clavicle as well as the spine and acromion of the scapula and inserts onto the deltoid tuberosity. The anterior fibers' action is to flex the arm at the glenohumoral joint (shoulder joint), the middle fibers cause abduction of the arm, and the posterior fibers act to extend the arm

Theres are 3 major flexors of the elbow and so they are synergistic to one another:

1. **Biceps brachii** – the name biceps means 2 heads and brachii refers to arm (see the body regions). Since this muscle has 2 heads, that means it has 2 points of origin, the coracoid process of the scapula and the edges of the glenoid fossa of the scapula. The biceps brachii inserts on the radial tuberosity.

2. **Brachialis** – is deep to the biceps brachii and is one of the synergists of the biceps brachii.
3. **Brachioradialis** – runs along the lateral antebrachium (forearm) and originates from the distal humerus to the distal radius. It is an elbow flexor as well and forms the lateral border of the antebrachial fossa (a common site for intravenous (IV) insertion in the hospital).

Since these 3 muscles work together synergistically to flex the elbow, they are sometimes referred to as the elbow flexor triumvirate. (I often refer to them as the 'triple B's').

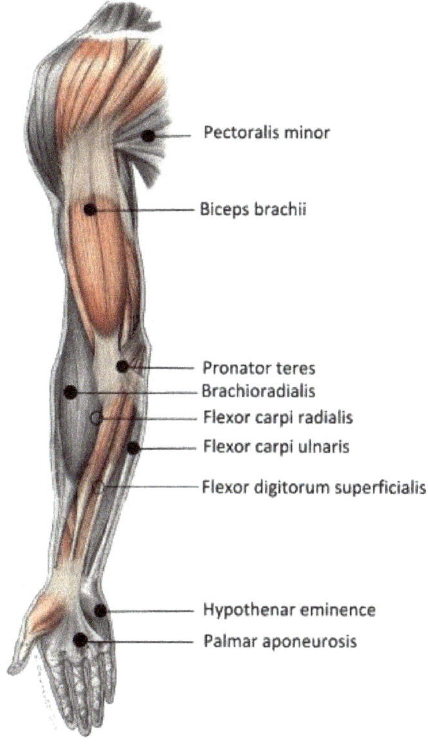

Below is an example of a flexed elbow in the human model. It is worth examining the skeletal model to identify the components involved in the formation of this hinge joint. Notice the concentric contraction of the biceps brachii, and the buldge deep (beneath) to it is the brachialis muscle assisting in the process.

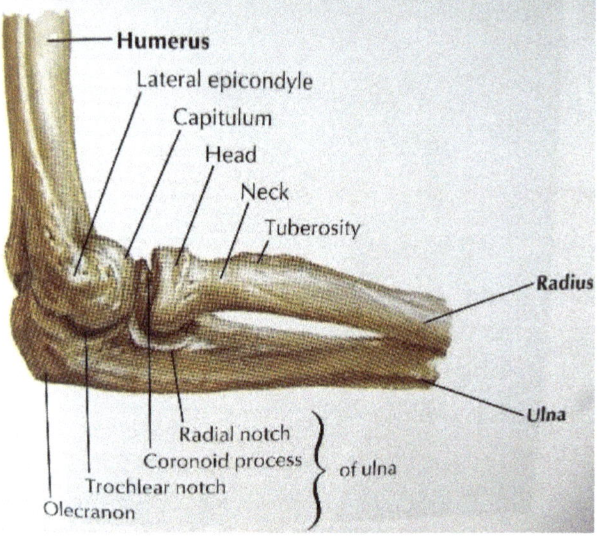

The triceps brachii is a 3-headed muscle located on the posterior aspect of the arm (brachium). It originates from the lateral scapula and the diaphysis of the humerus and inserts onto the olecranon process of the ulna. Its primary action is extension of the elbow. It is usually difficult to visualize the medial head even in a muscular person whereas the lateral and long heads can readily be seen.

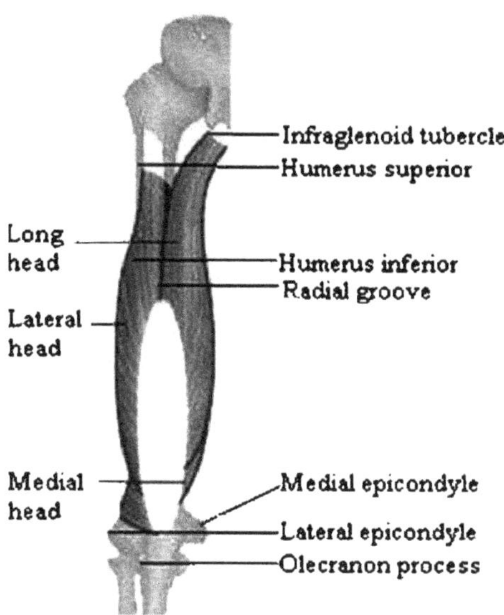

The forearm (antebrachium) muscles of importance are the brachioradialis (already mentioned as an elbow flexor), and the others are the two flexors and two extensors of the wrist. Simply keep in mind that when the two flexors work together, they are synergists and therefore flex the wrist and likewise for the extensors, when both extensors work together they extend the wrist.

- **The 2 wrist flexors are located ventrally in the forearm**:
  a. *Flexor carpi radialis* – runs along the lateral forearm on the thumb side, and obtained its name due to its proximity to the radius.
  b. *Flexor carpi ulnaris* – so named because of its proximity to the ulna (medial forearm – the side of the little finger when standing in the anatomical position)
- **The 2 wrist extensors are located dorsally in the forearm**:
  a. *Extensor carpi radialis longus* – besides being dorsally on the forearm, it is on the thumb side where the radius is found (lateral forearm). This muscle extends the wrist.
  b. *Extensor carpi ulnaris* – this muscle is also dorsally but medially located in the forearm.

Here in the illustration below, the left side shows extension of the wrist and on the right side, flexion of the wrist. As may be already understood, the wrist maintains its high degree of flexibility because of the carpal bones.

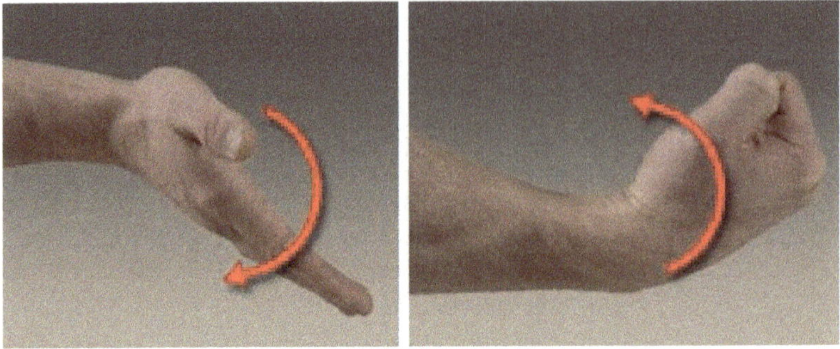

Anterior (picture on the left) and posterior (picture on the right) views of the forearm showing the ventrally located flexors of the wrist and the dorsally located extensors of the wrist respectively.

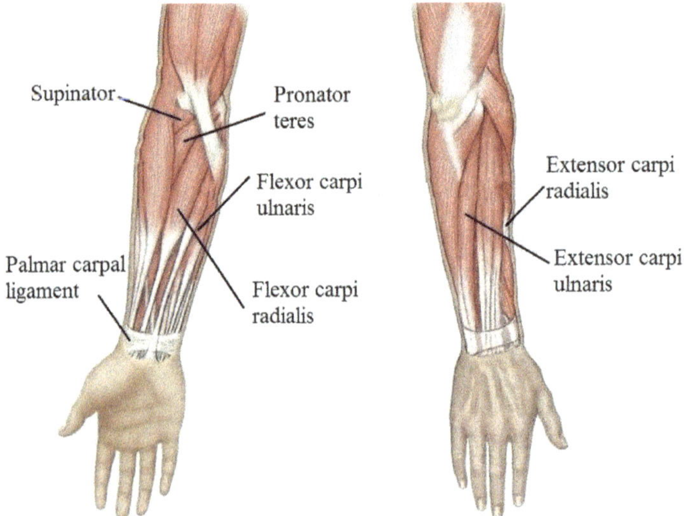

It should be noted that when the flexor carpi ulnaris and the extensor carpi ulnaris work together, their action leads to adduction of the wrist (as seen below on the left). And when the extensor carpi radialis longus and flexor carpi radialis work together, their action is then abduction of the wrist (bending of the wrist towards the thumb as seen on the right).

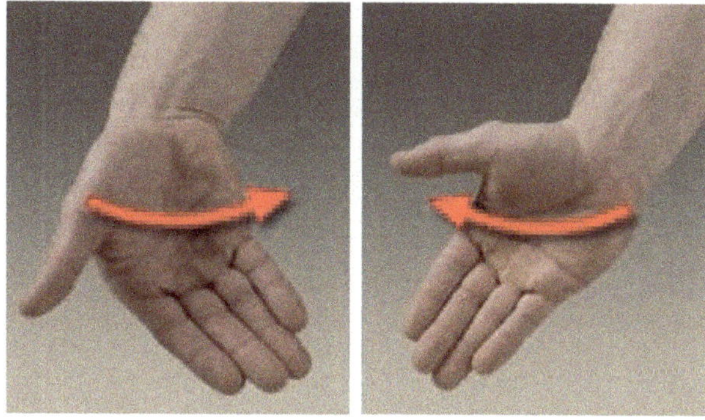

Notably as well is the small deep supinator muscle in the forearm supinates the forearm and the pronator teres sandwiched between the flexor carpi radialis and the brachioradialis muscles pronates the forearm (see illustration of forearm above).

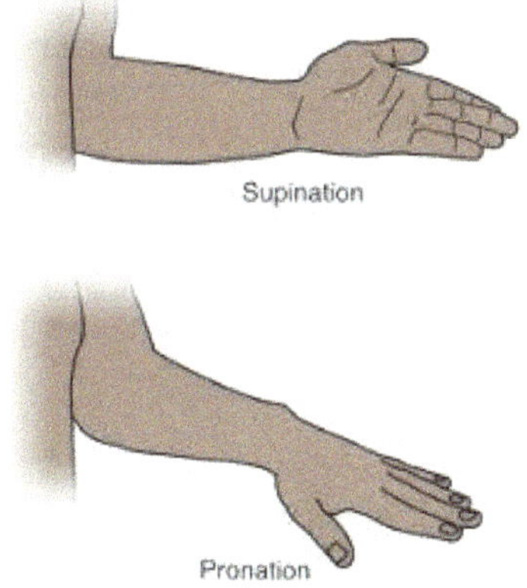

Muscles of the back

1. *Latissimus dorsi* – named so because it is wide (latissimus) and is located posteriorly (dorsi). It extends and adducts the humerus (arm) (think of rowing exercises at the gym).

Of note, we can see the splenius capitis, part of the sternocleidomastoid muscle, the trapezius and many other muscles already discussed as well.

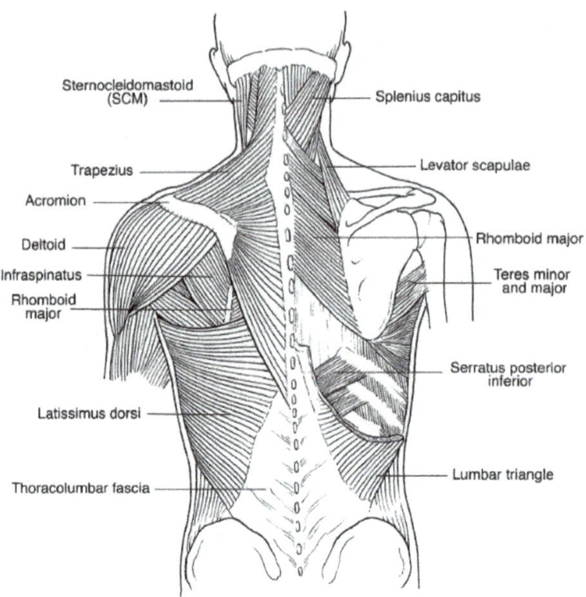

Quadratus lumborum (QL) is paired like the other muscles. It stabilizes the lumbar spine and is also involved in inspiration (inward breathing). Unilateral contraction results in flexion of the trunk, while bilateral contraction assists with extension of the trunk.

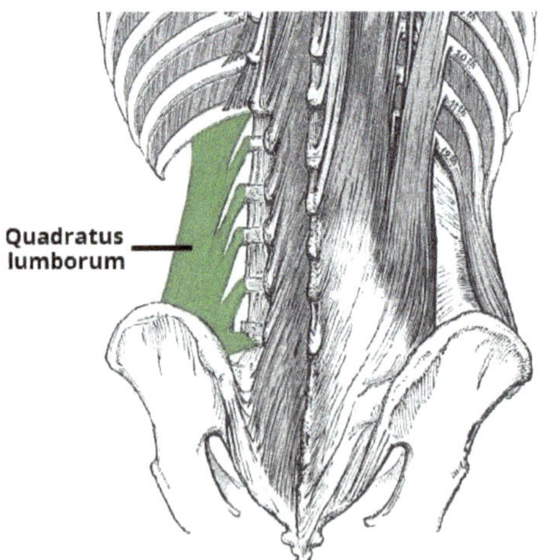

Erector spinae is a group of 3 muscles that are deep back muscles next to the spine. Their primary action is to extend the trunk, and therefore are antagonist to abdominal flexors (rectus bdominis, external obliques, etc…). Generally, the upper body muscles (including the upper extremities) located in the back function as extensors and the ventrally located muscles of the upper body are flexors. The opposite is true for the lower body muscles (lower extremities).

1. Spinalis
2. Longissimus
3. iliocostalis

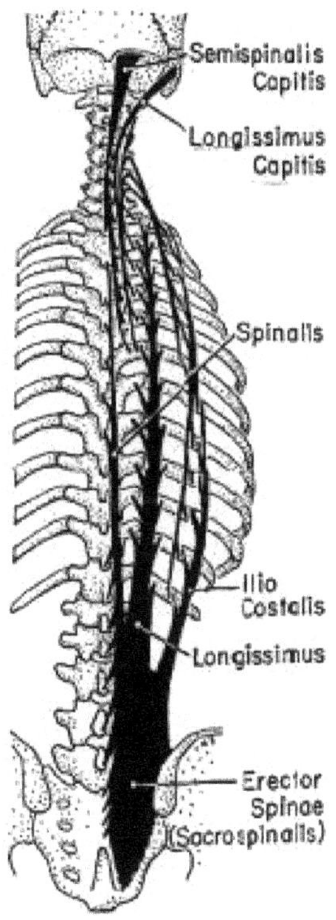

The rotator cuff **SITS** on the shoulder, a mnemonic device to remember this group of four muscles that stabilize the shoulder joint and perform a variety of movements:

S – subscapularis – located on anterior scapula, primary action is medial rotation and adduction of the humerus

I – infraspinatus – laterally rotates and abducts the humerus.

T - teres minor – also laterally rotates and abducts the humerus, therefore a synergist of the infraspinatus.

S - supraspinatus – abduction of the humerus (lateral movement of arm away from the body's midline).

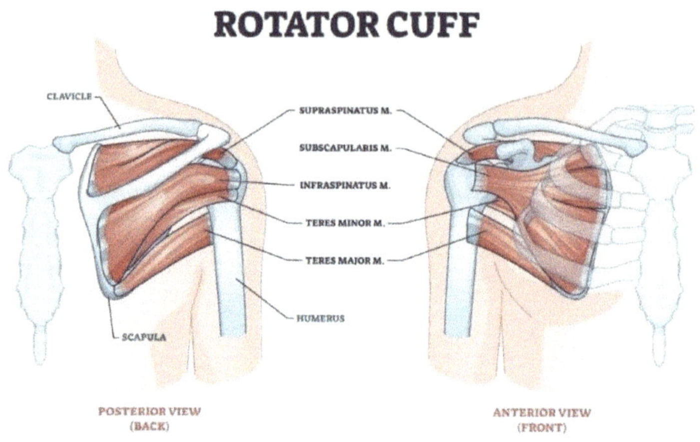

Know all the thigh adductors except adductor brevis and pectineus. The **iliopsoas** is a combination of the **psoas major** (the only lower extremity muscle attached to the vertebrae) and the **iliacus**. So, the iliopsoas muscle is a hip flexor (recall body movement terminology). The **sartorius** (sartor meaning tailor because of the cross-leg sitting position of tailors) flexes the hip, abducts and laterally rotates the thigh, and flexes the knee to boot.

Other muscles to consider is the Quadriceps femoris (known as 'the Quads'), a group of 4 muscles that collectively act to extend the knee:

1. Rectus femoris – also flexes the hip/thigh in addition to extending the knee joint (straightening of the lower extremity). It originates from the ilium and inserts on the patella (knee cap). It obtained its name due to its proximity to the femur and its vertical extension.

2. Vastus medialis – primary action is knee extension

3. Vastus lateralis – extension of the knee

4. Vastus intermedius – not visible here because it is deep to the rectus femoris. It also extends the knee.

Adduction of thigh

Muscles originate <u>medial</u> to hip joint

- Gracilis
- Adductor magnus
- Adductor longus
- Adductor brevis
- Pectineus

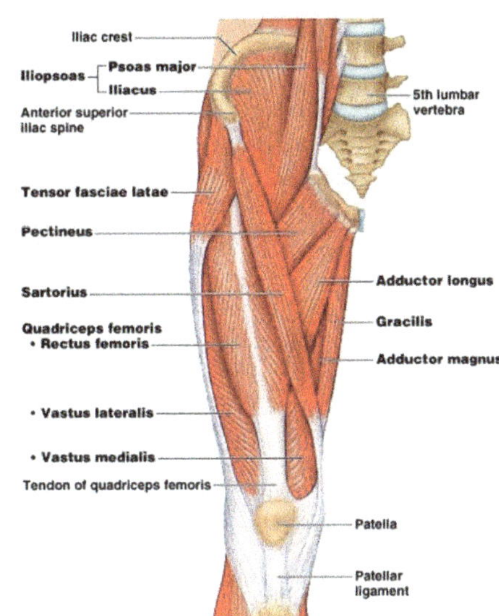

Anterior and posterior views of the thigh muscles – note the quadriceps femoris (vastus intermedius is member to the quadriceps but is not viewable due to its location deep to the rectus femoris).

The **hamstrings** is a group of 3 muscles in the posterior thigh in picture (b). Their primary action is **knee flexion** (bending of the knee). The 3 muscles that make up the Hamstrings are:

1. Biceps femoris – large muscle, lateral to the semitendinosus. Origin is from the ischial tuberosity of the hip bone and on the linea aspera of the femur. It inserts on the head of the fibula.
2. Semitendinosus – medial to the biceps femoris and superficial to the semimembranosus.
3. Semimembranosus – deep to the semitendinosus

Picture (a) is showing the quadriceps femoris excluding 2 muscles and a tendon, can you name the missing muscles and the tendon?

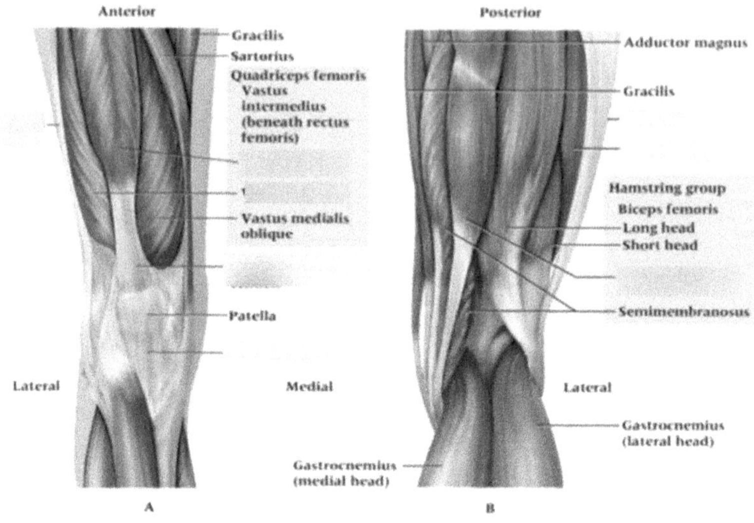

The hip flexors are:

1. Rectus femoris (remember it is part of the quadriceps femoris and also extends the knee joint straightening the lower extremity)
2. Tensor fasciae latae (TFL) – the TFL also abducts the thigh.
3. Iliopsoas

The gluteus maximus is pictured below, it extends the hip and laterally rotates it as well. It is used especially when getting up from a sitting position. Remember the other hip flexors as well. Deep to the gluteus maximus is a powerful hip abductor, the gluteus medius. The gluteus medius is a synergist of the TFL and the sartorius.

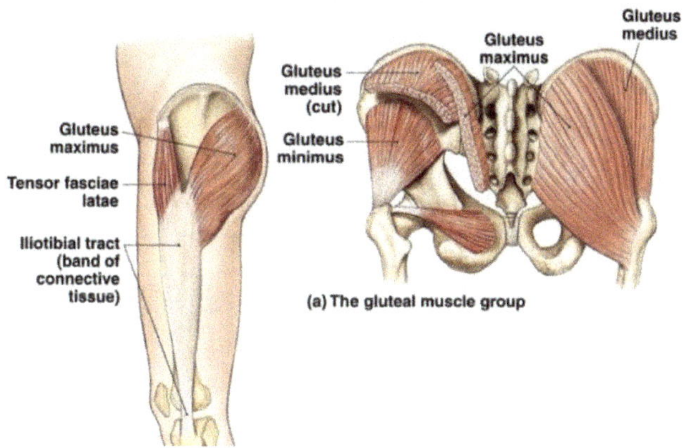

The hip abductors to know are : tensor fasciae latae, gluteus medius, and sartorius.

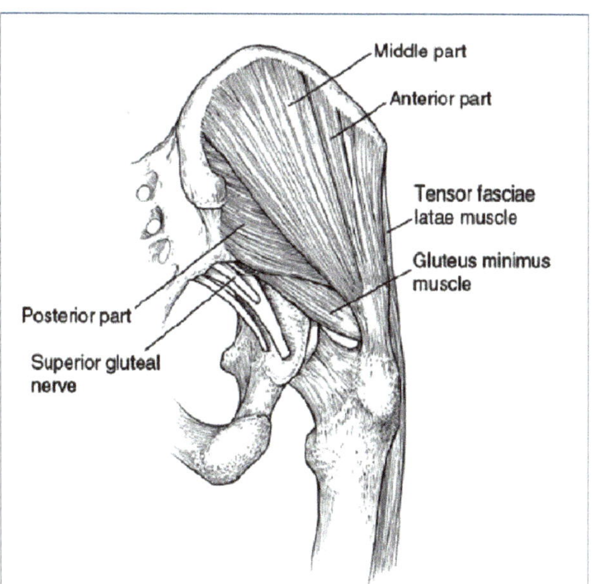

There are 3 muscles of interest involved in ankle movement, the **gastrocnemius** and **soleus** both are located in the posterior lower leg and the **tibialis anterior** slightly located in the lateral tibia in front of the lower leg. There are at least two movements of the ankle to know:

1. dorsiflexion (flexion of the ankle) – pointing toes superiorly, as in standing on the heels

2. plantarflexion (extension of the ankle) – standing on toes like a ballerina

Posterior and lateral views of the **gastrocnemius** and **soleus** muscles whose action is plantarflexion. The gastrocnemius originates from the medial and lateral condyles of the femur and inserts on the posterior calcaneus (heel bone) via the calcaneal tendon (Achilles tendon).

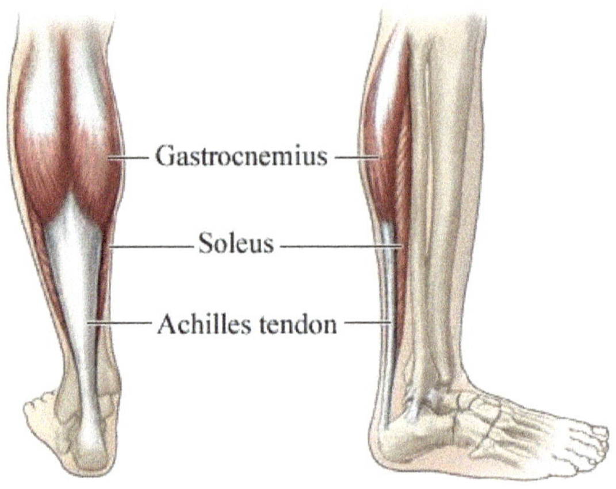

Below is the lateral view of the **tiabialis anterior**- its action is to dorsiflex the ankle. All three muscles can be view anteriorly depending on the angle of view. Note the lateral view both the gastrocnemius and the soleus.

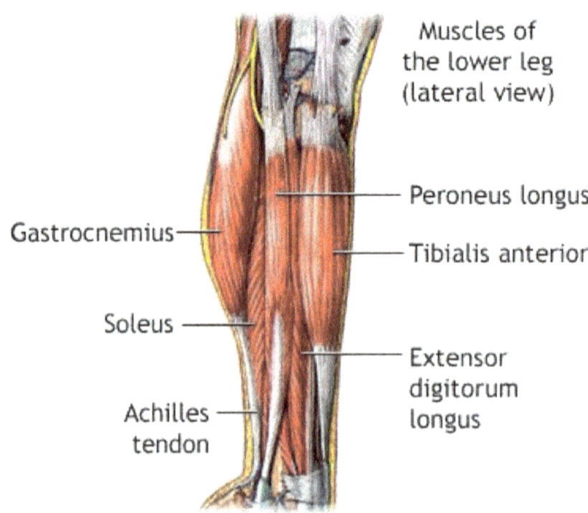

Anterior view of tibialis anterior and gastrocnemius

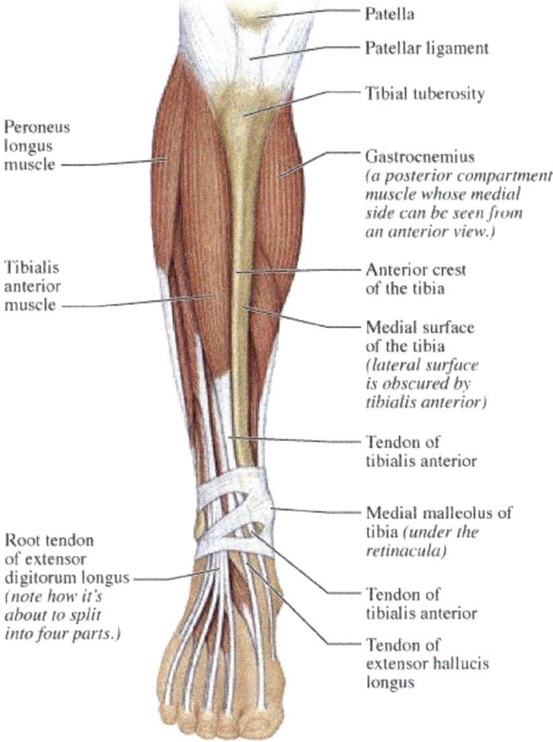

Practice identifying these muscles and their function

Muscles of the Head and Neck

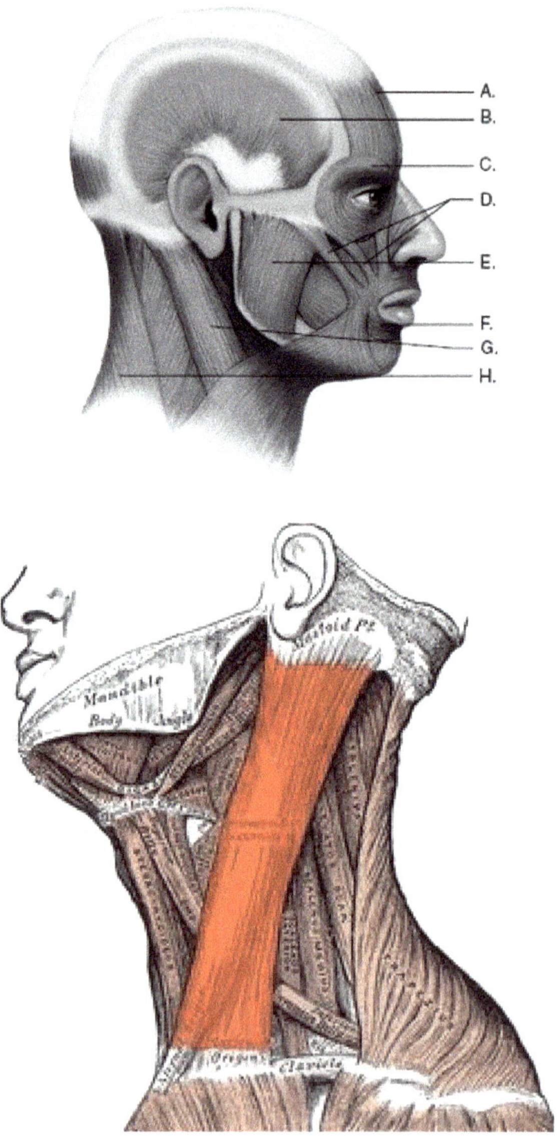

Which muscles allow her to turn her head in that position ? Be specific regarding the location (i.e., right or left) of these muscles.

Location of Superficial Muscles

A. Match these terms with the correct parts labeled in figure 7.3:

Adductors of thigh
Biceps brachii
Brachioradialis
Deltoid
External abdominal oblique
Flexors of the wrist and fingers
Pectoralis major
Quadriceps femoris
Rectus abdominis
Rectus femoris
Sartorius
Serratus anterior
Sternocleidomastoid
Tensor fasciae latae
Vastus lateralis
Vastus medialis

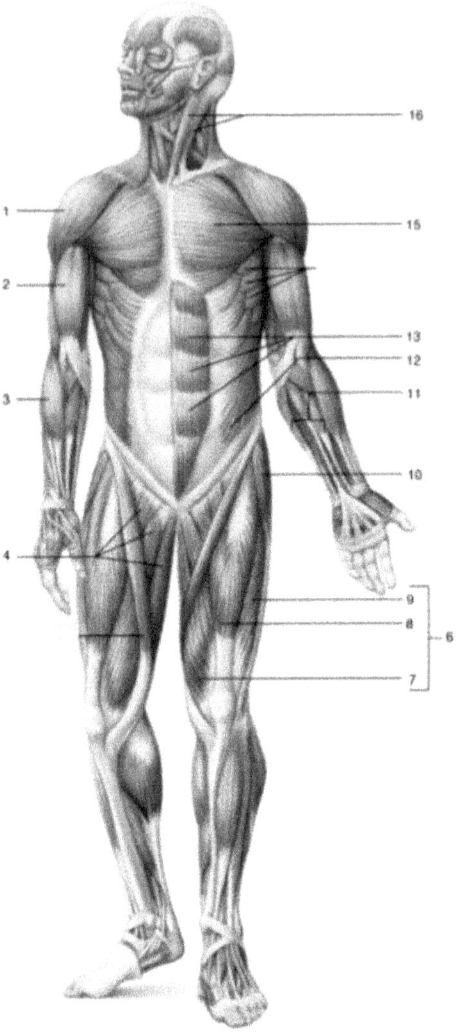

1. ______________________
2. ______________________
3. ______________________
4. ______________________
5. ______________________
6. ______________________
7. ______________________
8. ______________________
9. ______________________
10. ______________________
11. ______________________
12. ______________________
13. ______________________
14. ______________________
15. ______________________
16. ______________________

Part II
Human Anatomy and Physiology II

Chapter 1: The Endocrine System: An Overview

The endocrine and nervous systems regulate the activities of organs and systems through long-distance communication via chemical (i.e., molecules like hormones and neurotransmitters) signaling.

Both the endocrine and nervous systems use **chemical messengers** to communicate with their target cells and organs. The endocrine system secretes hormones into the blood, while the nervous system secretes neurotransmitters into synapses.

Hormones secreted by the endocrine glands travel through the bloodstream to their specific target cells and organs. Target cells contain receptors that bind specifically to the hormones that affect them.

Three types of hormones of the endocrine system:

a. **Steroid Hormones** are made from cholesterol, so they are lipid-based and hence lipophilic (fat-loving, fat-soluble), and like cholesterol, they are hydrophobic (not soluble in water, does not mix well with water). As a result, they can easily enter their target cells by diffusing through the phospholipid bilayer of the cell membrane (recall the composition and structure of the cell membrane from Biology of the Cell chapter in A&P I). They reach intracellular receptors and act as transcription factors. Examples of steroid hormones include **testosterone** (male characteristic hormone) and **estrogen** (female-characteristics hormone), both produced by the **gonads** (testes in males and ovaries in females, respectively).
b. **Amine Hormones** are produced from the modification of a single amino acid (amino acids bind together to form a protein molecule, thus known as the building blocks of proteins). In other words, an amine hormone is a derivative of an amino acid. Examples include **melatonin**, which is the

hormone secreted by the pineal gland to help regulate the sleep cycle (circadian rhythm) is derived from the amino acid, tryptophan; **epinephrine** (also called adrenalin), and **norepinephrine** (known as noradrenalin), which play a role in the fight-or-flight response are both secreted from the medulla of the adrenal gland are derived from the amino acid, Tyrosine.

c. **Peptide and Protein Hormones,** on the other hand, are derived from the linkage of multiple amino acids (i.e., a **peptide** defined as a chain of up to 50 amino acids, or **a protein** defined as a chain greater than 50 amino acids).

An example of a peptide hormone is **antidiuretic hormone** (ADH), a pituitary hormone made up of nine amino acids. ADH plays a role in water retention in the body, hence its importance in fluid balance. An example of a protein hormone is **insulin**, which consists of 51 amino acids produced by the pancreas to decrease blood sugar levels by promoting the uptake of glucose into body cells.

Steroid hormone (cholesterol, cortisol)

Cholesterol

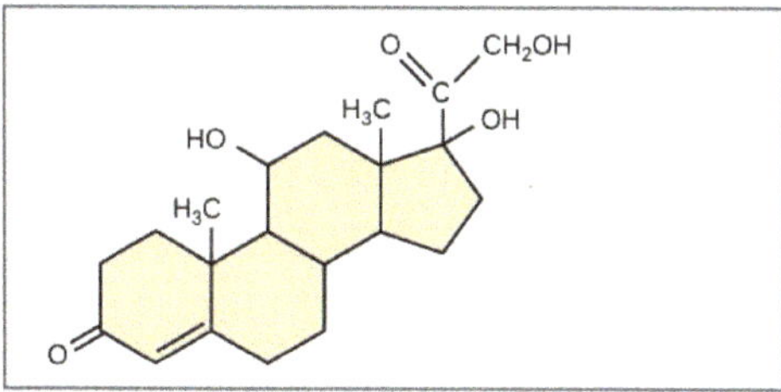

Cortisol

Amine hormone structure

Tyrosine Epinephrine Tryptophan Melatonin

(a) (b)

Peptide hormone structure

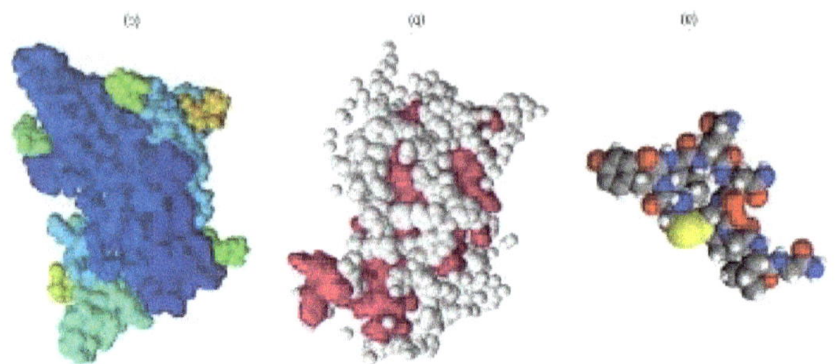

Major Glands of the Endocrine System

The four major glands of the endocrine system are the **pituitary**, **adrenal**, **thyroid**, and **parathyroid glands**. Other organs that have endocrine roles in addition to their other functions include the pancreas, stomach, small intestine, testes, and ovaries.

a. The **pituitary gland** is a small, round, pea-sized structure found within the sella turcica of the sphenoid bone. Anatomically, this gland is attached to the hypothalamus via a thin stem of nervous tissue called **infundibulum** (also called pituitary stalk). The pituitary gland plays an important role in the body and contains two lobes, the anterior lobe or **adenohypophysis** and the posterior lobe or **neurohypophysis** (please note anatomists may refer to these lobes as two distinct glands, anterior pituitary and posterior pituitary).
    - **The anterior pituitary** or **adenohypophysis** secretes various hormones, including growth hormone (**GH**), which stimulates bone and muscle growth, luteinizing hormone (**LH**) and follicle-stimulating hormone, which

stimulate endocrine activity in the testes and ovaries, respectively, and prolactin, which stimulates milk production in females.

- The **posterior lobe** or **neurohypophysis** stores and secretes hormones produced within the hypothalamus. Posterior pituitary hormones include **antidiuretic hormone (ADH)**, which decreases urine out by water retention in the body and increases blood pressure, and **oxytocin**, which plays a role in uterine contraction during childbirth.

Pituitary gland anatomy

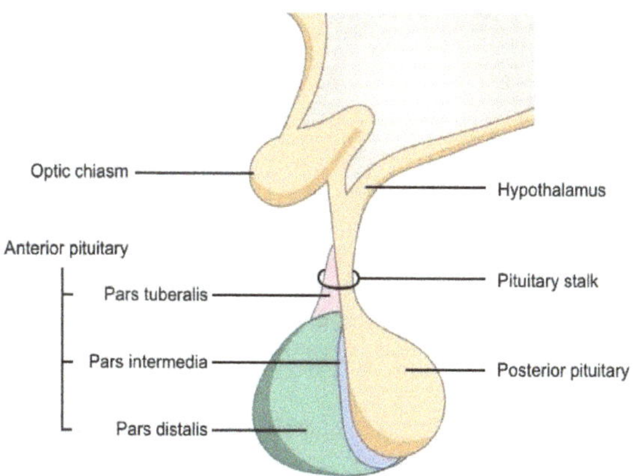

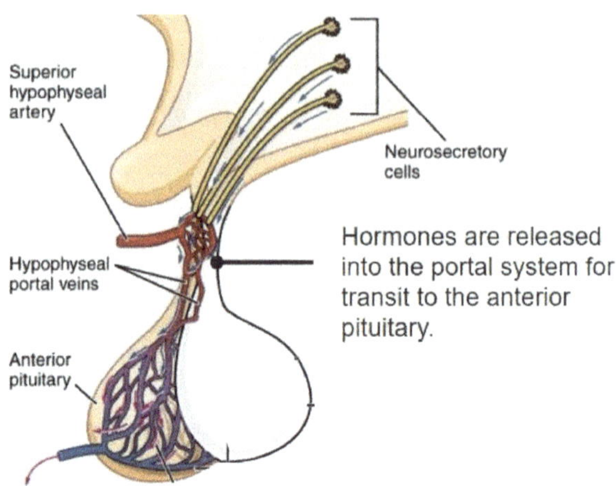

b. The **adrenal glands** or suprarenal renal glands are paired glands, each one sitting on top of a kidney. Each adrenal gland is encapsulated by a connective tissue **capsule** and held in place by adipose tissue all around the kidney. In cross-section, the adrenal gland has two primary regions: an external **adrenal cortex** and an internal **adrenal medulla**. The cortical region (adrenal cortex) secretes its own distinct group of steroid hormones collectively known as **corticoids.** Anatomically, the cortex is subdivided into three distinct layers of tissue:

- The **zona glomerulosa** produces and secretes **aldosterone**, a major mineralocorticoid due to its effects on the minerals in the blood, especially sodium and potassium (aldosterone promotes the retention of sodium and the excretion of potassium). It is secreted when blood volume is low, high potassium level, low sodium level, and low blood pressure (hypotension). So, it **regulates blood pressure** by increasing it to normal.
- The **zona fasciculata** is found deep in the zona glomerulosa and produces the stress hormone **cortisol**, a glucocorticoid, because of its role in glucose metabolism. In response to long-term stress, the hypothalamus

secretes corticotropic releasing hormone (CRH), which in turn triggers the release of adrenocorticotropic hormone (ACTH) by the pituitary. ACTH, in turn, causes the adrenal cortex to release glucocorticoids.

- The **zona reticularis** is the deepest/innermost zone of the adrenal cortex but superior to the adrenal medulla. This zone produces androgens, which are the sex hormone precursors that supplement the androgens from the gonads (testes and ovaries). During puberty and most of adulthood, androgens are produced by the gonads.
- The **adrenal medulla**, the central portion of the adrenal gland, is connected to the central nervous system and functions as a component of the sympathetic nervous system. When stimulated by the nerve signals from the sympathetic nervous system, the adrenal medulla secretes the **catecholamines (80% epinephrine and 20% norepinephrine)** into the bloodstream. Please note that these two compounds are considered **hormones** because of their effects on distant target cells or organs when released in the systemic circulation. On the other hand, these catecholamines are referred to as **neurotransmitters** because they can be secreted via exocytosis by nerve endings into synapses.

Adrenal gland, gross and microscopic anatomy

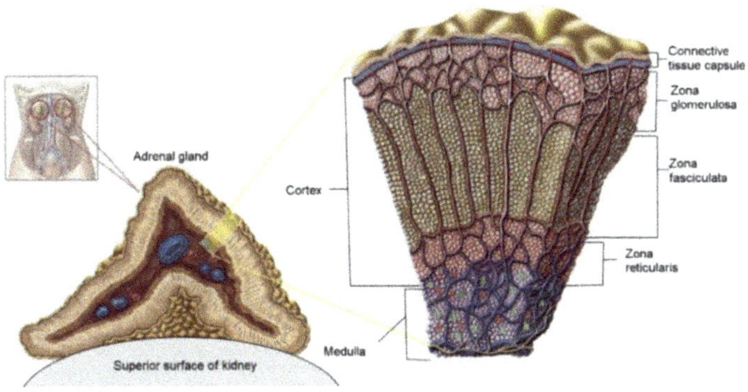

c. **The thyroid gland** is a butterfly-shaped gland that overlies the anterior and lateral surfaces of the trachea at the base of the cervical region (neck), inferior to the larynx. It consists of a pair of lobes connected by a thin piece of thyroid tissue called isthmus. The thyroid gland makes **thyroid hormones** (**triiodothyronine** (**T3**) and **thyroxine (T4)**, as well as **calcitonin**. The thyroid hormones stimulate cell metabolism, development, and growth and have many other functions. Calcitonin helps maintain homeostatic levels of blood calcium.

Under the microscope, the glandular tissue is composed mostly of thyroid follicles. The follicles are made of a ring of simple cuboidal epithelium made of **follicular cells** or **principal cells** that produce thyroglobulin, the precursor to the active form of the thyroid hormone (T3), which is then stored within the space of the follicle in a a sticky fluid is known as **a colloid.** Thyroid-stimulating hormone (TSH) from the anterior pituitary triggers the principal cells to take up thyroglobulin to convert it into the activated form of thyroid hormone (T3 and T4) and release it into the bloodstream.

Histology of the thyroid gland follicles showing **colloid** within the follicles and labeled C-cells, and principle cells.

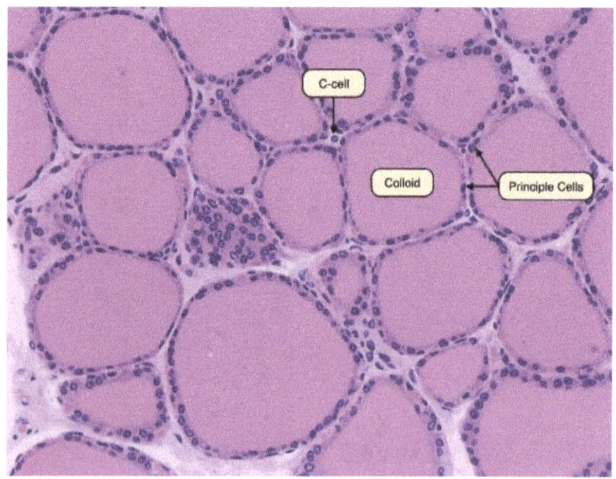

Thyroid gland location and anatomy (gross and microscopic) below

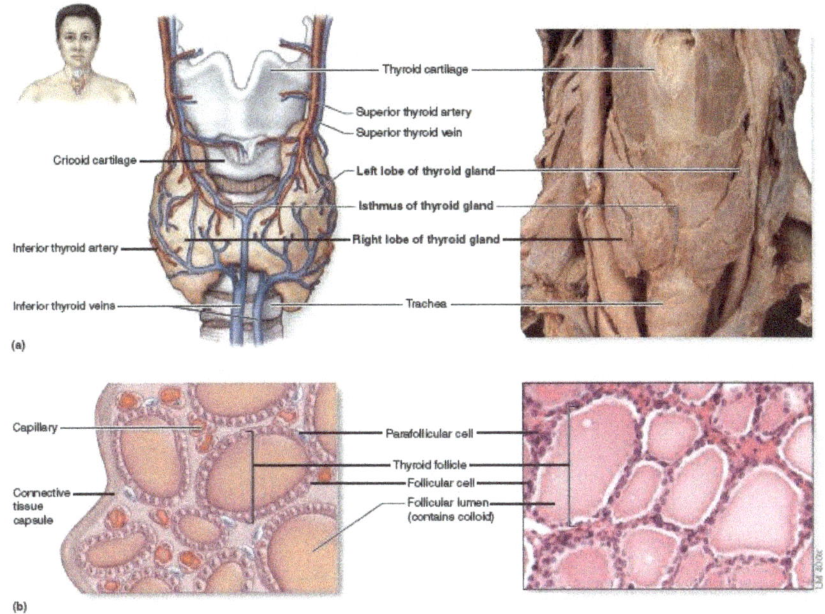

In between the follicles, there are blood vessels and **parafollicular cells** of **C cells**, which produce the hormone calcitonin when blood calcium levels are high. Calcitonin functions by inhibiting the activity of osteoclasts, thereby lowering calcium levels.

d. The **parathyroid glands** are tiny, round structures embedded in the posterior edges of the thyroid gland. Most individuals have four of these glands, though the number may vary in some people. The parathyroid glands produce **parathyroid hormone (PTH)**, a peptide hormone, and release it into the systemic circulation when blood calcium levels are low. PTH, therefore, works to **increase calcium levels** in the blood. It does so by increasing calcium reabsorption in the intestines (PTH initiates the production of the steroid hormone **calcitriol**, also known as active vitamin D or 1, 25 dihydrovitamin D. In turn, calcitriol helps dietary calcium reabsorption in the intestines), preventing calcium loss in the urine by substituting phosphate for calcium in the kidney, and by promoting the activity of osteoclasts and inhibiting osteoblasts activity. Please note that PTH is secreted in response to low calcium levels.

Microscopically, unlike the thyroid gland, the parathyroid glands have no follicles. Instead, they have smaller, darker clumps of cells known as **chief cells** that produce parathyroid (PTH). The parathyroid glands also contain fewer numerous, larger, paler cells known as **oxyphil** cells, the function of which is unknown.

Parathyroid glands

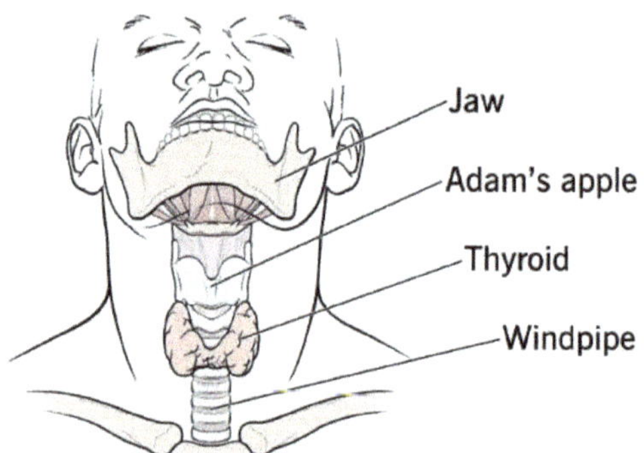

Front view of thyroid

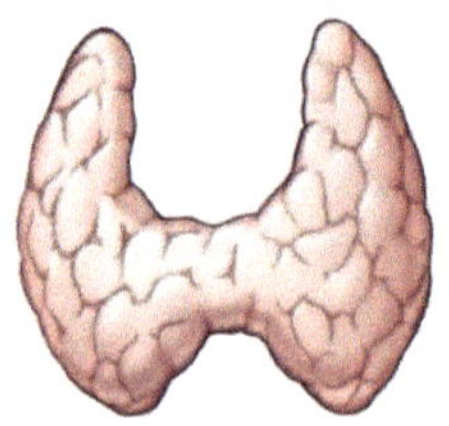

Back view of thyroid

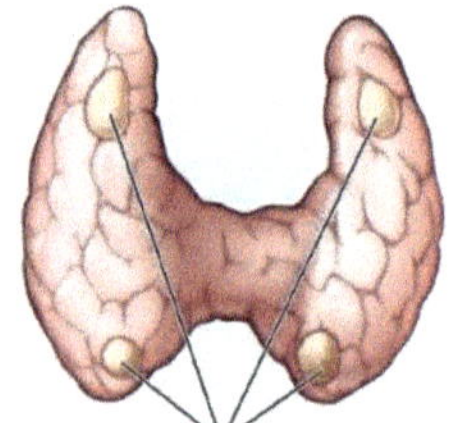

Parathyroid glands

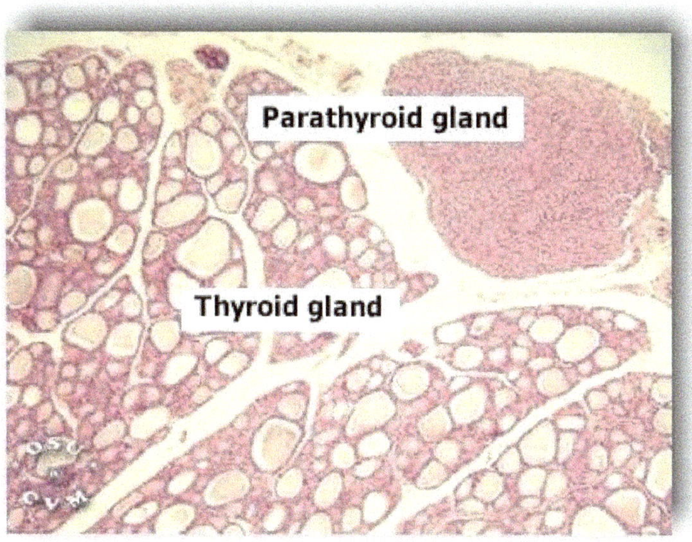

Practice questions

- Please examine the histology slide below. Can you differentiate the thyroid gland from the parathyroid gland?
- Can you identify the follicles, colloid, parafollicular cells, chief cells, and principles cells?
- What is the mechanism of action of calcitonin and PTH?

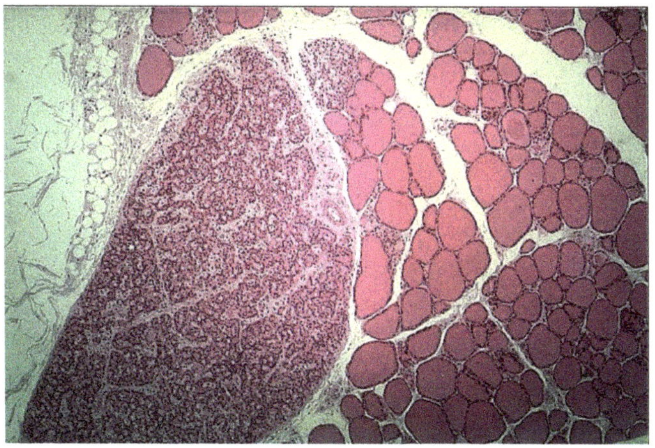

Chapter 2: Cardiovascular System: Blood

With this chapter, we begin our study of the cardiovascular (circulatory) system. This system consists of a fluid that circulates throughout the body (blood), a pump (the heart), and a series of tubes through which the blood flows (arteries, capillaries, and veins).

I. Functions and General Composition of Blood

Blood is a type of connective tissue. Cells (*i.e.,* **formed elements**) of the blood are located within a liquid matrix, which is known as the **plasma**. Formed elements of the blood include white blood cells (**leukocytes**), red blood cells (**erythrocytes**), and **platelets**.

The plasma and formed elements together constitute **whole blood** (see illustration below). Erythrocytes account for about half of the volume of whole blood. The percentage of blood volume occupied by erythrocytes is called the **hematocrit**. Nearly all of the remainder of the whole blood is plasma. Leukocytes and platelets account for less than 1% of blood volume. How is the hematocrit of a blood sample measured?

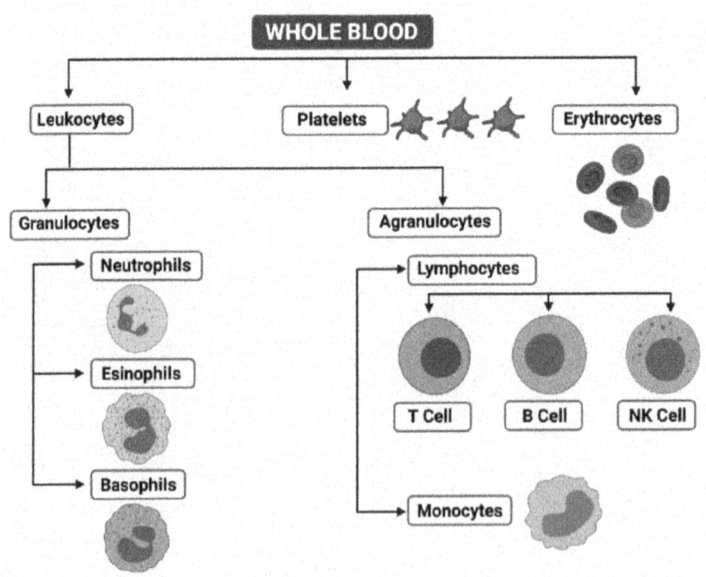

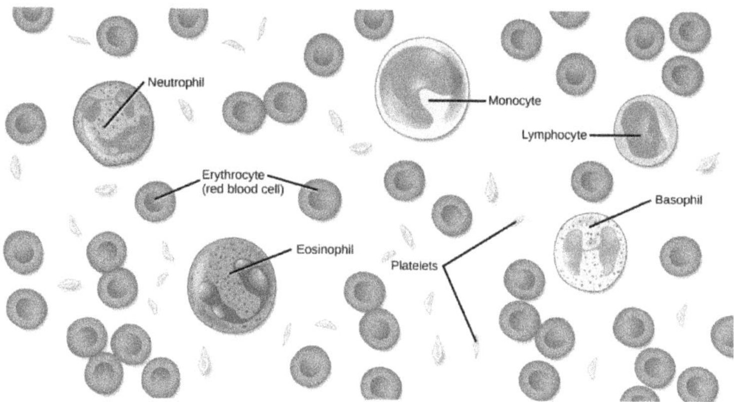

Functions of blood

A. Blood is important for the *transportation* of materials throughout the body. This includes the following functions:

   1. Delivery of oxygen and nutrients to the tissues of the body.
   2. Transport of carbon dioxide and other waste products from the tissues.
   3. Transport of hormones throughout the body.

B. Blood plays various roles in *the regulation* of homeostasis:

   1. Blood helps maintain body temperature by distributing heat.
   2. Various chemicals in the blood help to regulate pH in the body.
   3. Blood is important in regulating water and salt balance in the body.

C. Blood plays an important role in *the protection* of the body:

   1. Platelets and blood proteins help prevent blood loss.
   2. Leukocytes and other components of the immune system help defend the body against infection.

Physical characteristics of blood

Because blood contains formed elements and a variety of dissolved solutes, it is considerably more viscous (about five times more viscous) than pure water and more resistant to flow. Blood's pH is slightly alkaline, between 7.35 and 7.45. Would you consider blood with a pH of 7.25 to be acidic or alkaline?

The color of blood ranges from bright scarlet red (when well oxygenated) to dark red (when poorly oxygenated). A typical adult man's blood volume is 5 to 6 liters, and that of a typical adult woman is slightly less, about 4 to 5 liters.

II. Composition of Blood Plasma

Normal blood plasma contains over 100 different chemicals. However, water accounts for about 92% of the plasma volume. Proteins make up a significant fraction of the substances found in the plasma, approximately 8 g per 100 ml of plasma. Some of the more important plasma proteins are as follows:

Albumins account for about 60% of plasma proteins. These proteins are generally involved in transporting lipids within the bloodstream. Why do lipids need assistance for transport in the blood?

Globulins are important for transport and protection. Transport globulins carry thyroid hormones, metal ions, triglycerides, and steroids. **Immunoglobulins** are proteins made by the immune system to help attack foreign invaders.

Fibrinogen and **prothrombin** are used in the process of blood clotting. You will learn more about them later in this chapter.

Most plasma proteins are produced in the liver, and certain liver disorders can, therefore, alter the properties of the blood.

You should be generally familiar with the plasma solutes listed.

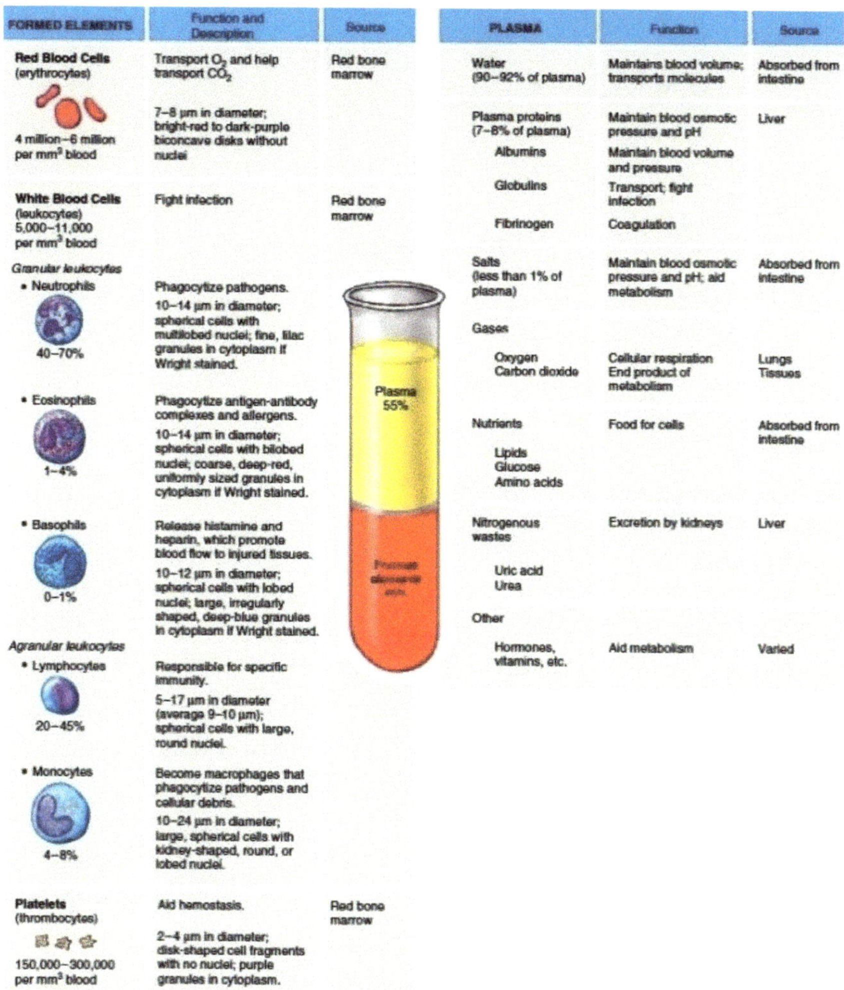

FORMED ELEMENTS	Function and Description	Source
Red Blood Cells (erythrocytes) 4 million–6 million per mm^3 blood	Transport O_2 and help transport CO_2. 7–8 µm in diameter; bright-red to dark-purple biconcave disks without nuclei	Red bone marrow
White Blood Cells (leukocytes) 5,000–11,000 per mm^3 blood	Fight infection	Red bone marrow
Granular leukocytes		
• Neutrophils 40–70%	Phagocytize pathogens. 10–14 µm in diameter; spherical cells with multilobed nuclei; fine, lilac granules in cytoplasm if Wright stained.	
• Eosinophils 1–4%	Phagocytize antigen-antibody complexes and allergens. 10–14 µm in diameter; spherical cells with bilobed nuclei; coarse, deep-red, uniformly sized granules in cytoplasm if Wright stained.	
• Basophils 0–1%	Release histamine and heparin, which promote blood flow to injured tissues. 10–12 µm in diameter; spherical cells with lobed nuclei; large, irregularly shaped, deep-blue granules in cytoplasm if Wright stained.	
Agranular leukocytes		
• Lymphocytes 20–45%	Responsible for specific immunity. 5–17 µm in diameter (average 9–10 µm); spherical cells with large, round nuclei.	
• Monocytes 4–8%	Become macrophages that phagocytize pathogens and cellular debris. 10–24 µm in diameter; large, spherical cells with kidney-shaped, round, or lobed nuclei.	
Platelets (thrombocytes) 150,000–300,000 per mm^3 blood	Aid hemostasis. 2–4 µm in diameter; disk-shaped cell fragments with no nuclei; purple granules in cytoplasm.	Red bone marrow

PLASMA	Function	Source
Water (90–92% of plasma)	Maintains blood volume; transports molecules	Absorbed from intestine
Plasma proteins (7–8% of plasma)	Maintain blood osmotic pressure and pH	Liver
Albumins	Maintain blood volume and pressure	
Globulins	Transport; fight infection	
Fibrinogen	Coagulation	
Salts (less than 1% of plasma)	Maintain blood osmotic pressure and pH; aid metabolism	Absorbed from intestine
Gases		
Oxygen	Cellular respiration	Lungs
Carbon dioxide	End product of metabolism	Tissues
Nutrients	Food for cells	Absorbed from intestine
Lipids, Glucose, Amino acids		
Nitrogenous wastes	Excretion by kidneys	Liver
Uric acid, Urea		
Other		
Hormones, vitamins, etc.	Aid metabolism	Varied

In the test tube above the thin filmy layer separating the straw-colored plasma and the erythrocytes below it is called the '*Buffy coat*' that contains the white blood cells and platelets.

Cell Types	Image	Diameter	Range (per µL)	Nucleus	Cytoplasm	Granules
Basophil		10-14 µm	0.5-1% of WBC 20-50	Bi-lobed or tri-lobed	Pale blue	Large purplish-black cytoplasmic
Eosinophil		10-14 µm	2-4% of WBC 100-400	Bi-lobed	Full of granules	Orange-red
Lymphocyte		5-17 µm	25-40% of WBC 1500-3000	Spherical or indented	Clear, Pale blue	-
Monocyte		14-24 µm	3-8% of WBC 100-700	U or kidney shaped	Gray-blue	Fine reddish (azurophil)
Neutrophil		10-12 µm	50-70% of WBC 3000-7000	2 to 5 segments or lobes	Pale blue-pink	Inconspicuous cytoplasmic
Erythrocytes		7-8 µm	4-6 million	Biconcave, anucleate disc; salmon-colored		
Thrombocytes		2-4 µm	150-500 thousand	Discoid cytoplasmic fragments containing granules; stain deep purple		

III. Formed Elements in the Blood

Hemopoiesis

Formed elements in the blood arise via the process of **hemopoiesis** (or hematopoiesis). In embryonic humans, this process occurs in the liver, spleen, thymus, and bone marrow. As humans mature into adults, bone marrow becomes the primary site of both red and white cell production.

Refer to the chart below to examine the steps in hematopoiesis. Also note that in the normoblast phase of erythropoiesis, the nucleus is ejected and the red cell is then without a nucleus (anucleate). Clinically, when a person is being treated for iron deficiency anemia (low iron levels and low red blood cell count causing weakness, easy fatiguability and increased heart rate), the patient is normally treated with iron sulfate tablets. To test whether the treatment is effective, the clinician will send a sample to the lab for a reticulocyte count. If the lab result shows an elevated reticulocyte count then it means the treatment is effective. If on the other hand, the reticulocyte count comes back low, then iron deficiency is not the cause of the anemia. Therefore, the cause of the anemia must be investigated since there exist other forms of anemia (megaloblastic anemia, anemia of chronic disease, sideroblastic anemia, thalassemia, etc…).

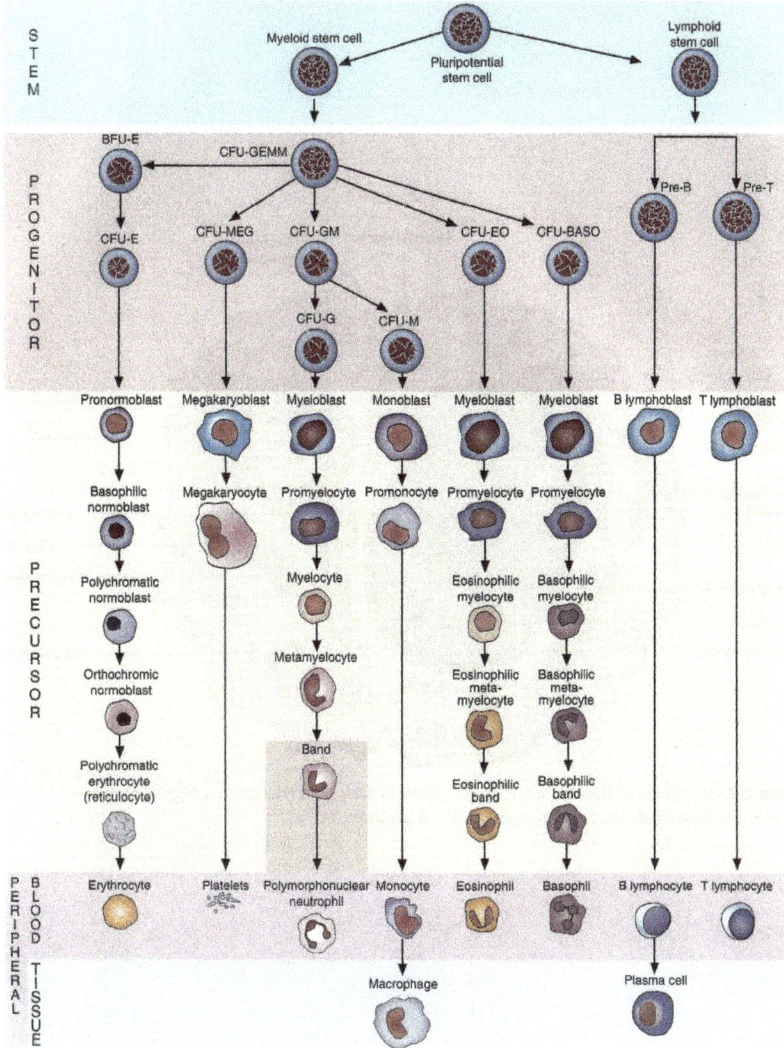

Erythrocytes

The most abundant cells in the blood are the erythrocytes. Mature erythrocytes are **anucleate** cells containing little more than a cytoskeleton and the protein, **hemoglobin**. Without a nucleus and other cellular machinery, the life span of a mature erythrocyte is short, about 120 days. How would you describe the shape of an erythrocyte?

Red Blood Cells (erythrocytes)

1. Function

erythrocyte as a bag for hemoglobin

- O_2 → transport, reactive oxygen species (ROS)
- CO_2 → transport, formation of HCO_3^-
- H^+ → transport, maintaining pH
  (35% of blood buffering capacity)

7 µm

Top View shows RBC to be circular

2. Structure

- large surface
  (for diffusion of gases)
- cytoskeletal proteins
  (for elasticity)
- membrane as an osmometer
  (Na^+/K^+-ATPase)

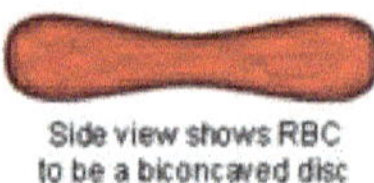

Side view shows RBC to be a biconcaved disc

Hemoglobin molecular structure and structural formula of heme.

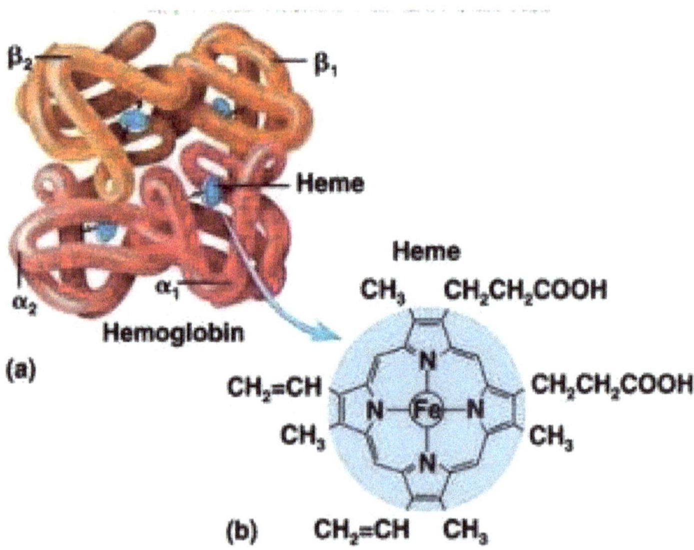

In a typical adult male, there are 5.1-5.8 million erythrocytes per mm^3 of whole blood; in females, the value is 4.3-5.2 million per mm^3. The normal range for the hematocrit of adult males is 4054%, and the normal range for females is 37-47%. What factors contribute to these differences between men and women?

The functional part of an erythrocyte is the hemoglobin molecule. Each molecule of hemoglobin contains four polypeptides (two "alpha" chains and two "beta" chains). Attached to each polypeptide is a structure called a **heme**. Each heme unit contains an iron atom (Fe), which can bind to a molecule of oxygen (O_2). Thus, each hemoglobin molecule can bind up to four molecules of oxygen. Hemoglobin bound to oxygen is referred to as **oxyhemoglobin**, and hemoglobin without any oxygen bound is referred to as **deoxyhemoglobin**.

Carbon dioxide can also bind directly to the polypeptides of hemoglobin. About 20% of the CO_2 in the blood is bound to hemoglobin.

Erythrocyte production in adults occurs primarily in the red bone marrow.

The production of erythrocytes is largely controlled by the peptide hormone **erythropoietin (EPO)**. **Hypoxia** in the kidneys and periphery largely stimulates EPO production. Why do certain athletes, such as cyclists use EPO?

What are some potentially life-threatening side effects of "blood doping"?

Know the specific steps of erythrocyte formation, they are the following steps: **Hemocytoblast** cells (stem cells) in the marrow divide into **myeloid stem cells**, which in turn give rise to erythrocytes and several types of leukocytes. Myeloid stem cells destined to become erythrocytes first become **proerythroblasts**, then **erythroblasts**, which begin the production of hemoglobin. The erythroblasts lose their nuclei to become **reticulocytes**, which enter the blood and become mature erythrocytes.

Cells in the liver, spleen, and bone marrow typically phagocytize dead and dying erythrocytes. These cells process hemoglobin, recycle the amino acids, and prepare the heme units for excretion. Free iron ions are toxic, and iron removed from hemoglobin needs to be stored safely. These iron ions are bound to the protein **transferrin** and transported through the bloodstream to the liver and spleen, where they are stored. Iron is stored in these organs bound to the proteins, **ferritin** and **hemosiderin**. Iron can then be transferred by transferrin to the bone marrow, where it is incorporated into new erythrocytes. The remainder of the heme unit is converted to **bilirubin**, a yellow pigment that is picked up by the liver, secreted into the small intestine, and converted to a brown pigment called **stercobilin**, which is excreted from the body.

Failure to process erythrocytes properly can result in a variety of clinical conditions:

Hemoglobinuria is an excess of hemoglobin in the urine, which results from the excess breakdown of erythrocytes in the blood without adequate processing; **jaundice** is an excess of bilirubin, which may result from the inability of the liver to excrete the products of heme breakdown.

Leukocytes

Leukocytes are much less numerous than erythrocytes, about 5,000-11,000 per mm^3. They are divided into five types. Review the functions and appearance of each type in your textbook and lab manual, and learn which of the five types are **granulocytes** and which are **agranulocytes**. I will get you started here:

A. **Neutrophils** are also referred to as *polymorphonuclear leukocytes* (PMNs) because of the shape of their nuclei. These are the most numerous leukocytes. They are phagocytic and partial to killing bacteria. The number of neutrophils in the body rises dramatically in response to acute bacterial infections.

B. **Eosinophils** respond to infections by multicellular parasites. They gather around these organisms and release digestive enzymes onto them.

C. **Basophils** are the least numerous of the leukocytes. They migrate to injury sites, where they release histamine and heparin.

D. **Monocytes** enter tissues and become phagocytic cells called **macrophages**. They are especially important in fighting chronic infections (*e.g.,* tuberculosis) and are partial to killing viruses and some bacteria.

E. **Lymphocytes** come in three types: **T cells**, **B cells**, and **natural killer cells**. They are primarily responsible for the body's specific immune response.

Like erythrocytes, leukocytes arise from **hemocytoblasts**. And also, like erythrocytes, neutrophils, eosinophils, basophils, and monocytes develop from **myeloid stem cells**. Lymphocytes arise from hemocytoblasts that differentiate into **lymphoid stem cells**. The life spans of leukocytes vary greatly: granulocytes typically live for just a few hours or days, while agranulocytes may live for just a few hours or many years.

Below are the five types of leukcocytes. Your task is to identify each one.

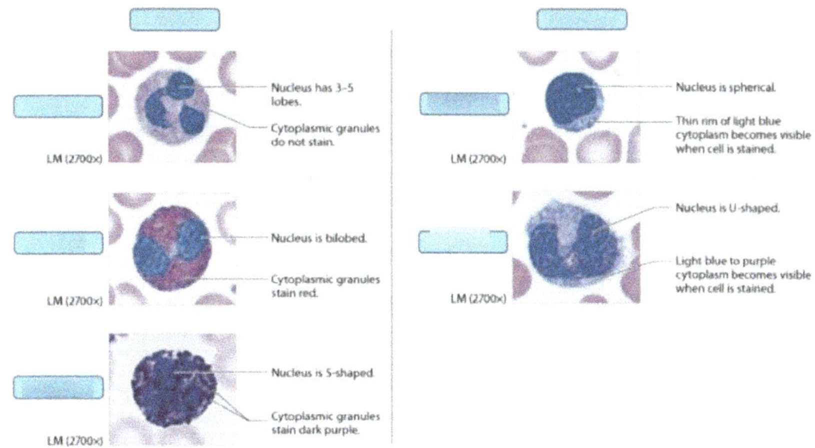

Use the blank spaces below to identify each labeled cell (WBC) with its corresponding function.

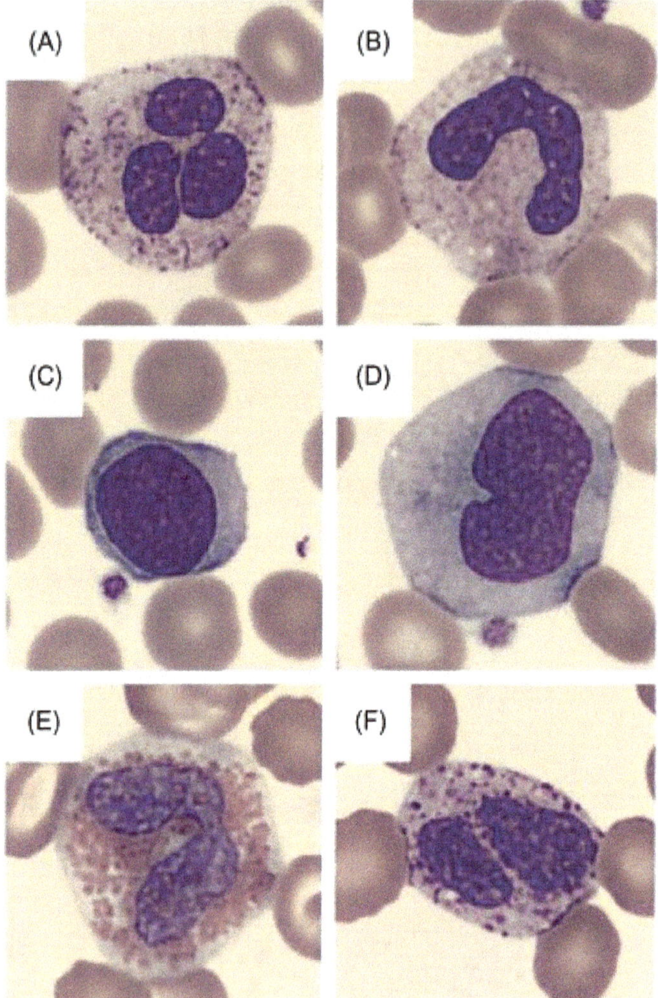

Platelets

Blood typically contains 150,000 to 400,000 platelets per mm^3. While most platelets circulate in the bloodstream, approximately one-third of the platelets in the body are held in the spleen as a reserve.

Platelets provide the following functions:

A. *Release of chemicals required for blood clotting.*
B. *The formation of a temporary patch in a damaged blood vessel.*

C. *Active contraction after clot formation to reduce the size of a break.*

Platelets form in the bone marrow from large cells called **megakaryocytes**. (Notice from the that, like erythrocytes and leukocytes, megakaryocytes arise from hemocytoblasts.) Megakaryocytes produce the various enzymes and other materials required by platelets and shed cytoplasm in membrane-enclosed packets, which are the platelets that enter the bloodstream.

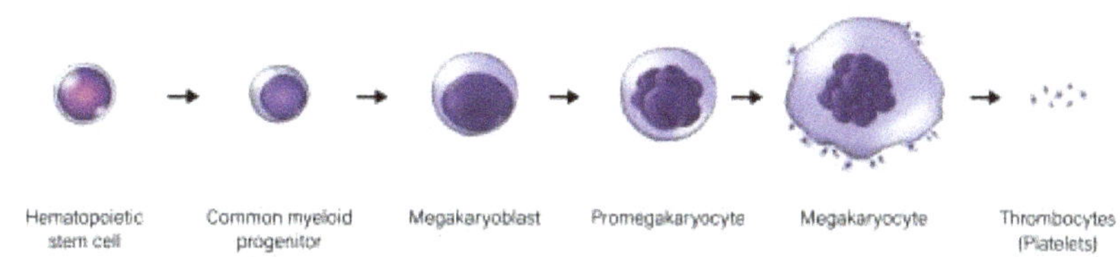

IV. Hemostasis

Clotting of the blood, which occurs to heal damaged blood vessels, is referred to as **hemostasis**. In addition to healing the vasculature, hemostasis establishes a framework for tissue repair. Hemostasis occurs in three phases: (1) vascular spasms, (2) platelet plug formation, and (3) coagulation.

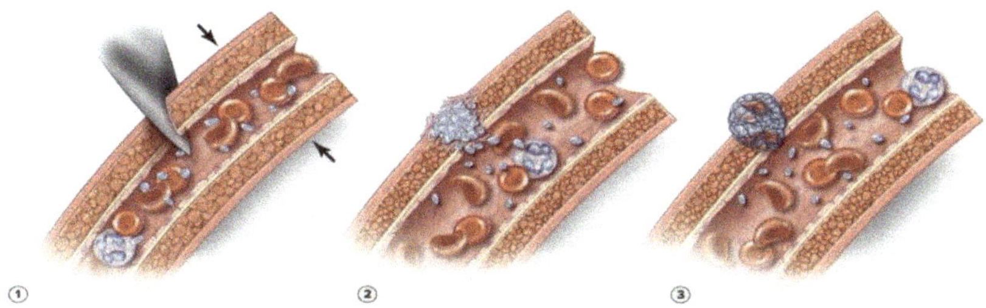

Steps of Hemostasis

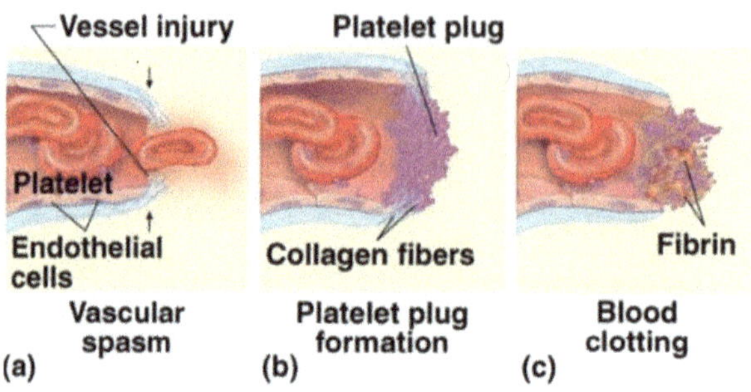

Vascular spasm

Damage to the blood vessel wall causes contraction of the smooth muscle cells of the vessel wall. This results in a "vascular spasm" that reduces or may stop the flow of blood through the vessel. This proccss may last for about 30 minutes. During this process, endothelial cells that line the inside of the blood vessel release a variety of chemicals. Some of these chemicals enhance contractions of the smooth muscle layer and stimulate the division of endothelial cells, smooth muscle fibers, and fibroblasts. In addition, the endothelial cell membranes become sticky.

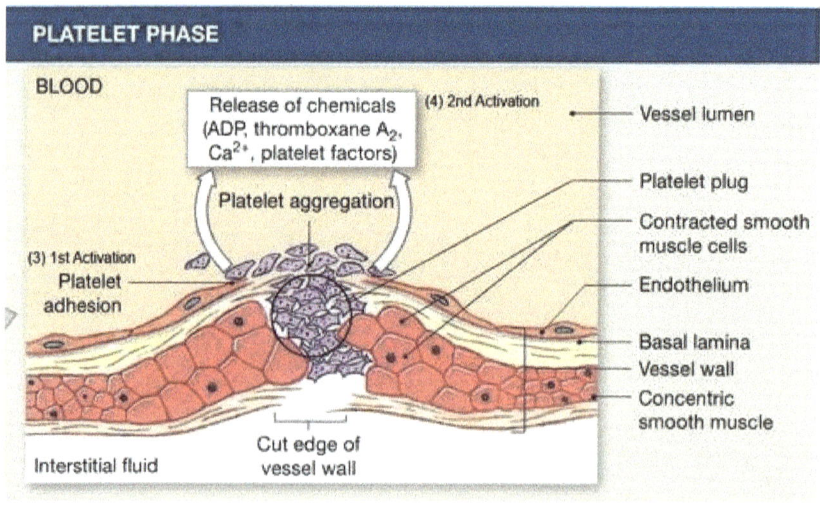

Platelet plug formation

Platelets stick to surfaces at the site of the injury (adhesion) and to each other (aggregation). This forms a plug that may close the wound if it is not very large. Adhesion and aggregation stimulate the platelets to release a variety of chemicals, including calcium ions and chemicals called **clotting factors**. These various chemicals promote adhesion and aggregation, vascular spasms, and blood clotting. You should see that hemostasis is enhanced by positive feedback loops.

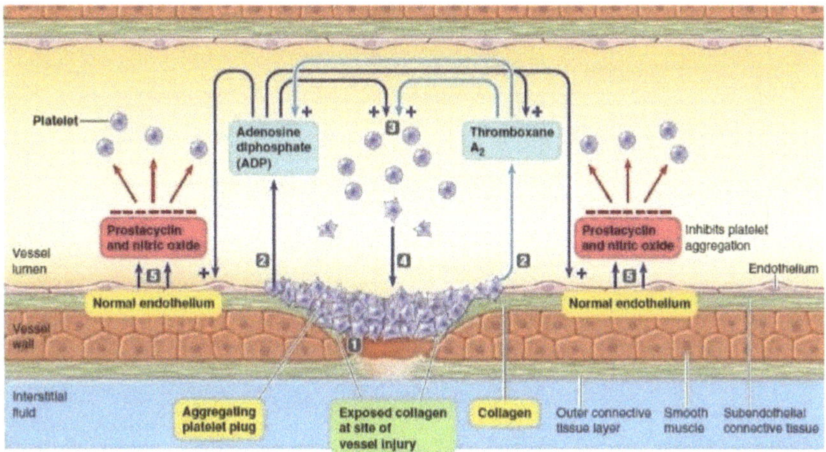

Coagulation

Coagulation, which is the true blood-clotting phase, begins 30 seconds or more after the injury. This is a complex process (see diagram below), which may follow two pathways: the **intrinsic pathway** results from damage to the inside of the vessel, and the **extrinsic pathway** is initiated by damage to tissue outside the vessel. Most of the chemicals needed for clot formation ("clotting factors") are proteins produced by the liver and require an adequate supply of vitamin K.

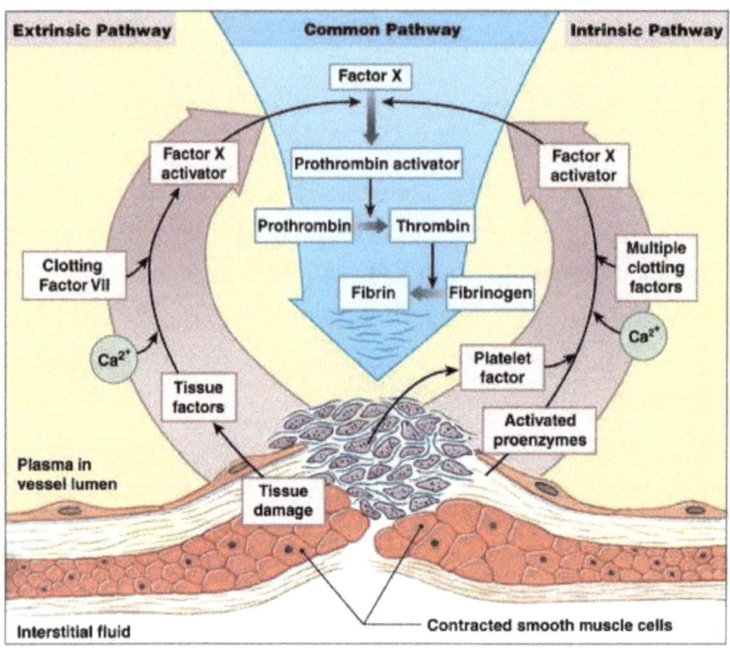

Coagulation cascade – intrinsic and extrinsic pathway

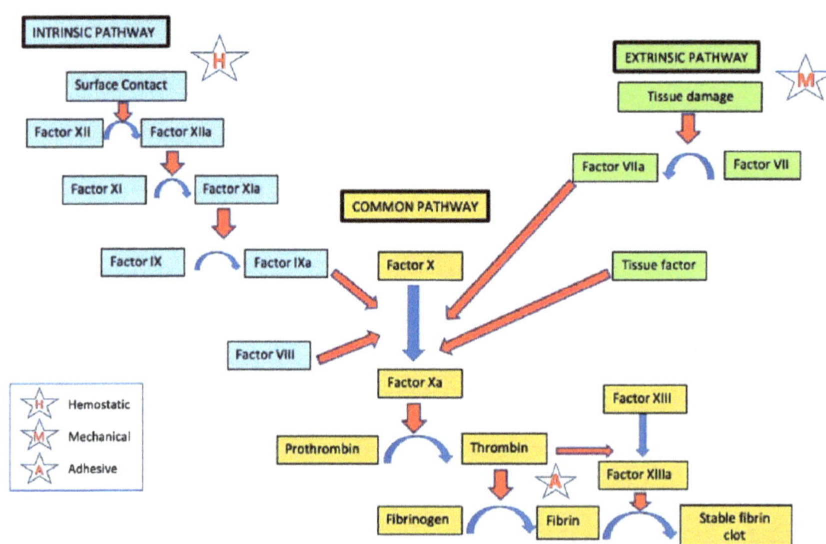

Both pathways lead to the formation of a substance called **prothrombin activator**, which converts the plasma protein **prothrombin** into an enzyme called **thrombin**. Thrombin catalyzes a reaction that causes **fibrinogen** molecules to link together into a mesh of protein called **fibrin**. Fibrin covers the platelet plug and traps additional cells and platelets in its fibrous network.

Elimination of the clot

Blood clots are stabilized by a process called **clot retraction**. Platelets contain the proteins actin and myosin, and (like muscle cells) they can contract. As clot retraction occurs, healing of the blood vessels also takes place, as **platelet-derived growth factor (PDGF)** promotes vessel repair. Eventually, the blood clot dissolves in a process called **fibrinolysis**.

Blood clotting is inhibited by various factors, including **antithrombin-III**, which specifically inhibits several clotting factors. **Heparin**, released by basophils and mast cells, accelerates the activation of antithrombin-III. Tubes used to collect blood samples typically contain either heparin or EDTA (a chemical that binds calcium ions). Why?

Practice by identifying the tagged formed elements in the illustration below.

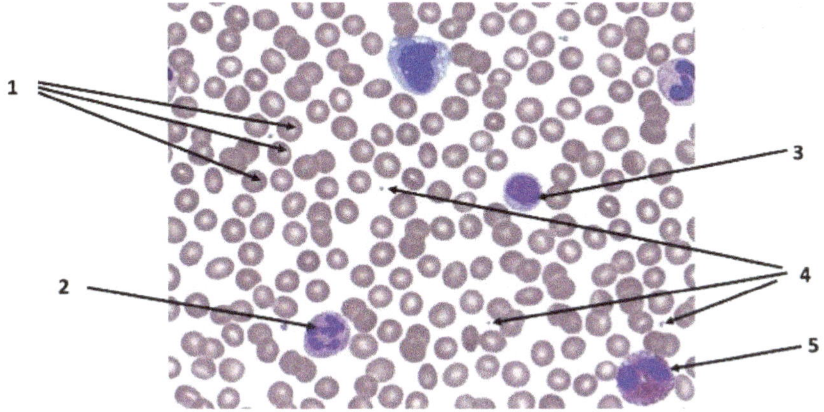

Identify the different types of leukocytes below. Note the ratio of WBC : RBC.

White Blood Cells or Leukocytes

WBC : RBC ratio = 1 : 700 – 1000

Main Function = **Fight infection**

There are **five** different types of leukocytes

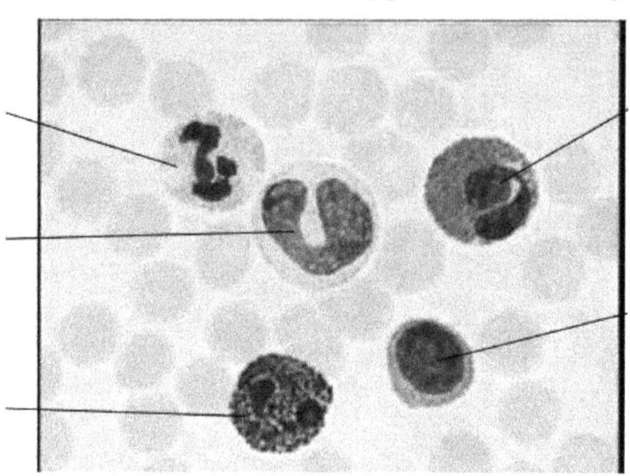

Chapter 3: Cardiovascular System

Heart

I. Location of the Heart and the Pericardium

Location and position of the heart

The heart lies in the thoracic cavity directly posterior to the sternum. More specifically, the heart is enclosed in the **pericardial cavity** in the anterior portion of the **mediastinum**. The superior end of the heart, to which are attached the "great veins and arteries" of the heart, is called the **base** (which may be a bit confusing). The inferior tip of the heart is called the **apex**. The heart is offset slightly to the left of the midline, and it lies at an angle to the longitudinal axis of the body, with the apex rotated slightly to the left.

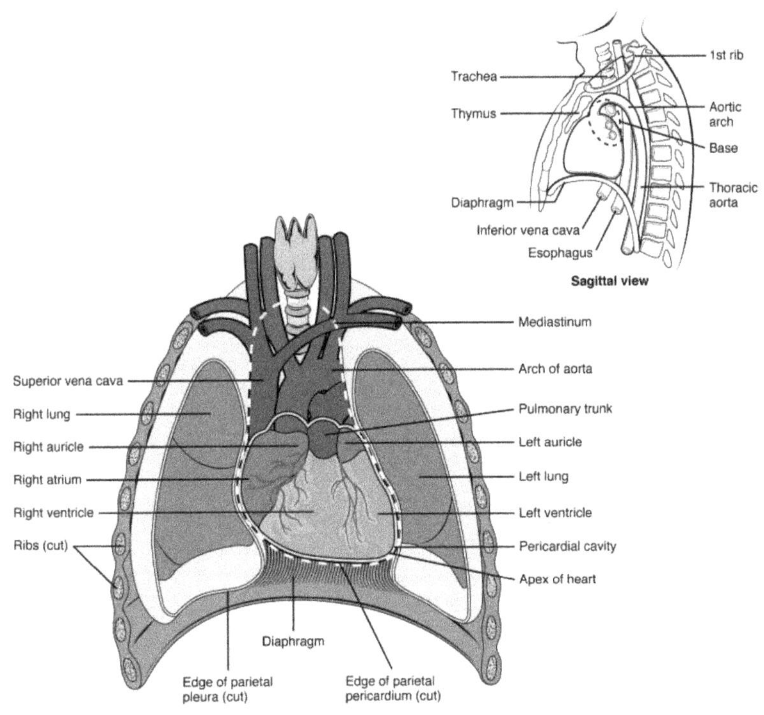

Characteristics of the pericardium

The heart is covered by a set of membranes called the **pericardium**. The pericardium has three layers (see image below):

A. The outermost layer is called the **fibrous pericardium**. It is made of fibrous connective tissue. It is tough and keeps a constant shape as it anchors the heart in place.

B. The next layer is called the **parietal pericardium**. It is directly attached to the fibrous pericardium. You should remember this and the next layer from the BIO 205 laboratory.

C. The innermost layer is called the **visceral pericardium**. This layer attaches directly to the heart and is also considered the outermost layer of the heart.

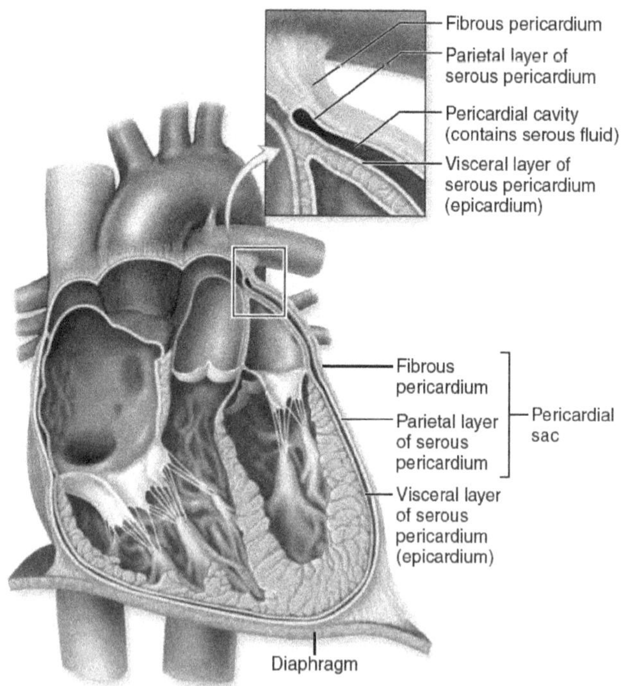

Pericardial fluid fills the space between the parietal and visceral layers. It acts like motor oil, lubricating the heart and reducing friction as it pumps.

II. Heart Anatomy

Layers of the heart wall

The heart itself is constructed in three layers (see image below):

A. The outermost layer of the heart is the **epicardium**. This is *the same structure* as the visceral pericardium.

B. The **myocardium** is the middle layer. It is the thickest layer of tissue in the heart and is made of cardiac muscle, blood vessels, and nerves.

C. The innermost layer is the **endocardium**, which is a thin layer of tissue that lines the chambers of the heart (simple squamous epithelium).

Layers of the ventricular wall of the heart (Epicardial layer with EAT = Epicardial Adipose Tissue)

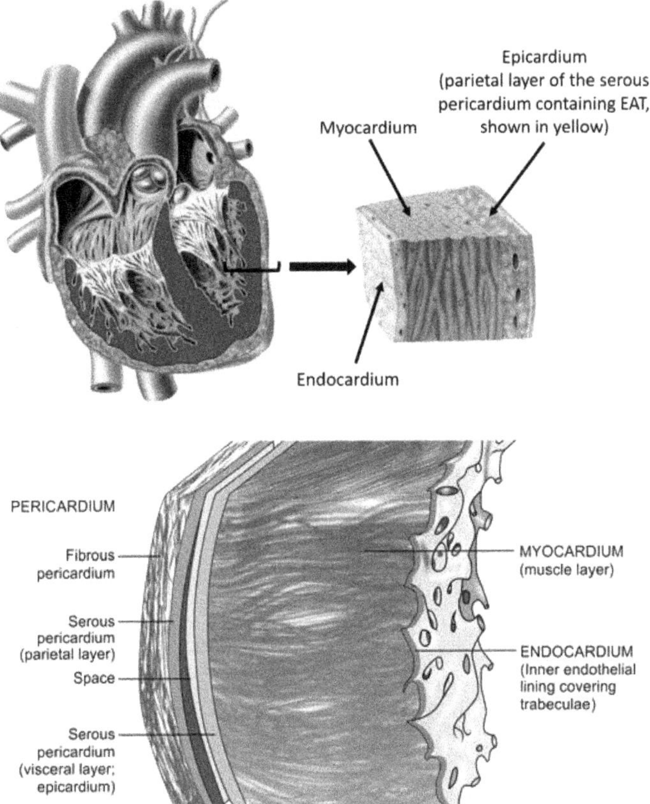

Heart chambers

The human heart has four chambers: the two smaller chambers are called **atria** and are located superiorly; the two larger chambers are called **ventricles** and are located inferiorly (see below). The interventricular septum separates the ventricles. A layer of connective tissue separates and electrically insulates the atria from the ventricles.

Superior and inferior venae cavae bring blood to the right atrium from the superior and inferior portions of the body, respectively. **Pulmonary veins** bring blood to the left atrium from the lungs. Blood exits the right and left ventricles via the **pulmonary trunk** and **aorta**, respectively.

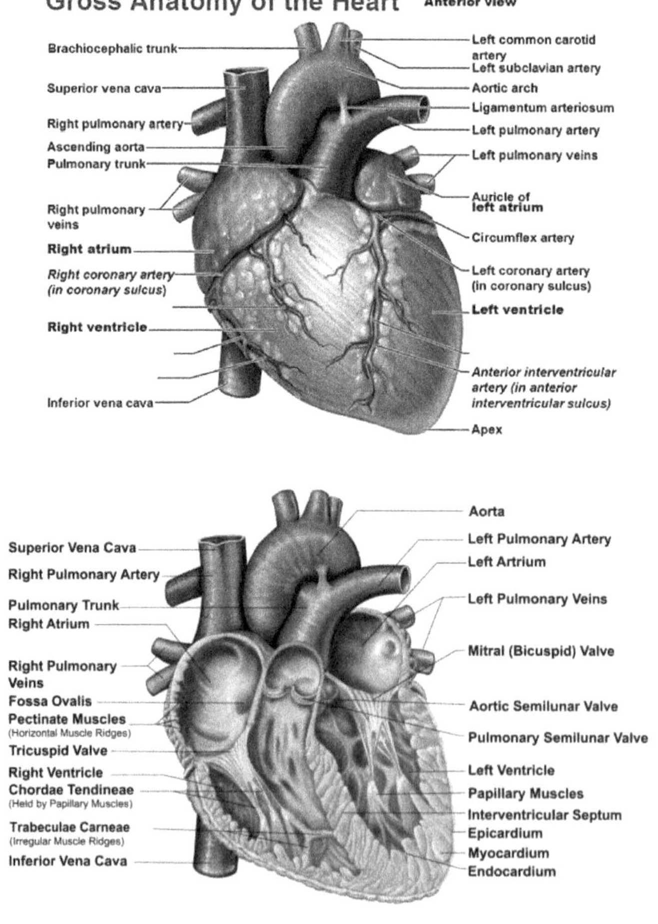

Heart (posterior view)

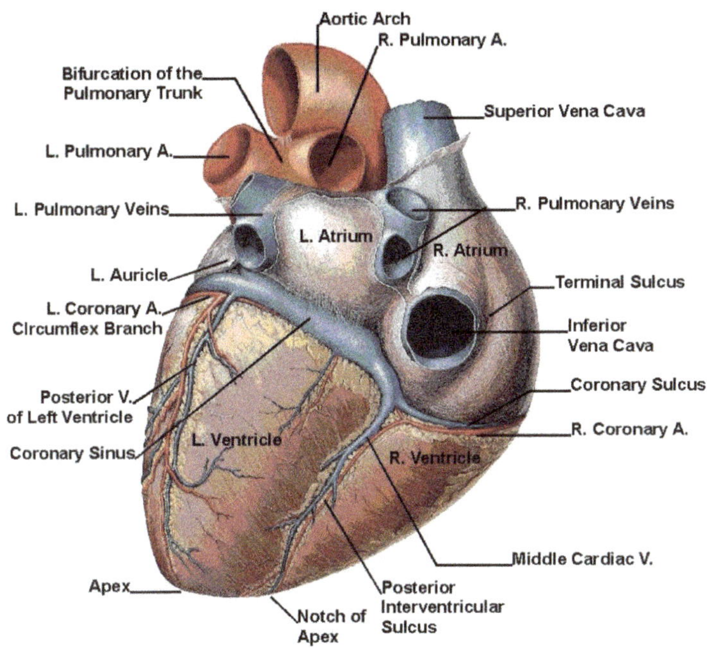

Heart valves

Blood from the right atrium is pumped into the right ventricle through the **right atrioventricular (AV) valve**. This valve is also known as the **tricuspid valve**, named for its three flaps (or cusps). When the right ventricle contracts, the tricuspid valve closes. Connective tissue fibers, called the **chordae tendineae,** prevent the valve from swinging back into the right atrium. From the right ventricle, blood is pumped through the **pulmonary semilunar valve** into the pulmonary trunk, which branches into **left and right pulmonary arteries** that travel separately to each lung.

Pulmonary veins lead from the lungs to the left atrium, which pumps blood through the **left AV valve** (also known as the **bicuspid valve** or **mitral valve**) into the left ventricle. When the left ventricle contracts, chordae tendineae prevent the left AV valve from swinging back into the left atrium. From the left ventricle, blood is pumped through the **aortic semilunar valve** into the **ascending aorta**, the **aortic arch**, and the **descending aorta**.

Anatomy students should be familiar with the four chambers of the heart and the direction in which blood flows through the heart. Thus , it is essential that students have a clear understanding of the functions of the valves (see image below). How would blood flow if the valves did not exist?

Atrioventricular valves and aortic and pulmonary valves (superior view – looking from above)

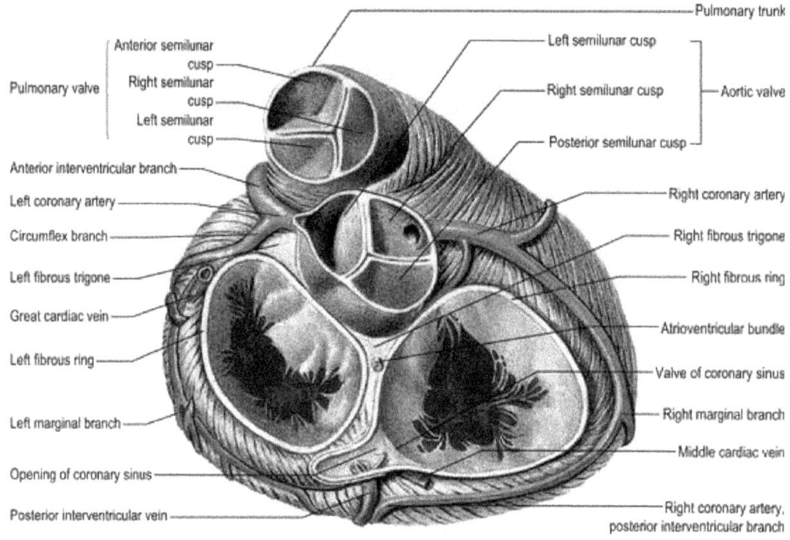

Superior view of heart valves during ventricular systole (ventricular contraction) below. (Note that the tricuspid and mitral valves are closed while the aortic and pulmonary valves are open when the ventricles contract to squeeze blood out of them).

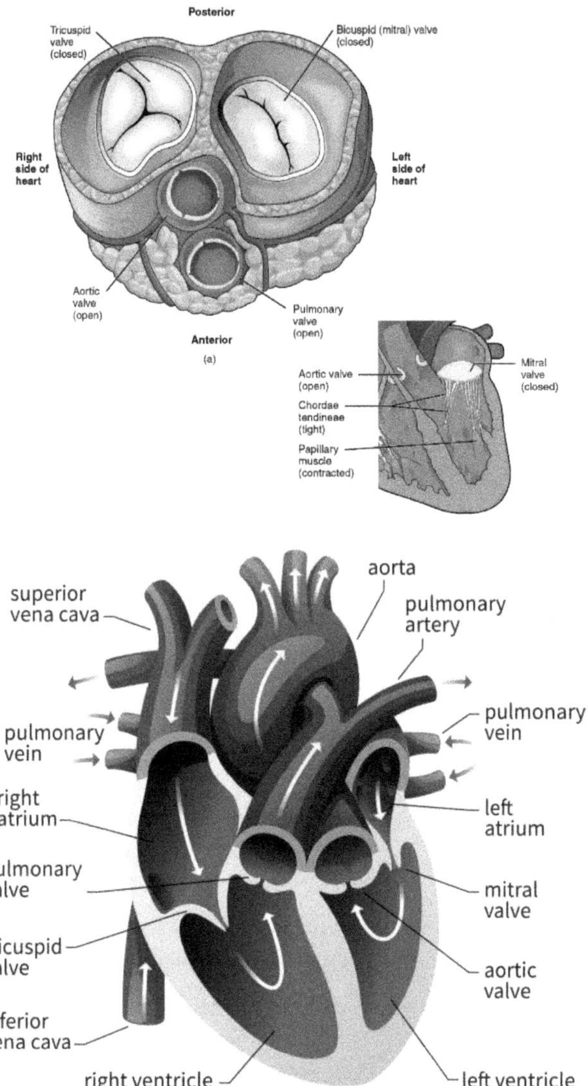

Although we will look at blood vessels in more detail in the next chapter, we can see at this time that the circulatory system can be divided functionally into two separate circuits: a **pulmonary circuit** and a **systemic circuit** (see diagram below). The pulmonary circuit involves the blood traveling to and from the lungs. Blood is pumped through the pulmonary circuit by the right side of the heart. The systemic circuit involves the blood traveling throughout the rest of the body. Blood is pumped through the systemic circuit by the left side of the heart.

Note the direction of flow to and from the heart in the illustration of the pulmonary and systemic circulation below. Which path is the longest? Which path is working against a greater pressure/resistance, and why?

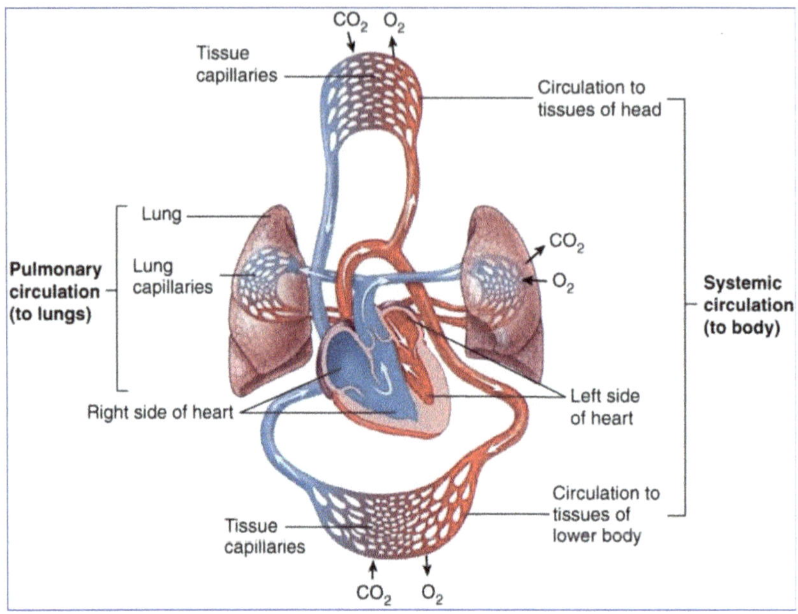

Microscopic structure of cardiac muscle

In many ways, cardiac muscle cells are similar in structure and function to skeletal muscle fibers. For example, both have actin and myosin arranged into sarcomeres that give the tissue a striated appearance. Some differences you should be aware of are as follows:

Whereas skeletal muscle fibers are very long and cylindrical, cardiac muscle cells tend to be short (hence, they are not called fibers), and they branch.

Structurally, skeletal muscle fibers are isolated from each other by their membranes. Although groups of skeletal muscle fibers may work together in a motor unit, each is innervated directly by a motor neuron and fires its own action potential. Cardiac muscle cells are structurally linked by structures called **intercalated discs**. These discs contain gap junctions that allow action potentials from cell to cell. The term **functional syncytium** is used to describe the fact that the branches and gap junctions link the entire myocardium into a single coordinated unit.

The action potential of a skeletal muscle fiber has a very short refractory period (about 1-2 ms), which contributes to its ability to enter a state of tetanus. Cardiac muscle cells have long refractory periods (about 250 ms), which normally prevents them from entering a state of tetanus. Why would you not want a cardiac muscle to enter tetanus?

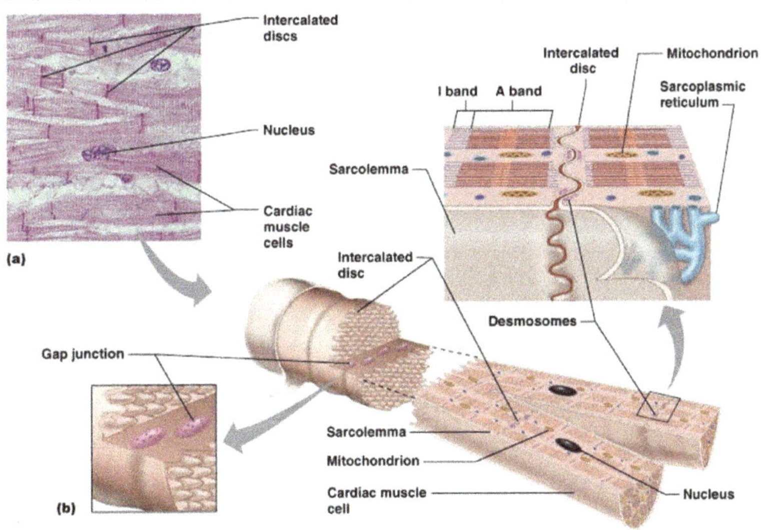

III. Coronary Vessels: Blood Supply of the Heart Wall

Coronary arteries

Because the heart works continuously, cardiac muscle cells require a constant supply of oxygen and nutrients. Branching off the aorta (just as it leaves the heart) are the **coronary arteries**, which supply the heart itself with blood (see image below).

Frontal view of hear showing the coronary arteries (on left) and coronary veins (on right)

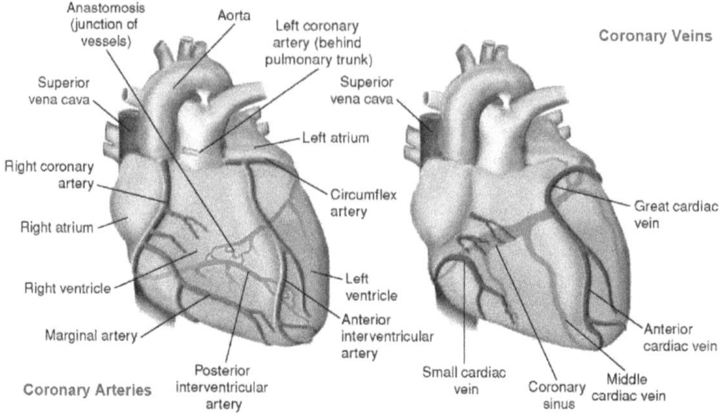

The **right coronary artery** supplies blood to the right atrium, portions of both ventricles, and portions of the heart's electrical conduction system. The **left coronary artery** supplies blood to the left atrium, portions of both ventricles, and the interventricular septum. Near the apex of the heart, branches of the right and left coronary arteries merge at junctions called **anastomoses**. Because the blood vessels merge, there is **collateral circulation** of the blood flow. This means that blood can reach much of the heart via different pathways. This collateral circulation may allow blood to flow to the entire heart even when an artery is blocked.

A heart attack typically results from a blockage of a coronary artery, which cuts off the supply of oxygen to part of the heart.

Coronary veins

Cardiac veins bring deoxygenated blood from the heart's capillaries to the **coronary sinus**, which dumps it into the right atrium.

IV. Anatomic Structures Controlling Heart Activity

Beating of the heart is a process that results from cooperative efforts of two types of cardiac muscle fibers: **Contractile cells** make up the bulk of the myocardium, and they are specialized to contract. Given the exceptions noted above, these cells function in a manner similar to skeletal muscle cells. **Autorhythmic cells** make up the conducting system of the heart and control the electrical activity of the heart. Unlike skeletal muscle cells and contractile cells, autorhythmic cells have unstable resting potentials that spontaneously move toward the threshold and lead to action potentials. This property is known as **automaticity**. Thus, autorhythmic cells will generate action potentials on their own. Hitting threshold leads to the opening of many voltage-gated Ca^{++} channels that allow an influx of Ca^{++}. Thus, the action potential of an autorhythmic cell is due to an influx of calcium rather than sodium.

The conduction system of the heart

The rate at which the heat beats is determined by both **intrinsic** and **extrinsic** control mechanisms. Intrinsic control is related to the property of automaticity described previously. Basically, the heart tends to beat at its own rhythm. However, the rate at which the heart beats can be affected by extrinsic (external) factors (remember the effects of the autonomic nervous system). First, we will examine the intrinsic control.

Cells of the conducting system cannot maintain stable resting potentials. After repolarization, the membrane potential inevitably drifts toward threshold and depolarization. The rate of spontaneous depolarization varies in different regions of the conducting system, but it is fastest in the **sinoatrial (SA) node**. Therefore, depolarization of the SA node sets the pace for the heart, and the SA node is referred to as the **pacemaker** of the heart. When fibers of the sinoatrial node contract, the electrical impulse spreads across both atria and causes the atria to contract.

The layer of connective tissue between the atria and ventricles prevents the direct spread of this impulse from the atria to the ventricles. A structure called the **atrioventricular (AV) node** carries the impulse to the ventricles but at a slower speed (). This creates a delay between the excitation of the atria and the excitation of the ventricles. The **atrioventricular bundle** (or **bundle of His**) and

Purkinje fibers then rapidly spread the impulse through the ventricles and cause the ventricles to contract. Why is it important to have a delay between the excitation of the atria and the excitation of the ventricles?

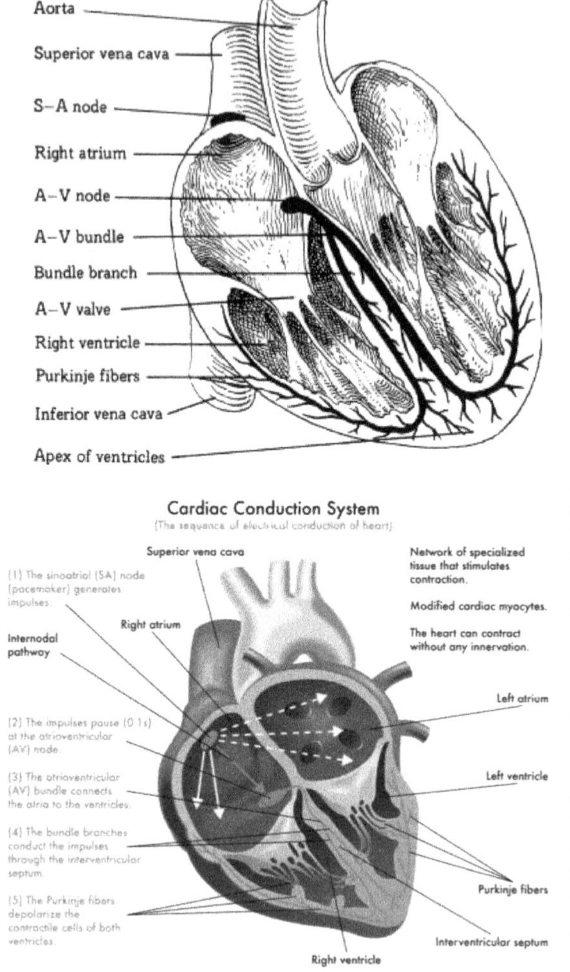

The electrical activity of the heart can be monitored with an instrument called an **electrocardiograph**. Abnormal function of the heart's conduction system can result in a number of serious clinical problems, generally characterized as **arrhythmias. Fibrillation** is the term used to describe rapid, uncoordinated contractions of cardiac muscle fibers. How does fibrillation affect the heart's ability to pump blood?

Electrocardiograh Electrocardiogram (reading of electrical activities in the heart on a strip of graph paper)

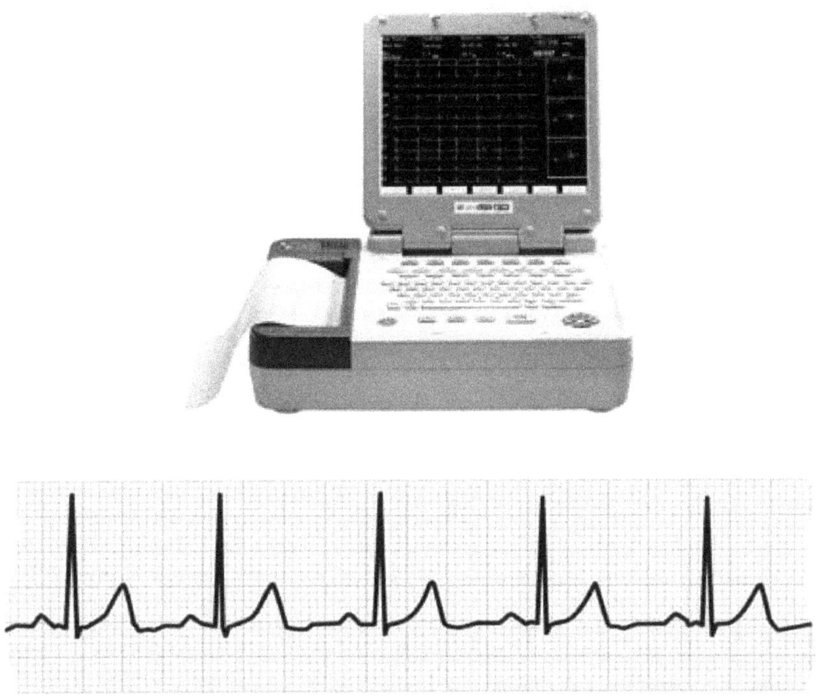

Each small square is 1mm in height and width, so voltage amplitude follows the vertical axis (wave height) and duration time in seconds follows the horizontal axis. There are five tiny squares for every large square, meaning if you count ten tiny squares (i.e., two large squares on the graph paper) high, that makes it ten (10) millimeters (mm) in height, which equates to one (1) millivolt (mV) in amplitude (1mV = 10 mm).

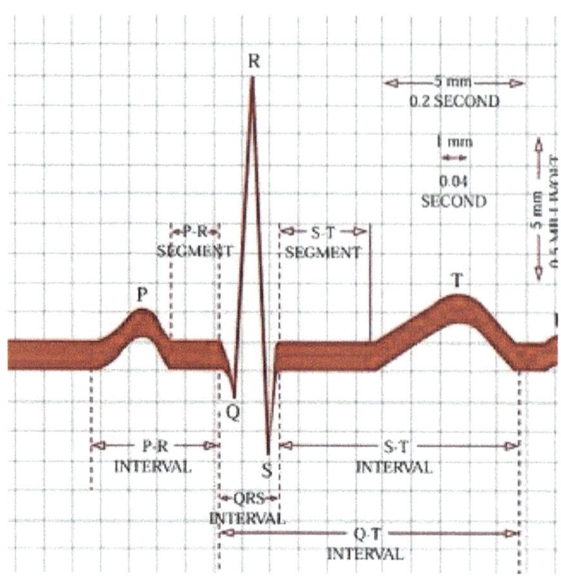

Calculating the voltage amplitude of a wave

Suppose you are asked to calculate the amplitude of a P wave. You would measure the height of the P wave from baseline to peak and multiply the P wave height by (1mV/10mm).

Example:

Measured P wave = 1mm (one tiny box above)

Multiply 1mm X (1mV/10mm)

Answer: 0.1mV

Calculating the duration of interval

Suppose you are asked to calculate the duration of the PR interval.

- Measure the width (number of tiny boxes) of the PR interval from the beginning of the P wave to the beginning of the QRS complex.
- Multiply PR interval width by (1s/25mm) **[1s/25mm means 1 second = 25 tiny boxes in the horizontal direction].**

Example:

Measured PR interval = 3mm

3mm X (1s/25mm)

Answer: 0.12 seconds

Calculating contraction rates

Suppose you are asked to calculate the atrial rate (i.e., the number of atrial contractions per minute).

- Measure the width of the PP interval from the beginning of one P wave to the beginning of the next P wave (1cycle).
- Divide 1500 by the PP interval width.

Example:

Measured PP interval = 30mm

Answer: 1500/30 = 50 atrial contractions/minute

Calculating ventricular rate is done using the same method for the exception of measuring the width of the RR interval instead of the PP interval.

Innervation of the heart

Although the heart has its own intrinsic rhythm, the heart rate can be adjusted extrinsically by control centers in the medulla oblongata. The **cardioacceleratory center** of the medulla sends signals to the heart via sympathetic pathways in spinal nerves T1-T5; these signals accelerate the heart rate, and they also increase contractility (the force with which the heart contracts). What neurotransmitter is released at the heart when the cardioacceleratory center is active?

The **cardioinhibitory center** of the medulla sends signals to the heart via parasympathetic pathways in the vagus nerves, slowing the heart rate. What neurotransmitter is released at the heart when the cardioinhibitory center is active?

Sympathetic and parasympathetic innervation of the heart and its effect on blood vessels below

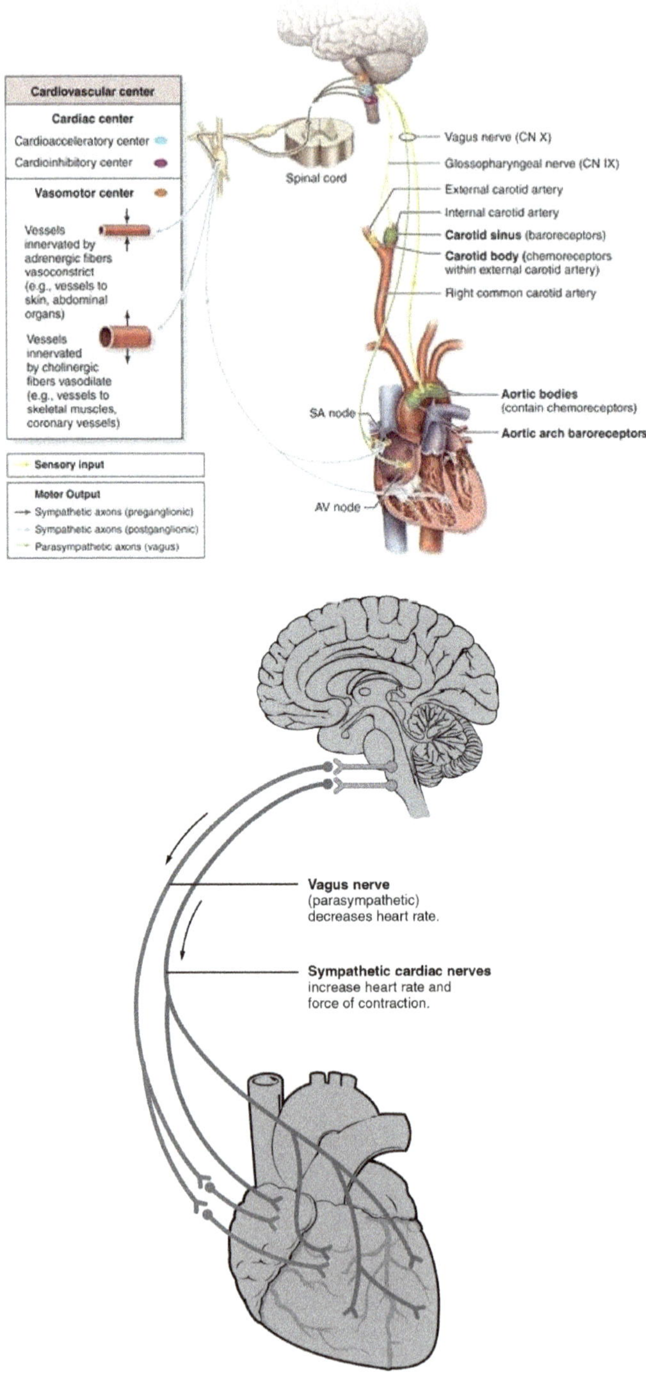

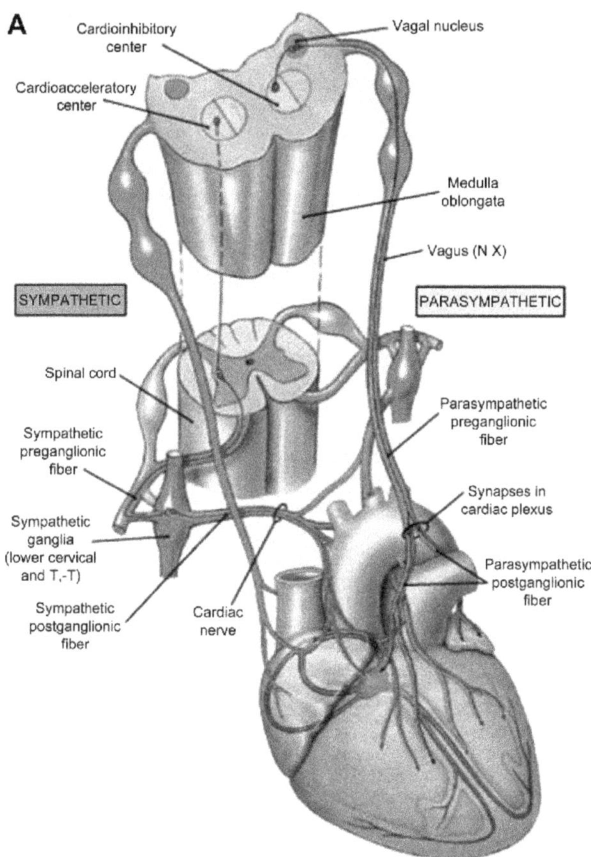

V. The Cardiac Cycle

The period from the beginning of one heartbeat to the beginning of the next is called the **cardiac cycle** (see diagram below). Contraction of a portion of the heart makes up a phase of the cardiac cycle called **systole**. Systole is followed by relaxation, called **diastole**. One key to understanding the cardiac cycle is understanding that blood will tend to flow from an area of high pressure to an area of low pressure.

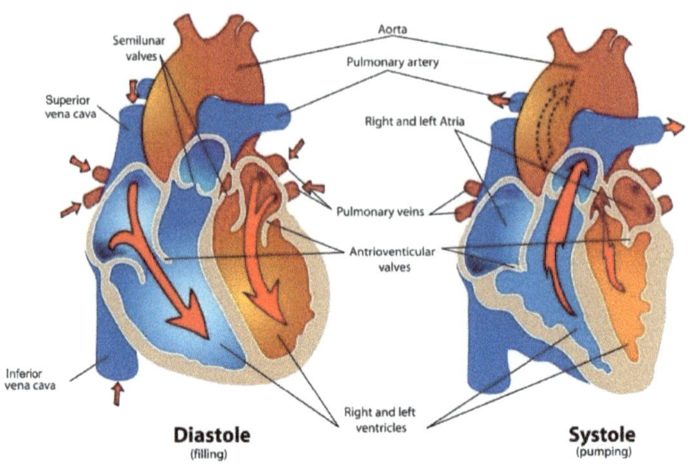

Events of the cardiac cycle

Because the beating of a heart is a cyclic process, we can start our description anywhere. We will begin our description at a point where all four chambers are in diastole, and the atria are about to begin contraction . . .

1. *Atrial systole.* The ventricles have already been filling with blood passively while the heart was relaxed. At the point when the ventricles are about 70% full and pressure has begun to build in the ventricles, the atria contract. Contraction of the atria is called **atrial systole**, and it lasts about 100 ms. This pumps additional blood into the ventricles as the pressure generated in the atria exceeds the pressure in the ventricles. The quantity of blood in each ventricle is maximal after atrial systole, and the volume of blood in the ventricles currently is referred to as the **end-diastolic volume (EDV)**. You should note that the atria do not do much in the way of filling the ventricles, they are already mostly full by the time the atria contract.

2. *Early ventricular systole.* As the atria relax after atrial systole, the ventricles begin to contract in a process called **ventricular systole**. As the pressure in the ventricles rises, the AVvalves shut, and for a time, all valves of the heart are closed. This phase is referred to as **isovolumetric contraction**.

3. *Late ventricular systole.* Eventually, the pressure inside the ventricles rises sufficiently to open the semilunar valves, and blood is ejected into the pulmonary trunk and the aorta. The ventricles eject a quantity of blood referred to as the **stroke volume (SV)**. SV is usually equal to about 60% of the EDV, and this percentage is referred to as the **ejection fraction**. The volume of blood in the heart at the end of ventricular systole is the **end-systolic volume (ESV)**.

4. *Early ventricular diastole.* Contraction of the ventricles is followed by **ventricular diastole**. As the ventricles begin to relax, pressure in the ventricles drops rapidly. At this point, pressure in the aorta exceeds that in the ventricles, and the semilunar valves close. For a brief period, all four valves are again closed at the same time. This phase is referred to as **isovolumetric relaxation**. The ventricles relax, but the volume does not change.

5. *Late ventricular diastole.* As the pressure in the ventricles drops further, the AV valves open, and blood flows passively from the atria into the ventricles. This brings us back to step 1.

The 4 phases of the cardiac cycle

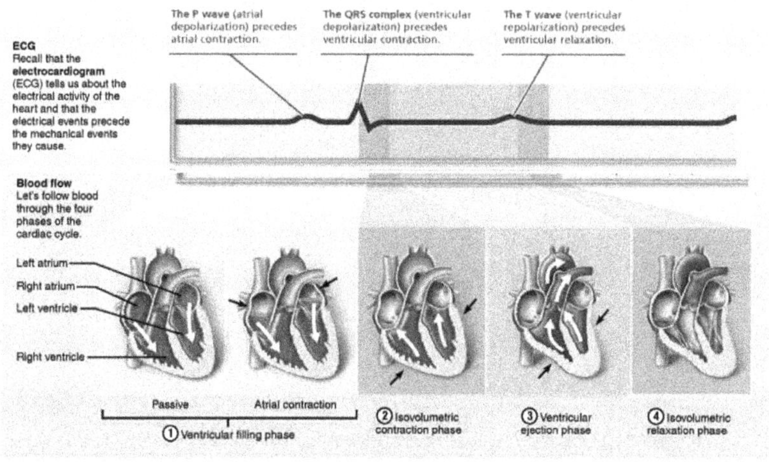

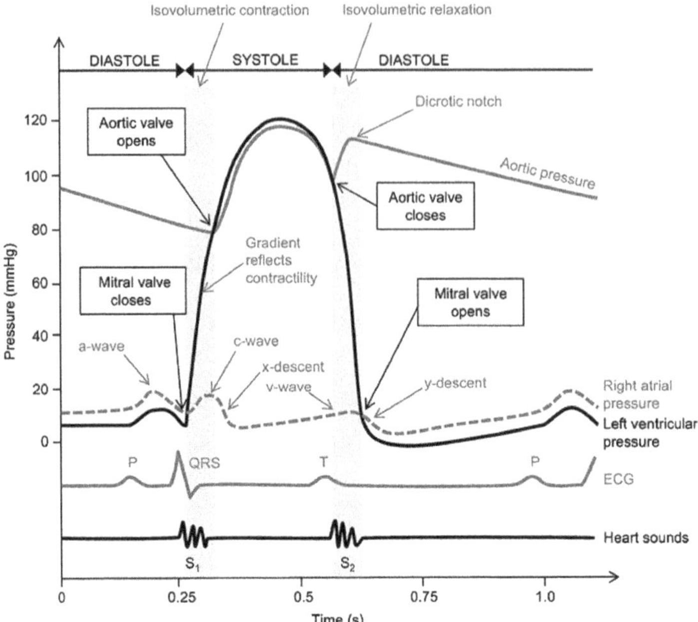

VI. Cardiac Output

Introduction to cardiac output

The volume of blood pumped per unit of time is the **cardiac output (CO)** of the heart:

CO (ml/min) = HR (beat/min) x SV (ml/beat)

where HR is the heart rate. Changes in either HR or SV will affect cardiac output.

Variables that influence cardiac output

Stroke volume is the difference between the EDV and the ESV:

SV = EDV - ESV Changes in either EDV or ESV will affect SV.

The EDV is affected largely by two factors:

A. **Filling time** is the duration of ventricular diastole. The longer the filling time, the greater the EDV.

B. **Venous return** is the rate of flow back to the heart from the body. The greater the rate of venous return, the greater the EDV.

The ESV is affected largely by three factors:

A. **Preload** is the degree to which the heart is stretched during ventricular diastole. The greater the stretching during preload, the greater the force with which the cardiac muscle will contract and the greater the SV will be. This principle was first described as the **Frank-Starling principle**. Increased preload leads to decreased ESV.

B. **Contractility** is the force produced during contraction independent of preload. Various factors may either increase contractility or decrease contractility. For example, the sympathetic nervous system increases contractility via the release of norepinephrine. Increased contractility leads to decreased ESV.

C. **Afterload** is the amount of force the ventricles must produce to eject blood through the semilunar valves. The most common cause of an increase in afterload is obstruction of the blood vessels, which can be particularly harmful if the myocardium has been damaged. An increase in the afterload increases the ESV.

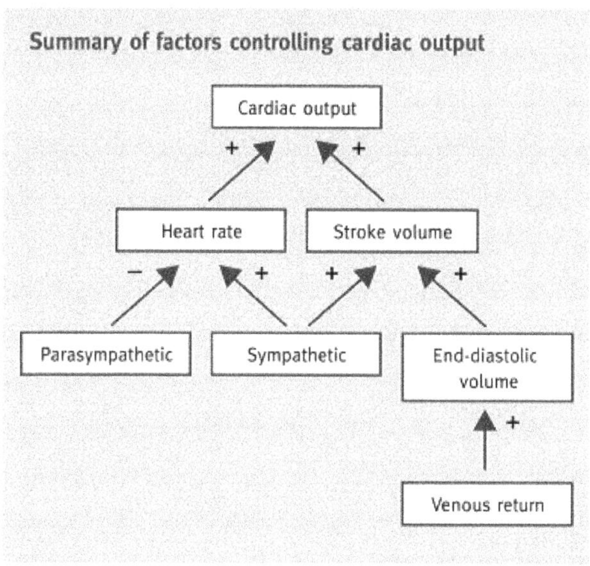

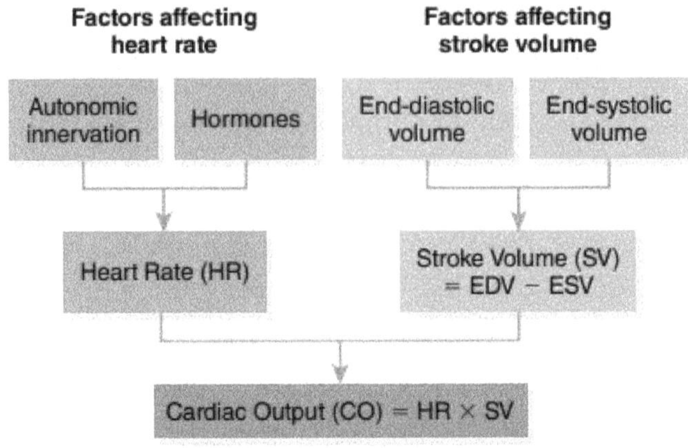

Factors affecting cardiac output.

A little bit more of Cardiophysiology…

Let's take a look at vital signs often referred to as 'vitals' in the clinical setting. They are very important in the assessment of an individual's overall health and are the first medical assessment performed at the doctor's office aside from the physical examination.

Pulse rate = Heart rate (number of times the heart beats per minute - bpm). Elevated **HR** (heart rate) is termed **tachycardia** (HR greater than 100 beats per minute) and slow HR is **bradycardia** (HR 65 beats per minute or less)

Pulse locations throughout the body (most common area (radial) to palpate for a pulse)

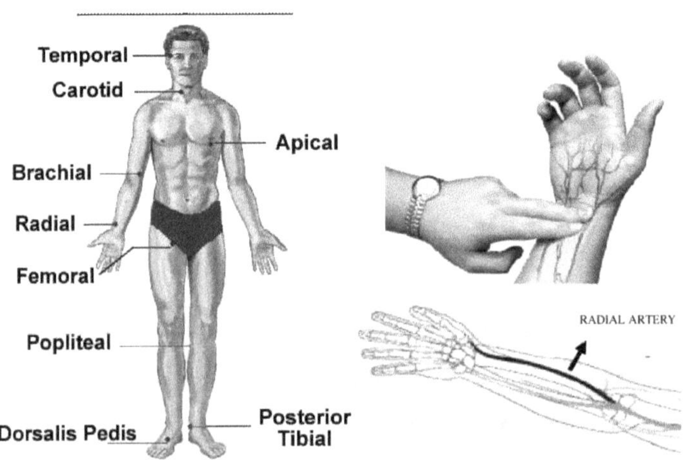

'Lub Dup' sounds of the heart represent the first and second heart sounds abbreviated as 'S1' and 'S2' respectively (see diagram below).

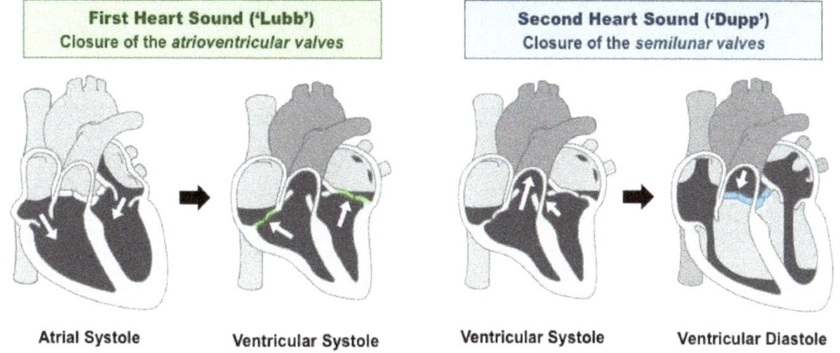

Introduction to chest auscultation using a stethoscope

Listening to the heart sounds allows the medical practitioner to detect a person's heart rate and rhythm, murmurs (blowing or swishing caused by turbulence blow flow through the valves and chambers), and pericardial disease. The murmurs are graded in intensity from I to VI (faint to loudest respectively).

This diagram describes the areas to auscultate

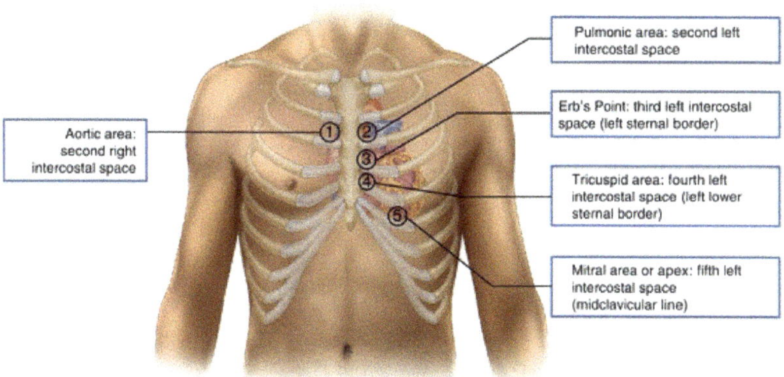

Quantitative analysis of blood pressure

By now, it can be said that we are quite familiar with the definition of blood pressure which is the measure of the hydraulic force exerted by blood flow upon the blood vessel walls. So, if necessary, review the definition of these parameters: **SBP** (systolic blood pressure), **DBP** (diastolic blood pressure), **MBP** (mean blood pressure), **PP** (pulse pressure), **SV** (stroke volume), **CO** (cardiac output), **ESV** (end systolic volume), **EDV** (end diastolic volume) and so on.

The following is a list of simple mathematical equations (discussed earlier) used to calculate **BP**, **HR** and other cardiophysiology parameters:

CO = HR X SV so, HR = CO / SV and SV = CO / HR

MBP = 2/3DBP + 1/3SBP

MBP = DBP + 1/3PP

PP = SBP - DBP (another way to calculate PP if the SV is available is to divide SV by 2: **PP = SV/2**).

There will be times when conversion from milliliters (mL) to liters (L) or vice versa may be necessary. The conversion of milliliters to liters follows: **1 L = 1,000 mL**, therefore, to convert 1mL to liter is:

L = 1mL / 1,000mL

Equally important is calculating the percentage of change in cardiac output, stroke volume, heart rate before and after exercise (keeping in mind that the vitals will increase after exercise), then the formulas follow:

$\% \Delta CO = [(CO_A - CO_B) / (CO_B)] \times 100$

$\% \Delta HR = [(HR_A - HR_B)] / (HR_B)] \times 100$

$\% \Delta SV = [(SV_A - SV_B)] / (SV_B)] \times 100$

Where CO_A with the subscript A to define cardiac output after exercise and CO_B with subscript B meaning cardiac output before exercise and likewise for percentage change in HR and SV. Note the units for each of the parameters, for example CO uses L/min (liters/minute), SV uses mL/beat (milliliters/beat), HR(bpm = beats per minute), and unit of pressure is mmHg (millimeters of mercury).

So, when reading a blood pressure measurement, the number will be recorded as a fraction: 120/80 mmHg (normal BP in a young adult).

Example: find the stroke volume if the cardiac output is 5 L/min and the heart rate (HR) is 78 bpm

Answer: 5 L must be converted into mL because the unit of SV is mL/beat which is the volume of blood ejected out the ventricle each time it contracts (like squeezing water out of a balloon). So, 5 L/min = 5,000 mL/min then 5,000 mL/min / 78 bpm = **64.1 mL/beat** (minutes are cancelled out of the calculation).

SV = 5,000 mL/min / 78 bpm = 64.1 mL/beat

Practice solving the following cardiac output problems

1. If CO = 6.4L/min, and SV = 73 ml/beat, what is the heart rate?
2. If Pulse = 110 beats/min and CO = 5.2L/min what is the SV?
3. If SV = 94mL/beat and SBP = 133mmHg, what is MBP?
4. At REST Pulse = 12 beats/10 seconds. BP = 110/70mmHg. What is CO?
    a. After exercise Pulse = 75 beats/45 sec, BP = 150/95mmHg. What is CO?
    b. What is the % change in HR from rest to post exercise?
5. If HR = 49 beats/min, CO = 5L/min, SBP = 105mmHg, what is DBP? (see answers on pages 416, and 417)

It is worth being familiar with the stages of embryonic development of the heart as represented in the diagrams below.

Looping of heart tubes according to stages of embryonic development

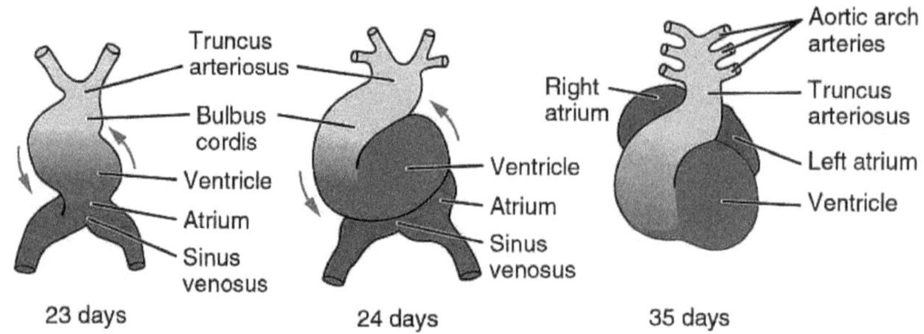

Embryonic heart tube during stages of development (longitudinal and cross-sectional views) below

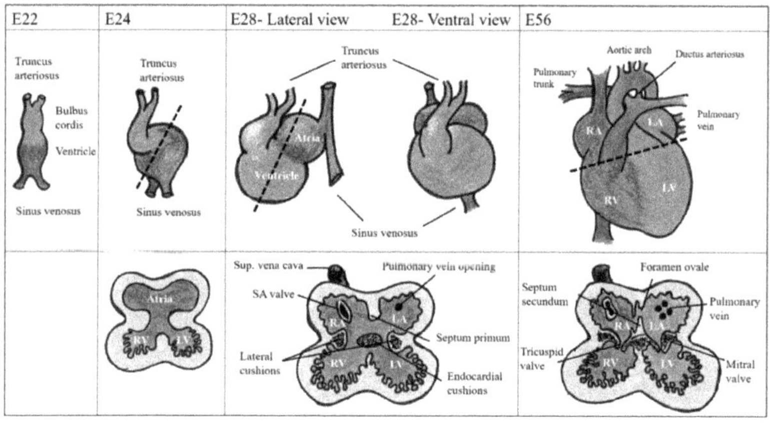

Chapter 4: Cardiovascular System: Vessels and Circulation

In this chapter, we examine blood vessels and the properties of flow through these vessels. Each blood vessel falls into one of three major categories: (1) **arteries** carry blood away from the heart, (2) **veins** carry blood toward the heart, and (3) **capillaries** allow diffusion between the blood and surrounding tissues. An adult human has more than ten billion capillaries.

I. Structure and Function of Blood Vessels

General structure of vessels

Arteries and veins are made of three layers of tissue, which surround a cavity called the **lumen** (see illustration below). These tissue layers provide strength and flexibility, but they are too thick to allow significant exchange of gases between the blood and surrounding tissues.

A. The **tunica intima** is the innermost layer. It consists of endothelial tissue that lines the blood vessel and an underlying layer of areolar connective tissue with lots of elastic fibers. The endothelial tissue provides a slick surface to minimize friction with the flowing blood.

B. The **tunica media** is the middle layer. It is made of concentric rings of smooth muscle in a framework of loose connective tissue. Collagen fibers bind this layer with the inner and outer layers. This is the thickest layer in arteries.

C. The **tunica externa** is the outermost layer. It is a layer of fibrous connective tissue that adds strength and support to the vessel. This is generally the thickest layer in veins. Connective tissue fibers of this layer blend into surrounding tissue to anchor the vessels.

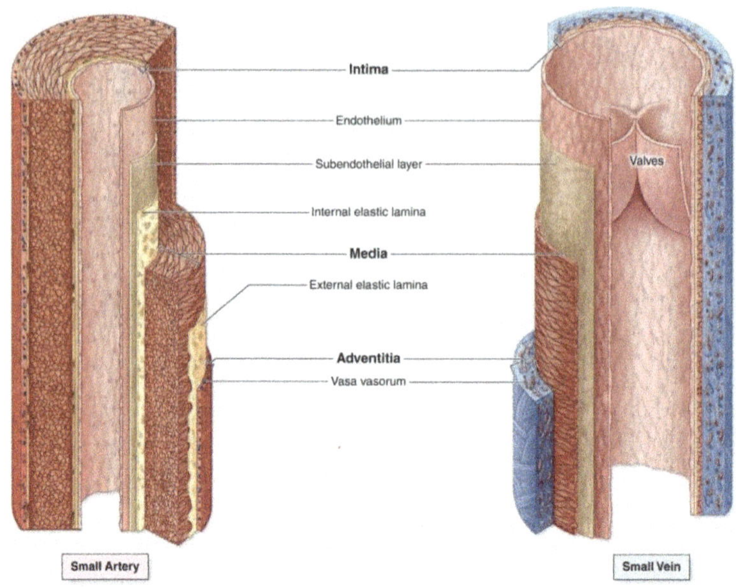

Arteries

Elastic fibers and thick muscular walls give the arteries both **elasticity** and **contractility**. Elasticity allows these vessels to respond to and dampen changes in blood pressure generated by the pumping of the heart. Contractility allows **vasoconstriction** and **vasodilation** in response to signals from the nervous system. Refer back to the chapter on the general effects of the parasympathetic and sympathetic nervous systems on blood vessels.

In traveling from the heart to the capillaries, blood flows through three types of arteries with somewhat different properties:

A. **Elastic arteries** are large vessels designed to transport large volumes of blood to major regions of the body. The tunica media of these vessels contains large proportions of elastic fibers and a smaller proportion of smooth muscle. In response to ventricular systole, these vessels expand to accommodate blood from the heart and dampen the increase in pressure. As blood moves through the vessels and the ventricles enter the diastole, these vessels contract and dampen the drop in pressure.

B. **Muscular arteries** distribute blood throughout the muscles and organs. These vessels contain a higher proportion of muscle than the elastic arteries.

C. **Arterioles** are the smallest of the arteries; some contain just one layer of smooth muscle cells. They are largely responsible for regulating the amount of blood that flows to a particular set of capillaries. They do this by contracting or relaxing to restrict or increase the amount of blood flowing through them. Contraction or relaxation changes the **resistance** to flow through the vessel.

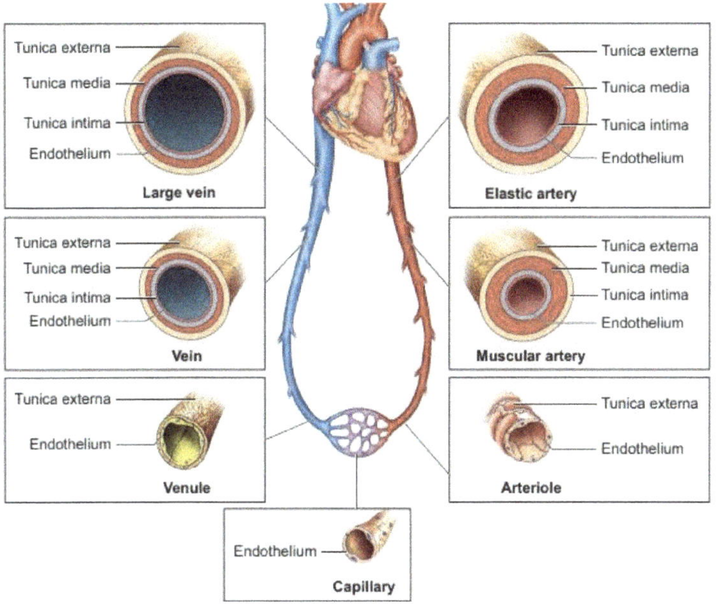

Structure of Blood vessels

Structural and histological comparison of artery and vein. Note the difference between the size of the layers of both types of vessels as well as the size of the lumen. Can you tell which vessel is more flexible and why?

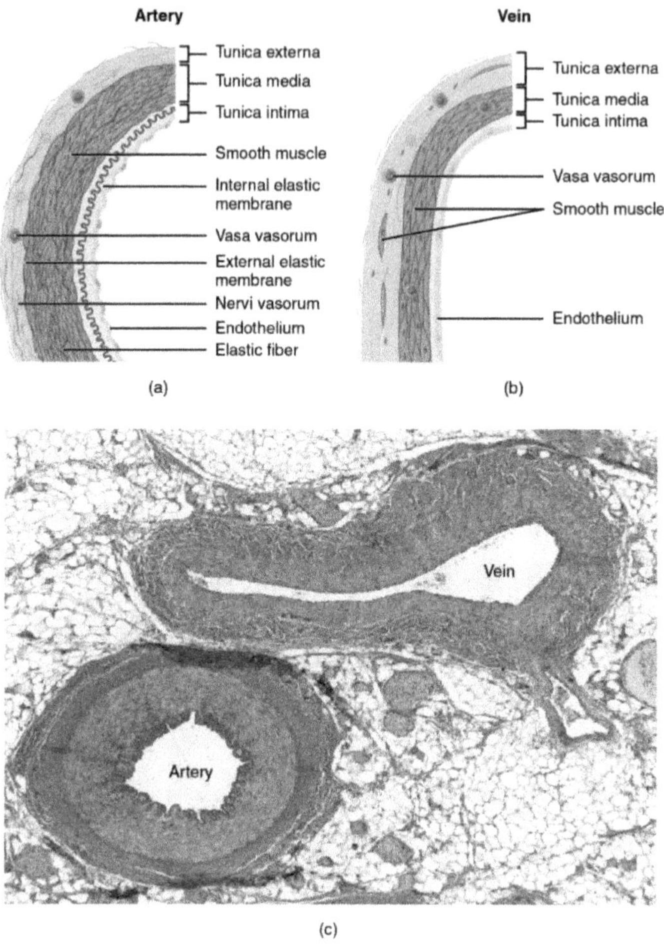

Capillaries

Capillaries have very thin walls (one cell layer thick) in order to allow efficient gas exchange between the blood and surrounding tissues. Neither tunica media nor tunica externa is present.

Within the body are three slightly different types of capillaries:

A. **Continuous capillaries** are the most common capillaries. They are particularly abundant in the skin and muscles. As their name suggests, the endothelial cells of these capillaries form a continuous lining. The endothelial cells are held together with tight junctions, but there are some gaps (**intercellular clefts**) that allow small amounts of fluid to pass between cells. An exception to this is found in the capillaries of the brain, in which tight junctions form a complete barrier to the movement of fluids between cells. Recall the blood-brain barrier discussed in the nervous system section (Part I).

B. **Fenestrated capillaries** contain endothelial cells that have pores. These pores make the fenestrated capillaries considerably more permeable to fluids than the continuous capillaries. The fenestrated capillaries are typically found where active absorption of materials into the blood is required, such as in the small intestine and endocrine organs. They are also found in the kidneys where filtration occurs.

C. **Sinusoids** have large lumens, very few tight junctions between the endothelial cells, and fenestrations. Thus, they tend to be especially leaky, allowing even large molecules and cells to pass between the blood and surrounding tissues. These capillaries are often used by formed elements to move in and out of the blood vessels. Sinusoidal capillaries are found primarily in the liver, bone marrow, spleen, and other lymphoid organs

Different Types of Capillaries

Continuous	Fenestrated	Discontinuous
Continuous ring of endothelial cells surrounded by a continuous basement membrane. Found in most tissues.	Highly permeable to water and solutes In tissues that specialize in fluid exchange – kidneys, exocrine glands, choroid plexus	Large junctions and discontinuities Highly permeable to plasma proteins.In organs where RBC and WBC need to migrate between blood and tissue e.g. bone marrow. Also liver – proteins cross membranes

Capillaries function as parts of interconnected networks referred to as **capillary beds**. **Precapillary sphincters** regulate the flow of blood through various capillaries. **Thoroughfare channels** are direct connections between arterioles and venules that allow blood to bypass most of the capillaries in a capillary bed. Blood flow must be carefully regulated to the parts of the body that need it. A human does not have enough blood in his or her body to fill up all of the capillaries at once.

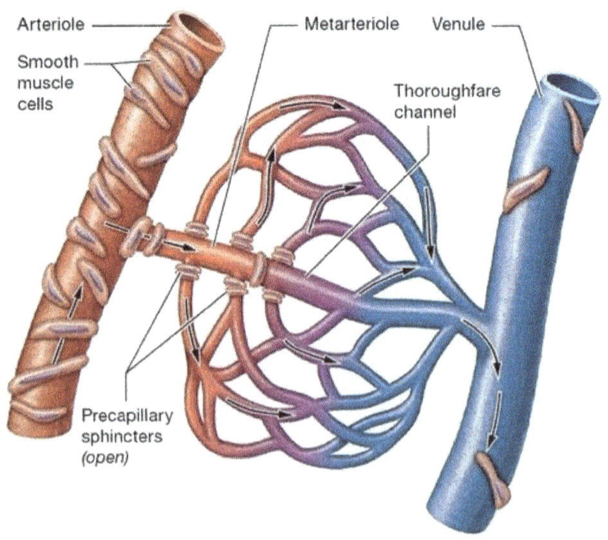

(a) Blood flow through a capillary bed when precapillary sphincters are open

Capillary Beds

- **<u>Precapillary Sphincter</u>** - surrounds the root of each true capillary and acts as a valve to regulate the blood flow into the capillary.
  - When the sphincters are *open*, blood flows through the capillaries and takes part in *exchanges* with tissue cells.
  - When the sphincters are *closed*, blood flows through the shunts and *bypasses the tissue cells.*

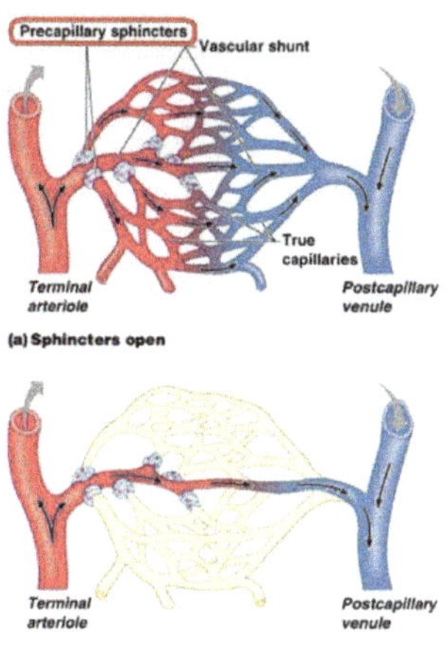

Veins

Veins collect blood from the tissues and return it to the heart. Their walls are thinner than those of arteries because of lower blood pressure. **Venules** are the smallest veins, and the tunica media is generally absent in all but the largest venules.

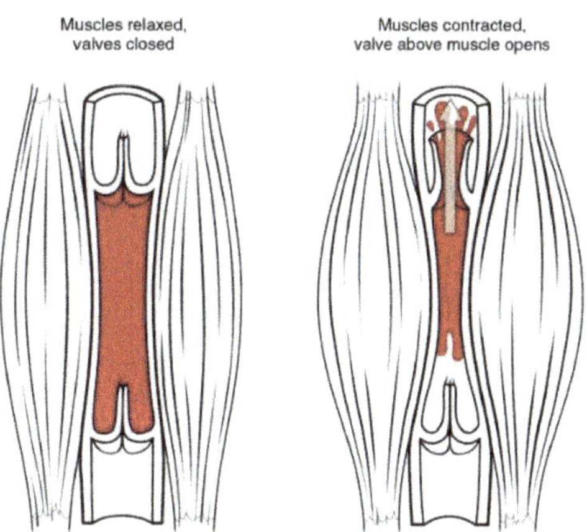

The blood pressure in veins may be sufficiently low to not overcome the force of gravity. Therefore, most veins contain valves to help prevent the backflow of blood. How does contraction of certain skeletal muscles (*e.g.,* in the legs) work with valves to enhance **venous return**?

II. Capillary Exchange

Diffusion and vesicular transport

One of the major functions of capillaries is the diffusive exchange of respiratory gases, oxygen, and carbon dioxide. When we examine the respiratory system, we will study this in detail.

Bulk flow

Bulk flow is the movement of a fluid and any dissolved substances in the fluid down a pressure gradient. When considering the bulk flow of blood and other fluids in the body, two types of pressure should be considered:

1. **Hydrostatic pressure** is the physical pressure exerted by a fluid on its surroundings. As we examine the exchange between the blood in a capillary and the interstitial fluid surrounding the capillary, we will consider blood hydrostatic pressure (HP_b) and interstitial fluid hydrostatic pressure (HP_{if}). The HP_{if} is generally so low that we can consider it zero.
2. **Colloid osmotic pressure** is created by the presence of large molecules (mainly proteins) that cannot diffuse across the capillary wall. These molecules attract water and create osmotic pressure. We will consider blood colloid osmotic pressure (COP_b) and interstitial fluid colloid osmotic pressure (COP_{if}).

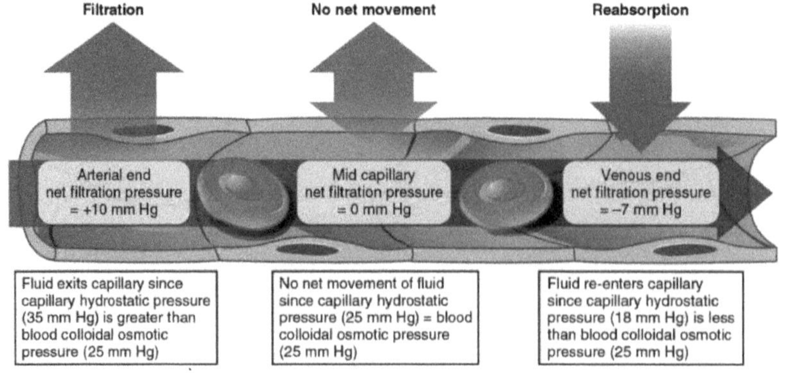

Net filtration pressure

Hydrostatic pressure is higher in arteries than in veins. Thus, hydrostatic pressure is greater at the arterial end of a capillary than at the venous end. Along the entire length of the capillary, HP_b is generally greater than HP_{if}, which is approximately zero. This means that hydrostatic pressure tends to drive fluid out of the capillaries, especially at the arterial end.

The concentration of molecules that create colloid osmotic pressure tends to be greater in the blood than in the interstitial spaces. Thus, COP_b is generally greater than COP_{if}. This means that colloid osmotic pressure tends to draw fluid into the capillaries.

At the arterial end of a capillary bed, the net HP typically exceeds the net COP, and there is a net loss of fluid from the capillaries at the arterial end. At the venous end of a capillary bed, the net COP typically exceeds the net HP, and there is a net gain of fluid into the capillaries at the venous end.

Overall, there tends to be a net loss of fluid from the capillaries. We will see later that this fluid is absorbed into the lymphatic system.

III. Blood Pressure, Resistance, and Total Blood Flow

Blood pressure

Blood pressure is a measure of the hydraulic force exerted upon blood vessel walls. It is recorded in millimeters of mercury (mm Hg). All blood vessels exhibit a blood pressure, but arterial pressure is the most frequently measured and the most useful. Arterial pressure is not constant, it varies with the stages of the cardiac cycle.

The graph below shows the approximate blood pressure as blood flows through the circulatory system. Pressure is highest at the aorta and lowest at the venae cavae. Thus, blood expelled by the left ventricle flows down its pressure gradient until it reaches the right atrium.

Pressure varies significantly with the beating of the heart until blood reaches the arterioles.

Pressure is highest in the arteries after the ventricles contract and expel blood into the arteries. This peak in pressure is what we call **systolic blood pressure** (approximately 120 mm Hg in healthy adults). Elastic arteries expand to absorb some of the pressure. As the ventricles enter the diastole, the pressure drops to what we call the **diastolic blood pressure** (approximately 70 to 80 mm Hg in healthy adults). Elastic arteries contract to keep the pressure from dropping too low.

The difference between systolic pressure and diastolic pressure is called **pulse pressure**, and this difference is what is felt when someone measures a pulse. Be aware of the difference between "pulse pressure" and "pulse"! The former is a measurement of pressure; the latter is a measure of heart *rate*. Look at the graph below and explain why the pulse cannot be felt in a vein.

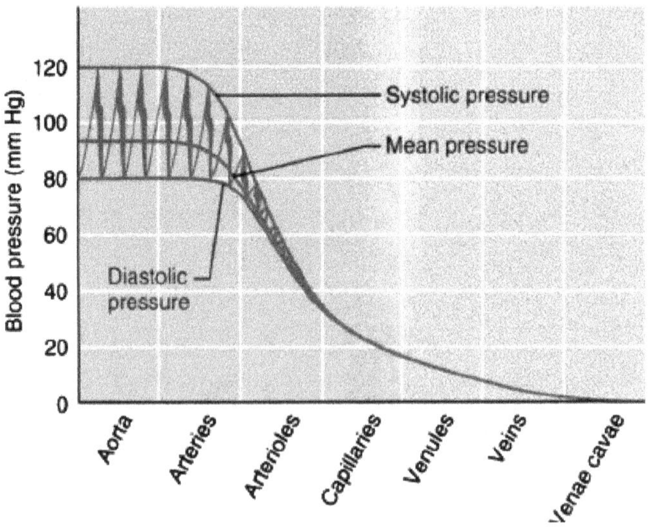

Resistance

Blood flow through the blood vessels can either be laminar or turbulent. When this laminar (uninterrupted) flow encounters any physical change within the lumen of the blood vessel like an obstruction for instance, then blood flow is now disturbed and thus becomes turbulent (see illustration below). This obstruction causing a change in the diameter of the blood vessel can be compounded by friction between blood and the vessel walls and that contributes to what we call resistane to blood flow. So, change in the diameter of the blood vessel is one of several factors contributing to resistance.

Laminar flow on left side vs. turbulent flow on right side caused by obstruction

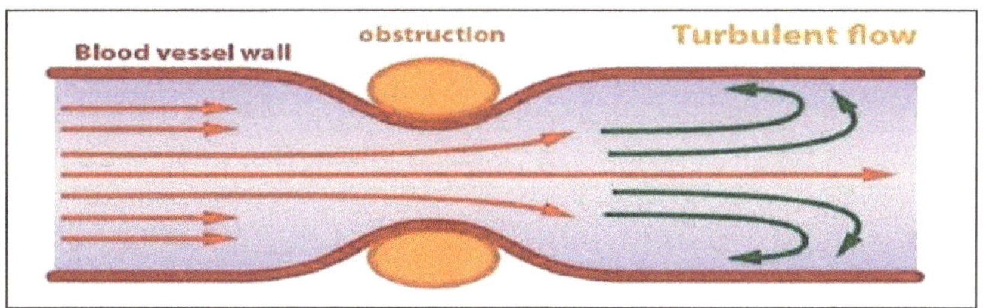

Resistance to the flow of blood through a blood vessel is determined by three factors: (1) the viscosity of the blood, (2) the length of the blood vessel, and (3) the diameter of the blood vessel. Understand how a change in each of these factors increases or decreases the resistance. In the space below, give at least one example of how these factors may change naturally in the human body.

Viscosity–

Vessel length–

Vessel diameter–

When considering vessel diameter, be aware that resistance within a vessel is inversely proportional to the fourth power of the radius ($R \alpha 1/r^4$). What effect does decreasing the radius by a factor of two have on resistance?

Peripheral Resistance

Blood cells and plasma encounter resistance when they contact blood vessel walls.

- If resistance ⬆, then more pressure is needed to keep blood moving. Three main sources of peripheral resistance:
  - blood vessel diameter
  - blood viscosity
  - total vessel length

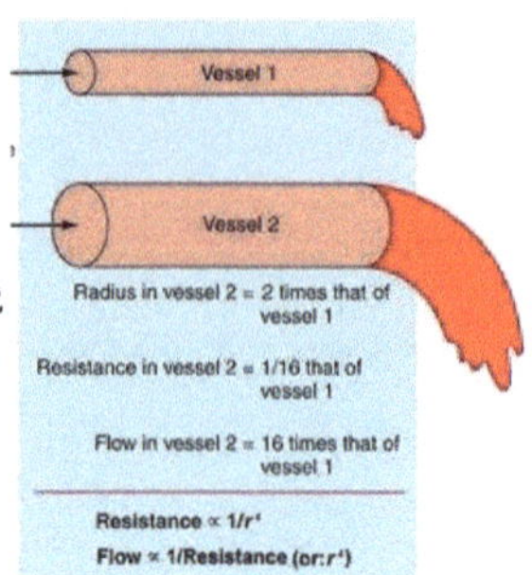

$F = \Delta P/R$

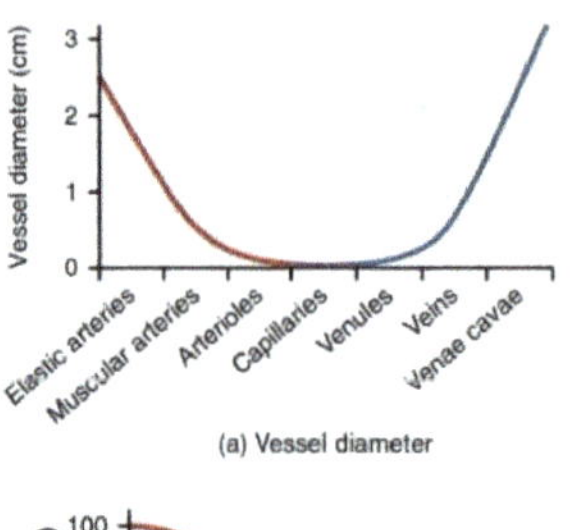

(a) Vessel diameter

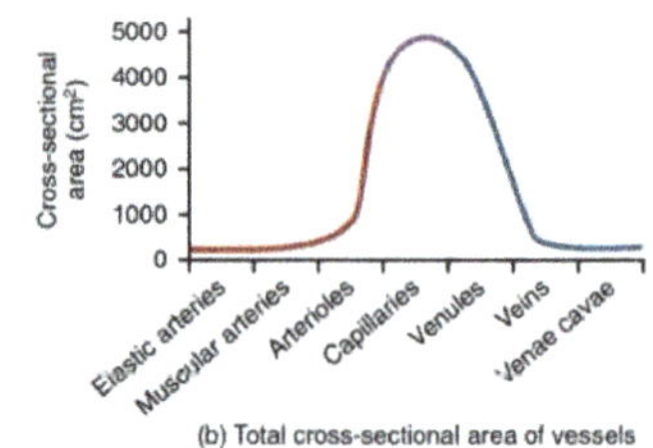

(b) Total cross-sectional area of vessels

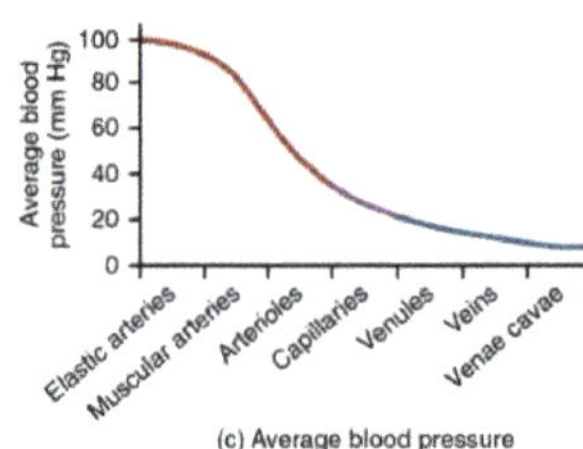

(c) Average blood pressure

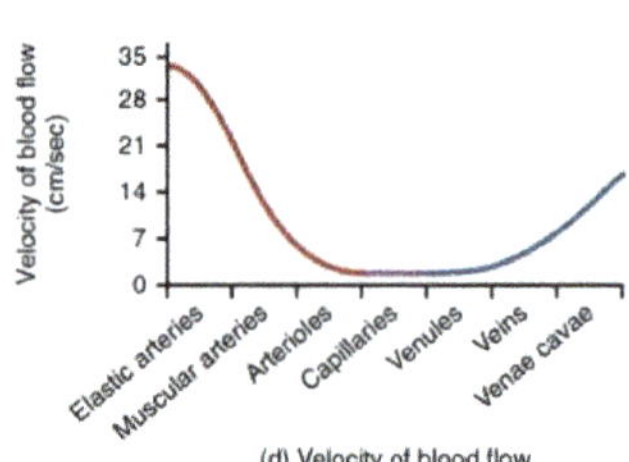

(d) Velocity of blood flow

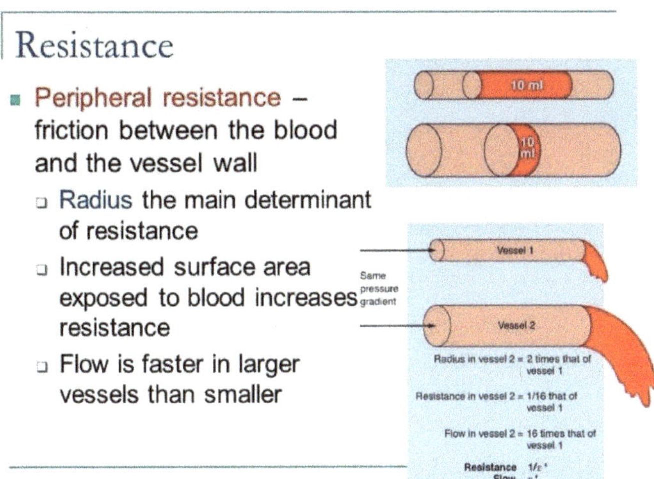

How does this relate to atherosclerosis?

The relationship of blood flow to blood pressure gradients and resistance

A fluid (like blood) will flow through a tube (like a blood vessel) as a result of differences in pressure at different points in the tube. The fluid always flows in the direction from high pressure to low pressure (*i.e.,* down a pressure gradient). The relationship between flow (F) through a tube and the pressure gradient (ΔP = change in pressure) is given by the following equation:

$F = \Delta P/R$ (ΔP = higher pressure - lower pressure, see illustration below)

where R is the resistance to flow.

When studying the circulatory system, we can say that flow through the entire system is equivalent to cardiac output (CO); the difference in pressure is approximately equal to the mean blood pressure (BP); and resistance comes primarily from friction in the small vessels of the periphery (total peripheral resistance, TPR). Thus, the equation given above becomes the following equation:

CO = BP/TPR

Diagram below showing difference in pressure determines direction of flow and factors affecting pressure

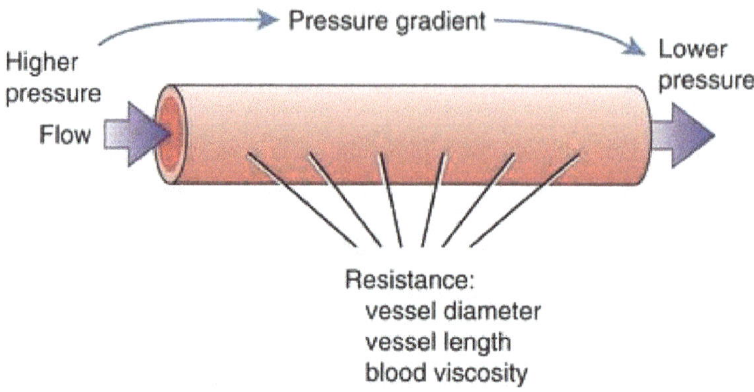

IV. Regulation of Blood Pressure and Blood Flow

Because flow depends upon blood pressure, it is important to keep blood pressure high enough to maintain sufficient transport of gases and nutrients. However, a blood pressure that is too high may result in hemorrhaging or aneurysms. The body has both short- and long-term mechanisms to regulate blood pressure. By rearranging the equation given above, you can see that blood pressure can be altered via changes in cardiac output and total peripheral resistance:

$$BP = CO \times TPR$$

Neural regulation of blood pressure

Short-term regulation of blood pressure is mediated primarily by the nervous system. Neural control is generally accomplished via reflex arcs that involve baroreceptors or chemoreceptors, the **vasomotor center** of the medulla, and sympathetic efferent pathways. The effectors of these reflex arcs are vascular smooth muscles, primarily of the arterioles.

The vasomotor center constantly sends stimulation along these sympathetic pathways, leading to a constant level of constriction, called **vasomotor tone**. **Baroreceptors** sense increased pressure in the blood vessels as the vessels are stretched. They are located in the carotid arteries, the aortic arch, and most other larger arteries of the neck and thorax. Stimulation of the baroreceptors sends afferent signals along the glossopharyngeal nerves that inhibit the vasomotor center. What effect does this have on the blood vessels and blood pressure?

Afferent impulses from the baroreceptors also inhibit the cardioacceleratory center and stimulate parasympathetic activity. The baroreceptor system functions to protect against harmful short-term changes in blood pressure, but it is rather ineffective in protecting against long-term changes. Which cranial nerve will carry parasympathetic signals to the heart?

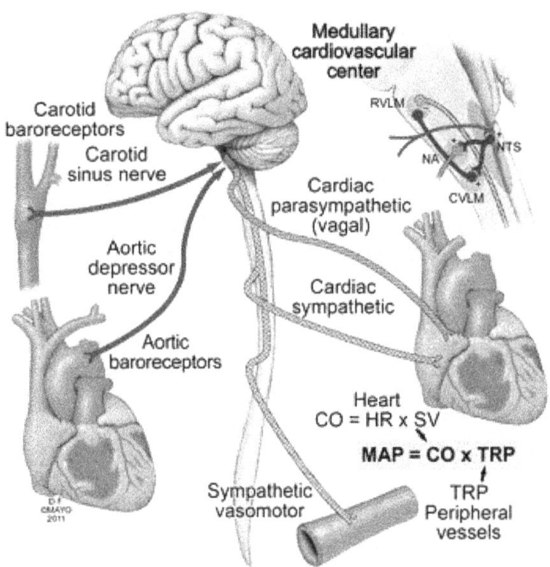

Chemoreceptors in the aortic arch and large arteries of the neck monitor the concentration of CO_2 in the blood. As the level of CO_2 rises, the chemoreceptors stimulate the cardioacceleratory and vasomotor centers. You should be able to out the effect on blood pressure and how it relates to the change in CO_2. A similar effect is caused by decreasing levels of oxygen in the blood.

Baroreceptors and chemoreceptors work via reflex pathways. Higher brain centers can also affect blood pressure. For example, the hypothalamus can cause a strong rise in blood pressure during the fight-or-flight response. The cerebrum can also affect blood pressure.

Start thinking about taking one of my exams, and your blood pressure may rise!

Hormonal regulation of blood pressure

Long-term regulation is accomplished primarily by the kidneys, and it involves changes in blood volume. The kidneys act to regulate blood volume both directly and indirectly.

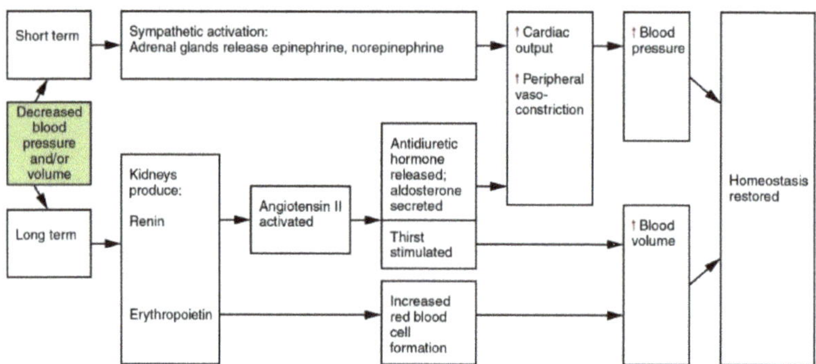

The **direct renal mechanism** of blood pressure regulation is fairly simple and intuitive: When blood pressure is high, the amount of filtration of fluid into the kidneys naturally increases, and more urine is formed. Blood volume decreases, and blood pressure decreases. When blood pressure is low, less fluid is filtered into the kidneys, and fluids are conserved. Combined with an intake of fluids, this results in an increase in blood volume and blood pressure. (Note that this is not the result of hormones).

The **indirect renal mechanism** of blood pressure regulation involves various chemicals that act to elevate blood pressure in response to a decline in blood pressure. A drop in blood pressure causes the kidneys to release the enzyme, **renin**, into the blood. Renin leads to the production of the chemical, **angiotensin II**, which is a strong vasoconstrictor. What effect does systemic vasoconstriction have on total peripheral resistance and blood pressure?

Hormonal regulation of blood pressure (renin-angiotensin system) below

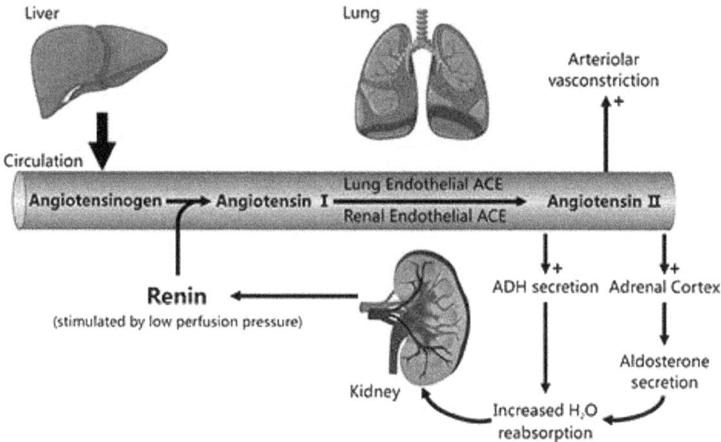

Angiotensin II also stimulates the release of the hormone, **aldosterone**, from the adrenal cortex. Aldosterone promotes the retention of sodium by the kidneys and the release of ADH by the pituitary gland. What effect does sodium retention have on water retention?

What is the effect of ADH?

What is the effect of aldosterone on blood pressure?

Autoregulation

We have just discussed how the nervous and endocrine systems can generally affect cardiac output and systemic blood pressure. It is also important that the body be able to specifically regulate blood flow to individual organs based on their changing demands for blood. To some extent, this is accomplished by **autoregulation**, in which local conditions affect the flow to a particular organ.

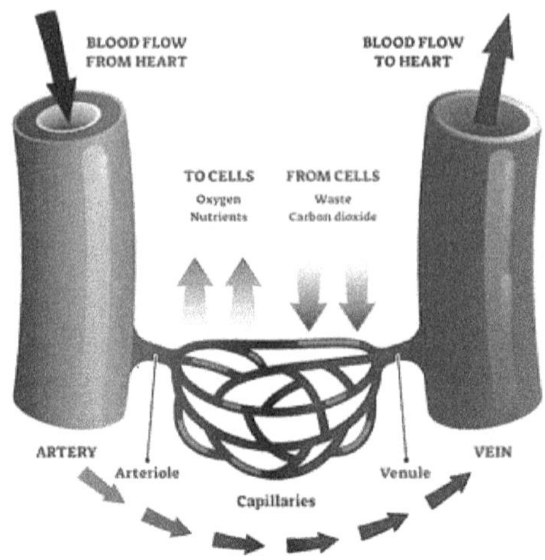

Declining nutrient levels in an organ stimulate vasodilation of nearby arterioles and relaxation of precapillary sphincters. This, obviously, allows more blood and more nutrients into the organ. Is this an example of positive or negative feedback?

Arterioles are generally able to respond automatically to changes in blood pressure. The stretch of vascular smooth muscle tends to cause vasoconstriction.

How can this protect an organ? Reductions in stretch cause increased tone and vasodilation. How can this help maintain homeostasis?

V. Velocity of Blood Flow

Blood flow through tissues is called **perfusion**. Each organ in the body typically has a small number of arteries bringing blood to it. Once these arteries enter the organ, they rapidly branch into arterioles and capillaries. Proper delivery of oxygen and other nutrients to the organ depends on properly matching the organ's demand for blood to the supply of blood to the organ.

The graph below shows that as blood travels through the body, the **velocity** of blood flow is inversely proportional to the cross-sectional area of blood vessels being filled. Although capillaries are smaller than arteries and veins, the immense number of capillaries means that as blood flows from arteries to arterioles and then to capillaries, the total cross-sectional area increases (and velocity decreases). This is beneficial as the slow movement of blood through capillaries allows adequate time for the exchange to occur. As capillaries merge to form venules and veins, the cross-sectional area decreases, and flow increases as blood returns to the heart.

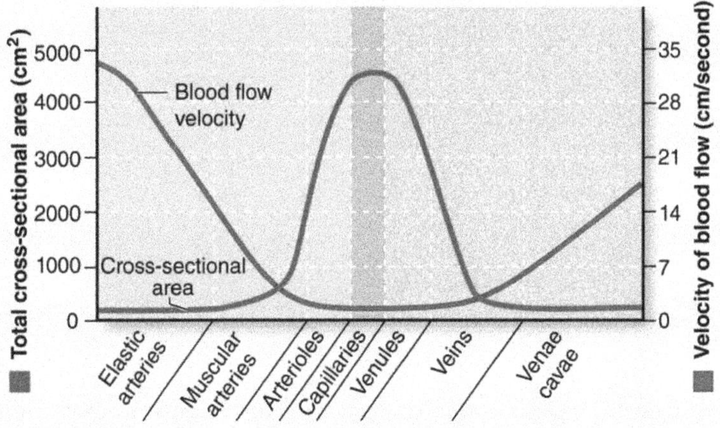

ARTERIAL SYSTEM (ANTERIOR VIEW)

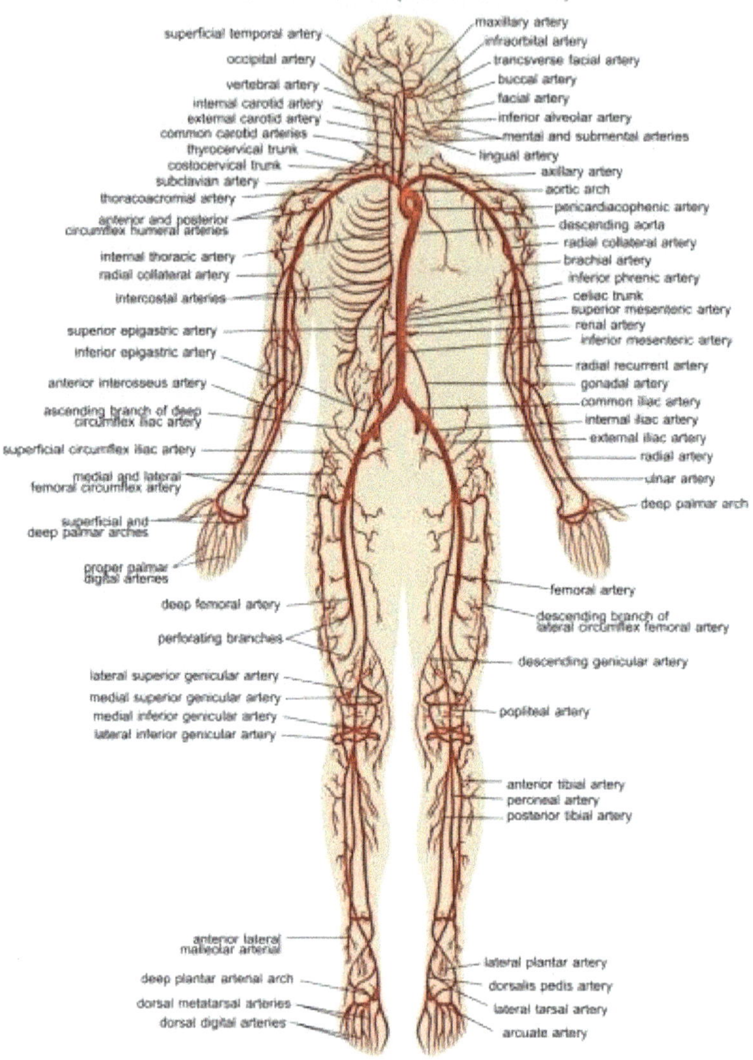

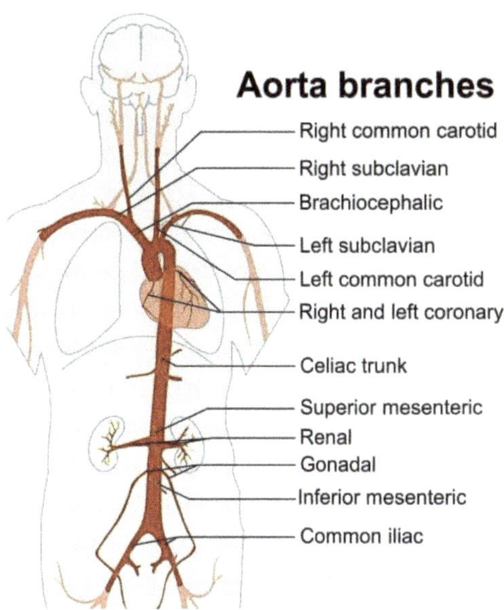

Questions

- **Name the three (3) single abdominal arteries, (*hint*: start with the first one that emerges from the abdominal aorta)**
- **Which arteries branching from the aorta are asymmetrical?**

Thoracic and abdominal arteries. Name the missing labeled vessels in this illustration.

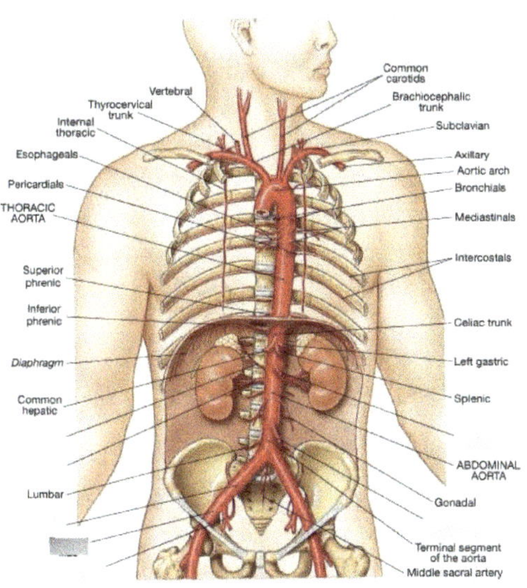

major arteries of right upper and lower extremities

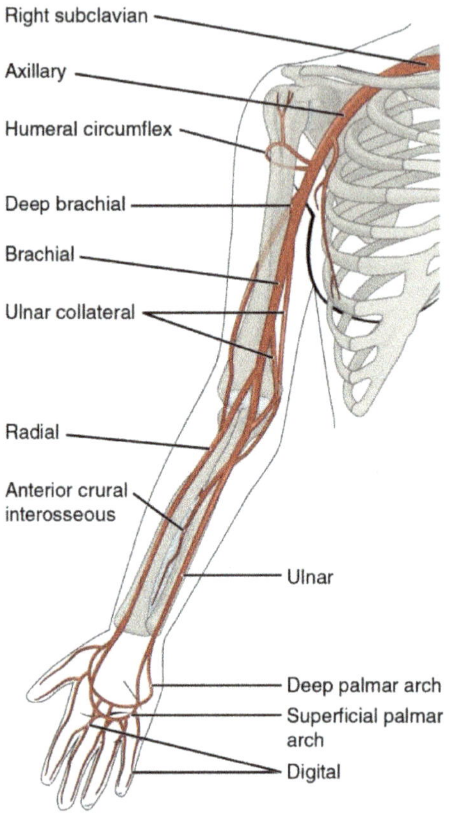

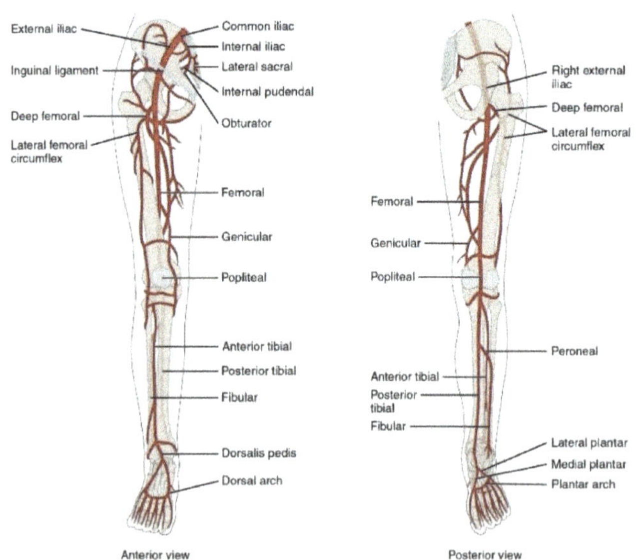

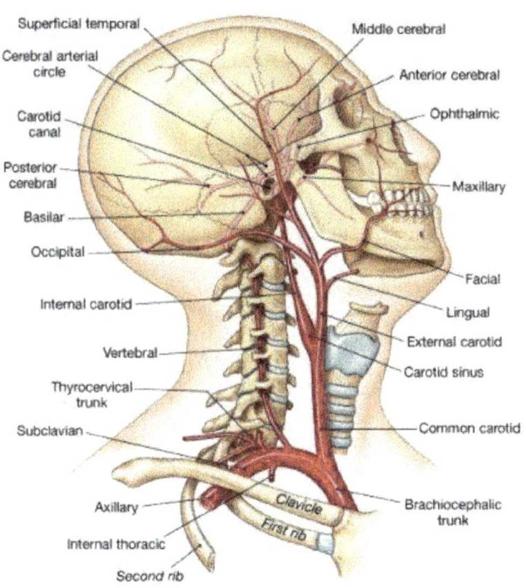

The internal carotid artery enters the skull through the carotid canal in the temporal bone and branches out into three arteries: the anterior cerebral supplies the frontal and parietal lobes of the cerebrum; middle cerebral which supplies the lateral cerebrum; and the ophthalmic artery supplying the eye. It should be noted that the anterior and middle cerebral arteries are joined to the posterior cerebral arteries to form the cerebral arterial circle (**circle of Willis**).

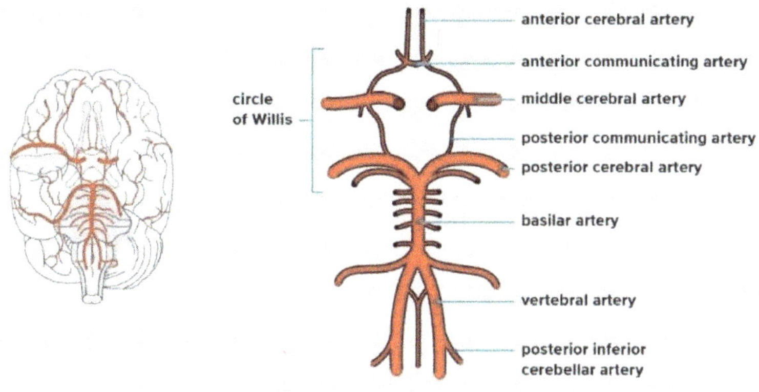

Major veins of the head and neck

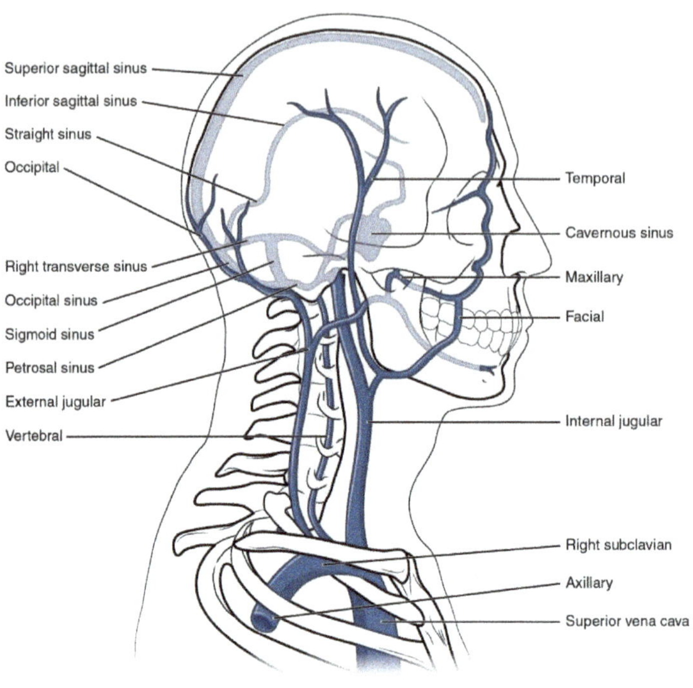

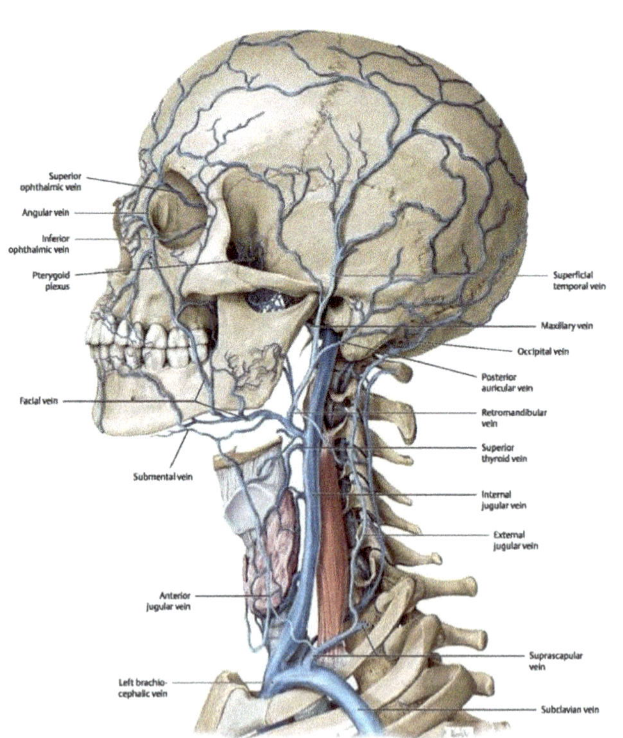

Major veins of upper extremities

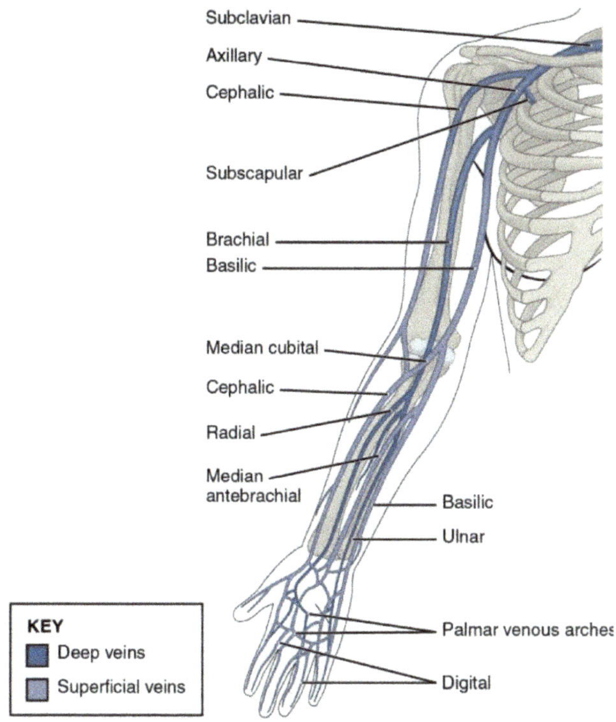

Now would be a good time to test your knowledge of the circle of Willis below. Remember, don't rush through this, make it fun.

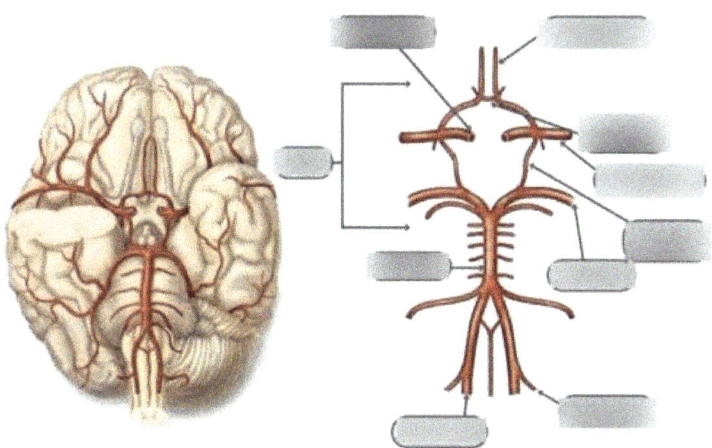

What is the path of a red blood cell (RBC) from the IVC (inferior vena cava) to the left common carotid artery, naming all blood vessels and heart valves that this RBC passes through?

Chapter 5: Lymphatic System

The lymphatic system serves two major functions for the body:

1. Assist the body in defense against pathogens (bacteria, viruses, protozoa), toxins, cancer cells, etc. Lymphatic tissues and organs transport and house phagocytic cells and lymphocytes to aid in the body's defense.
2. Transport fluid from the interstitial spaces to the bloodstream. Hydrostatic and osmotic pressures result in the movement of fluid out of the circulatory system at the arterial ends of the capillaries. Much of the fluid is reabsorbed at the venous ends of the capillaries, but the balance must also be returned to the bloodstream, or the blood volume will drop. The lymphatic system absorbs fluid from the interstitial compartment and later returns to the venous circulation.

Notice that both of these functions involve supporting other body systems, including the immune and cardiovascular systems.

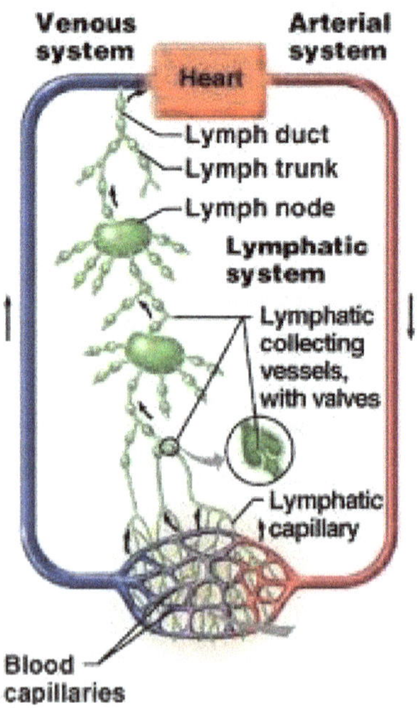

I. Lymph and Lymph Vessels

Lymph and lymphatic capillaries

You should already be aware that the body contains interstitial fluid in the spaces between and around cells. You should also be aware that throughout most of the body, fluid from the blood leaks through capillaries into the interstitial fluid. A slight pressure gradient drives interstitial fluid into capillaries of the lymphatic system, which are found throughout most of the body. Once fluid enters these vessels, it is known as **lymph**. Over the course of a day, about three liters of fluid are absorbed into lymph capillaries. The main ingredient of lymph is, of course, water. Lymph also contains small solutes (such as sodium and potassium) and a small amount of protein. Leukocytes and foreign particles are also found in the lymph.

Lymph capillaries are found throughout the body in most places where capillary beds are found. Notable exceptions are in the bones, bone marrow, and central nervous system. Lymph capillaries are more permeable than blood capillaries, and they easily take in fluid and particles as large as proteins. Pressure in the interstitial fluid is generally somewhat higher than pressure in the lymph capillaries, and this pressure opens flaps on the lymph capillaries to allow fluid and particles into the lymph capillaries. Should pressure inside the lymph capillaries exceed that of the interstitial fluid, the flaps shut to prevent backflow out of the lymph capillaries. When tissue becomes inflamed, interstitial pressure may rise sufficiently to force particles such as viruses and bacteria into the lymph capillaries.

Blind end lymphatic capillary

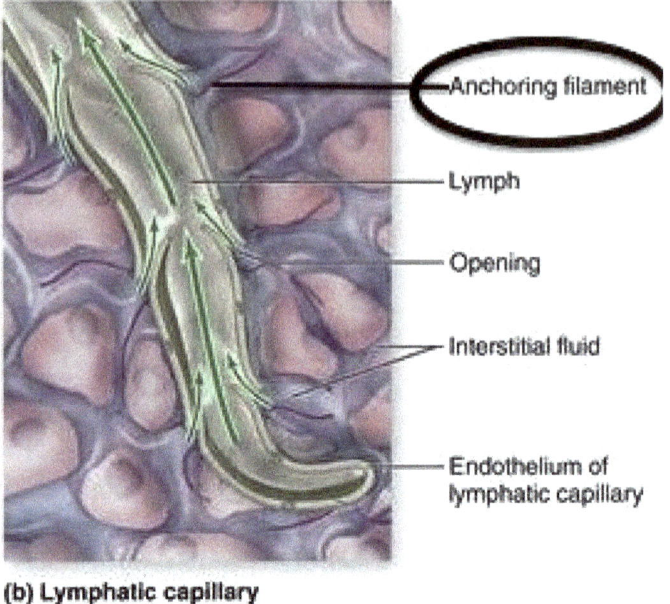

(b) Lymphatic capillary

Key features of lymphatic capillary:

- Absent in avascular structures (i.e., cartilage), brain and spinal cord, splenic pulp and bone marrow
- Blind end
- Single layer of overlapping endothelial cells more permeable than that of blood capillaries
- Combined to form lymphatic vessels which resemble veins with thin walls and more valves
- Lymph flows through lymph nodes towards large veins above the heart (i.e., lymph empties into bloodstream)

Lacteals are lymph capillaries found in the villi of the small intestine. They absorb lipids and lipid-soluble vitamins (i.e., A, D, E, and K) from the intestine and transport them to the bloodstream.

INTESTINAL VILLI

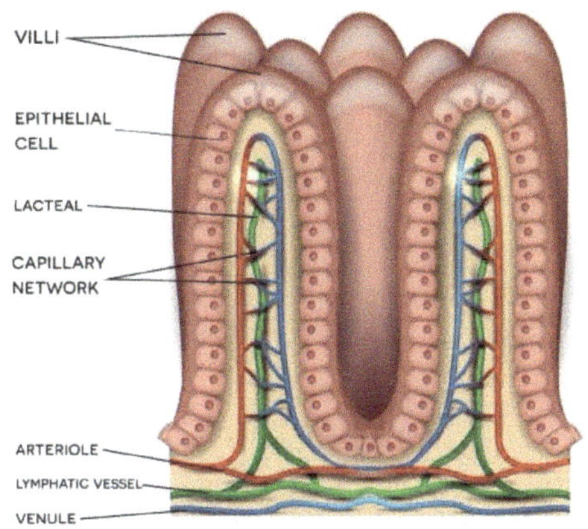

Lymphatic vessels, trunks, and ducts

From lymph capillaries, lymph travels toward the heart via increasingly larger vessels.

Lymphatic vessels are similar in structure to veins; they have the same three tunics as veins, and they have valves. Why are valves critically important to the lymphatic vessels?

Lymphatic vessels converge to form **lymphatic trunks**, which drain major body regions, such as the head, the arms, the abdominal organs, and the legs. The lymphatic trunks eventually lead to one of two large **lymphatic ducts** (see illustration below): The **right lymphatic duct** drains the right arm, the right side of the head, and the right side of the thorax. The right lymphatic duct empties lymph into the venous circulation at the right subclavian vein. The **thoracic duct** drains the rest of the body. The thoracic duct empties lymph into the venous circulation at the left subclavian vein.

Diagram below showing lymphatic drainage. Note that the right lymphatic duct drains into the right subclavian vein and the thoracic duct drains into the left subclavian vein. Also, on closer look the right lymphatic duct drains the right side of the chest, the right side of the head and the right upper extremity. Whereas the thoracic duct drains the rest of the body.

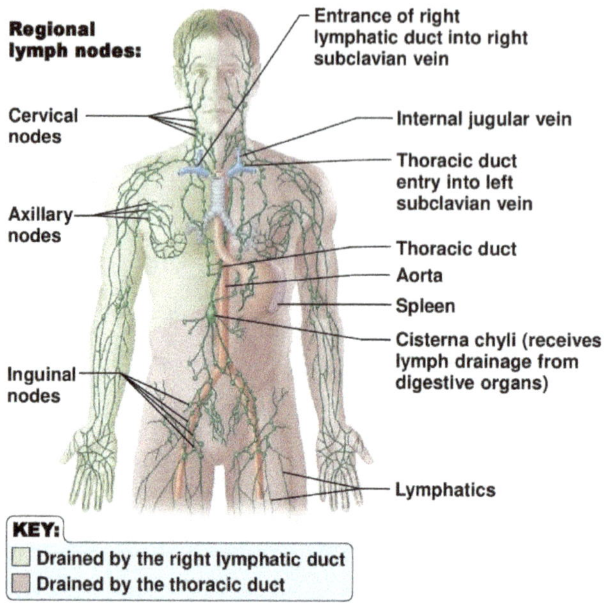

Flow through the lymphatic system occurs at a much lower rate than flow through the circulatory system. Movement of fluid is accomplished primarily by contraction of skeletal muscles and changes in pressure of the thoracic cavity (caused by breathing), combined with assistance from the valves. Compare the flow rate through the lymphatic system (about three liters of fluid per day) with the flow rate through the cardiovascular system.

II. Overview of Lymphatic Tissue and Organs

In addition to lymphatic vessels and lymph, the lymphatic system contains clusters of tissue and organs. The various tissues and organs are categorized as either primary or secondary lymphatic structures.

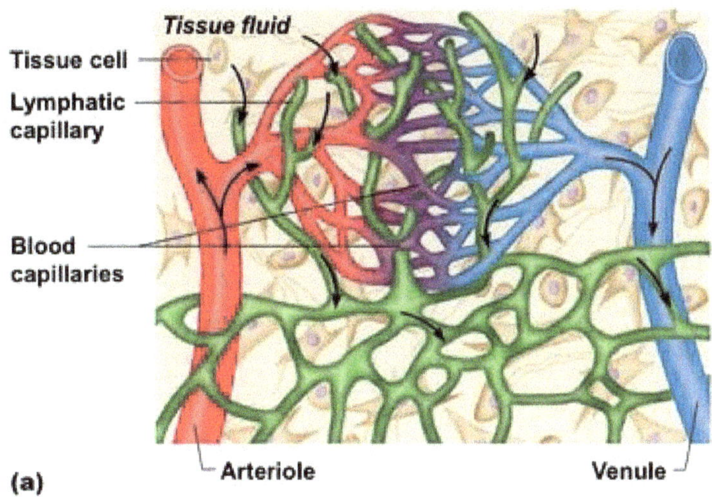

Primary lymphatic structures are involved in the formation and maturation of lymphocytes.

Secondary lymphatic structures are not involved in lymphocyte maturation. They are sites where immune responses are initiated.

Primary and secondary lymphoid organs and their functions in chart below

Lymphoid organ	Lymphoid organ type	Description	Function
Red bone marrow	Primary	In adults, this is diffusely located in the central portions of flat bones such as the ribs, sternum, pelvis and vertebrae, with smaller amounts found in the epiphyses of some of the larger, long bones, such as the femur	To act as the major haemopoietic tissue, producing all the formed elements of blood including erythrocytes (red blood cells), leukocytes (white blood cells) and platelets (thromobocytes)
Thymus gland	Primary	A small, butterfly-shaped gland located on the superior surface of the heart (Fig 1)	Plays a key role in programming the immune system to recognise 'self', and provides a site for the maturation of T-lymphocytes
Spleen	Secondary	The largest lymphoid organ, the spleen has a lobular structure and is located on the left-hand side of the abdomen, between the ninth and 11th ribs	Helps trap foreign material and remove damaged and aged erythrocytes, while also acting as a reservoir for erythrocytes, leukocytes and platelets
Lymph nodes	Secondary	Small, bean-shaped lymphoid organs found in major clusters, including the cervical nodes (neck), axial nodes (armpits) and inguinal nodes (groin regions)	Primarily filter lymph, trapping foreign material such as bacterial and viral particles

III. Primary Lymphatic Structures

Red bone marrow

Red bone marrow tissue is considered a primary lymphatic structure. As you learned in the previous sections (and in A&P I), it is the site of hemopoiesis. Both B and T lymphocytes are formed in the red bone marrow (see histology slide below).

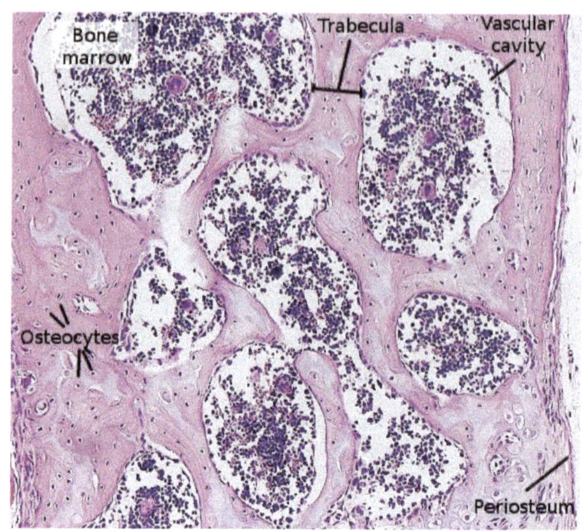

Thymus

The **thymus** is located in the superior portion of the mediastinum. After T-lymphocytes are formed in the red bone marrow, they must travel to the thymus to properly mature. The "T" in the phrase "T-lymphocyte" or "T cell" comes from the fact that these cells mature in the thymus. The thymus grows throughout childhood and slowly atrophies through adult life. This atrophy may explain, to some degree, why older persons are often more susceptible to infection.

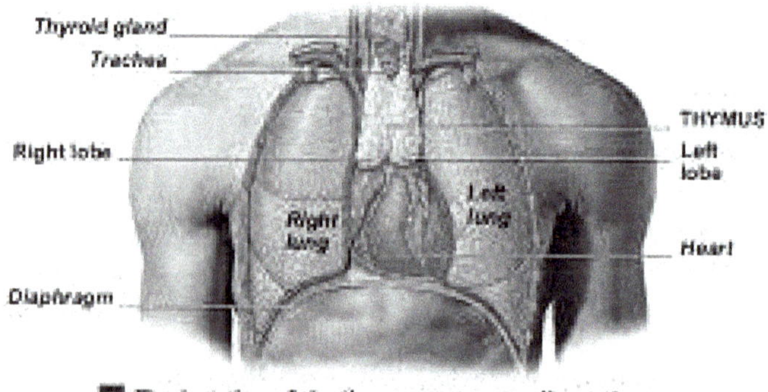

The location of the thymus on gross dissection; note the relationship to other organs in the chest

Child thymus (prepuberty) – highly active

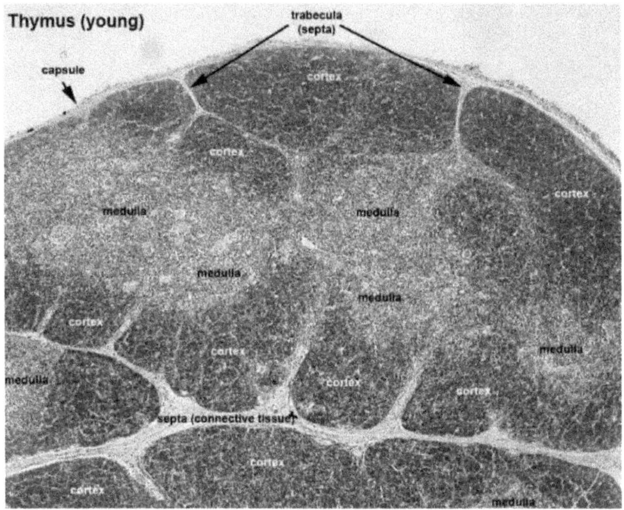

Child (a) thymus compared to and adult (b) thymus consisting mostly of fatty tissue and is at this point vestigial (biologically non-functioning)

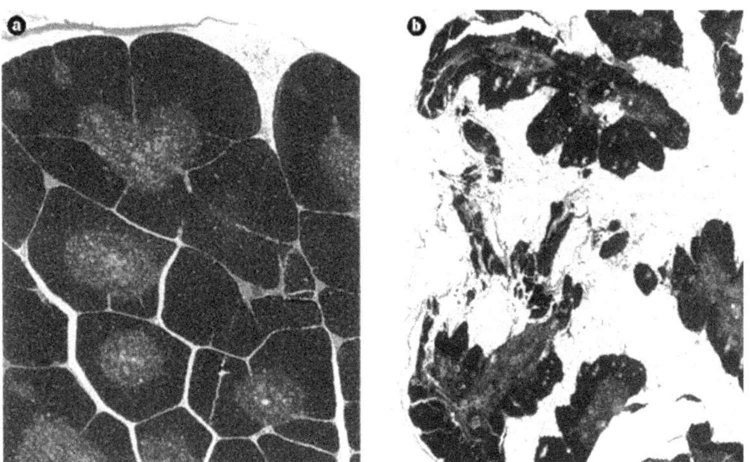

IV. Secondary Lymphatic Structures

Secondary lymphatic structures house lymphocytes and other cells of the immune system. They are mainly composed of reticular connective tissue. There are four key cell types in the lymphatic system:

1. **Lymphocytes** direct the specific immune response against specific antigens.

2. **Macrophages** that phagocytize pathogens.
3. **Dendritic cells** present antigens to T-lymphocytes.
4. **Reticular cells** produce a fibrous network in the lymphoid organs (structural frame).

Secondary lymphatic structures provide an environment for the proliferation of lymphocytes. They also provide strategically placed surveillance points for lymphocytes and macrophages to look for pathogens. Pathogens that enter the body are likely to invade the interstitial spaces or blood and end up in the lymphatic system.

A secondary lymphatic structure that is enclosed by a capsule is considered an organ; these include the spleen and lymph nodes. Other secondary lymphatic structures have incomplete capsules or lack capsules altogether; these structures are not regarded as proper organs.

Lymph nodes

Lymph nodes are relatively small, round structures found along the lymphatic vessels. They act to filter the lymph as it travels through the vessels. While they are found throughout most of the body, lymph nodes are concentrated in the inguinal, axillary, and cervical regions of the body (see picture above). Phagocytic cells in the nodes remove microorganisms and debris from the lymph, and lymphocytes monitor the lymph for specific antigens that may require a specific immune response.

Lymph nodes are generally less than 1" in length. Each node is surrounded by an **outer capsule** made of connective tissue. Extensions of the capsule, called **trabeculae**, extend into the lymph node and make compartments within the node. The outer layer of the lymph node, or **cortex**, contains numerous **lymphatic nodules** (discussed in more detail later). The inner portion of the node is called the **medulla**.

Multiple **afferent lymphatic vessels** bring lymph to each lymph node. The lymph makes its way to the medulla, where lymph flows into **lymph sinuses**, which are spanned by networks of reticular fibers. These networks provide an environment in which many macrophages and lymphocytes reside, as they monitor the lymph for invaders. One side of each node has an indented area, called the **hilum**. At the hilum, an **efferent lymphatic vessel** allows lymph to exit the lymph node.

Structures of a lymph node below

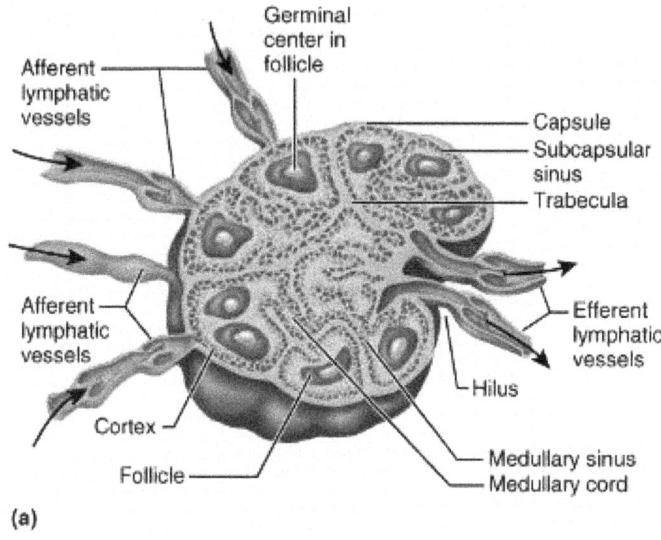

Histology slide of lymph node

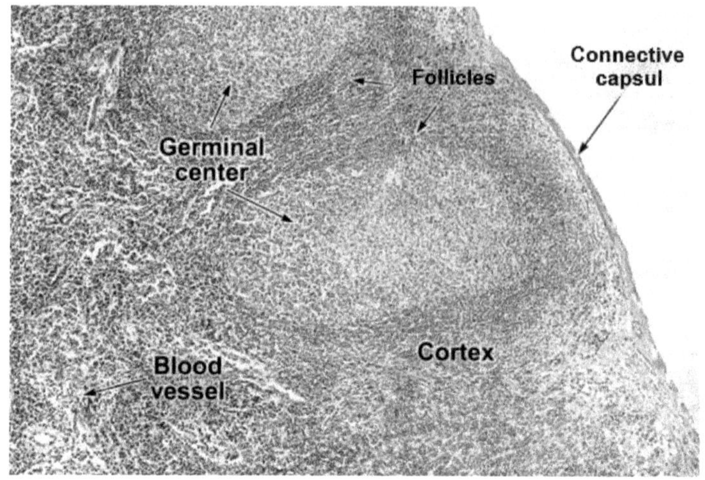

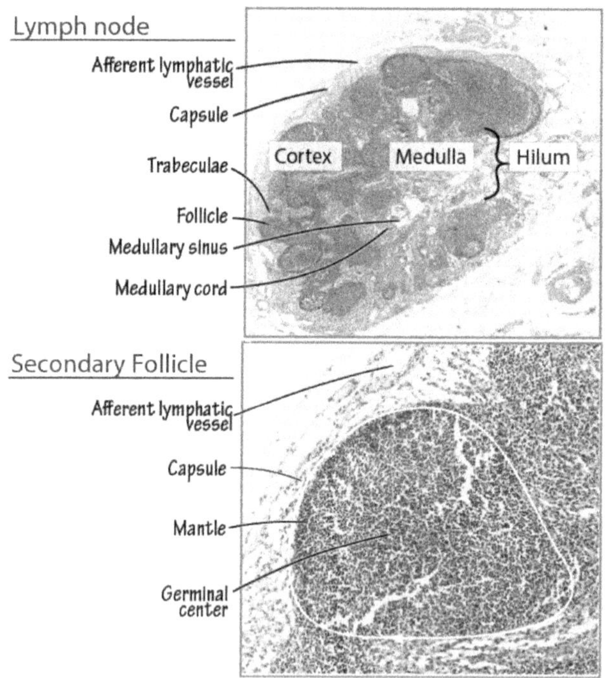

Spleen

The **spleen** is located below the diaphragm and curls around the left margin of the stomach. It receives a rich supply of blood and is generally deep red in color. The functions of the spleen are as follows:

1. Like the lymph nodes, the spleen is a site of lymphocyte proliferation and immune system surveillance. Unlike the lymph nodes, the spleen monitors the blood rather than the lymph.

2. The spleen aids in removing from the bloodstream and breaking down aged and defective erythrocytes and platelets. Iron is stored in the spleen and released when needed by other parts of the body.

3. The spleen is a storage site for platelets.

Areas of the spleen that consist primarily of lymphocytes are called **white pulp**. These areas are the primary sites for the spleen's immune system functions. The rest of the spleen has a high density of erythrocytes and makes up the **red pulp**. Arteries and veins enter and exit the spleen at the **hilum**. While the spleen has efferent lymphatic vessels, it lacks afferent lymphatic vessels.

The **capsule** of the spleen is quite delicate, and injury to the spleen can be detrimental because of the potential for extensive internal bleeding. However, a damaged spleen can generally be removed without serious complications since other lymphoid organs and the liver can handle its functions.

Gross anatomy of the spleen (largest lymphoid organ) and its vascularization

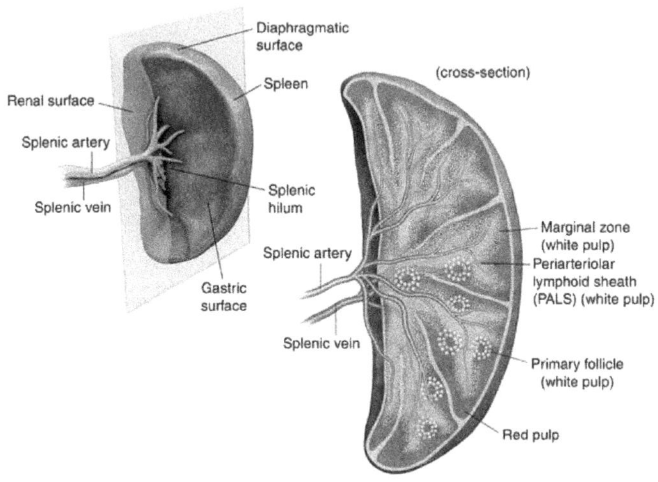

Sectioned spleen

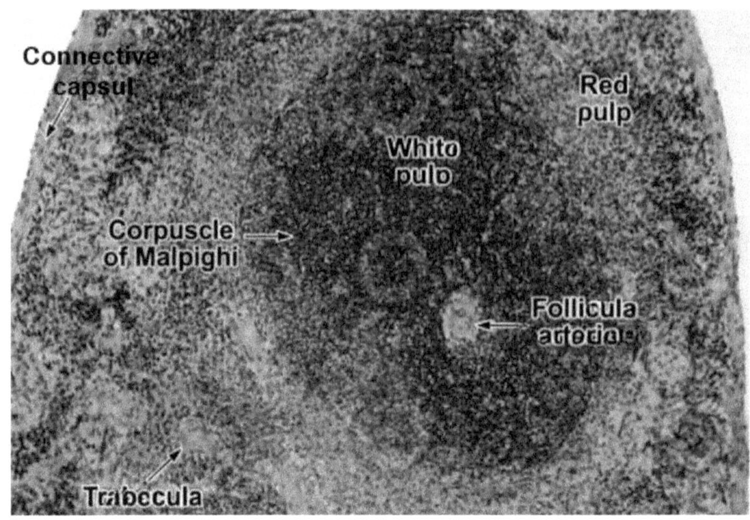

Spleen

- Functional tissue:
- White pulp
  - Lymphocytes around central arteries
  - Branches of trabecular arteries
  - Immune response to antigens
  - Produces both B & T lymphocytes

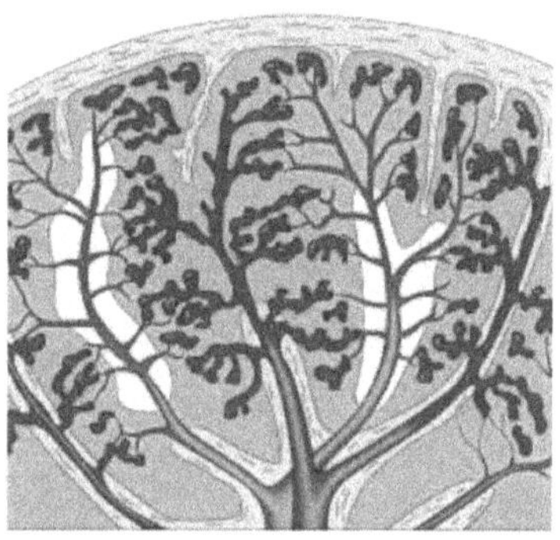

Tonsils

The **tonsils** are found in the walls of the pharynx. They are not completely surrounded by a capsule, so they are not considered true lymphatic organs. There are three main tonsils: the **palatine**, **lingual**, and **pharyngeal tonsil**s (also called the adenoids). Located in the pharynx, they monitor the entrance to both the respiratory and digestive systems.

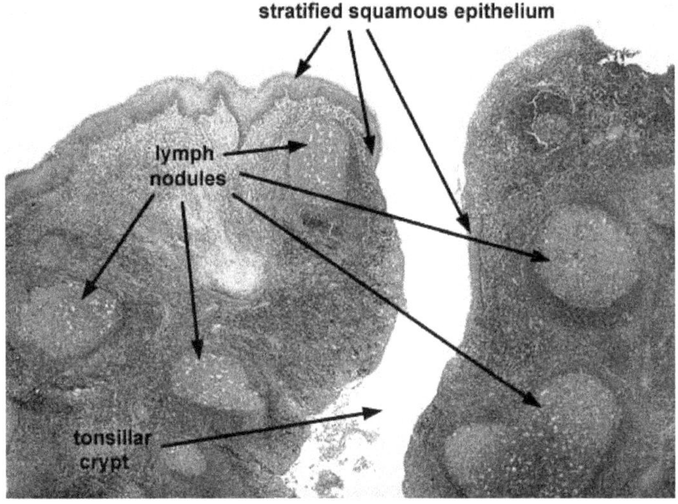

Lymphatic nodules and MALT

Lymphatic nodules (or *follicles*) are dense, spherical bodies of lymphoid tissue. Groups of nodules make up **diffuse lymphatic tissue** found throughout most body organs.

Nodules are found in the lymphoid organs, including the lymph nodes, spleen, and tonsils. Specific collections of lymphatic follicles are also found in other places in the body: **Peyer's patches** are aggregates of follicles found in the wall of the small intestine. The **appendix** also contains many follicles.

The term **MALT (mucosa-associated lymphatic tissue)** refers to the various lymphoid tissues and organs associated with the digestive and respiratory tracts (including the Peyer's patches, lymph nodes, etc.).

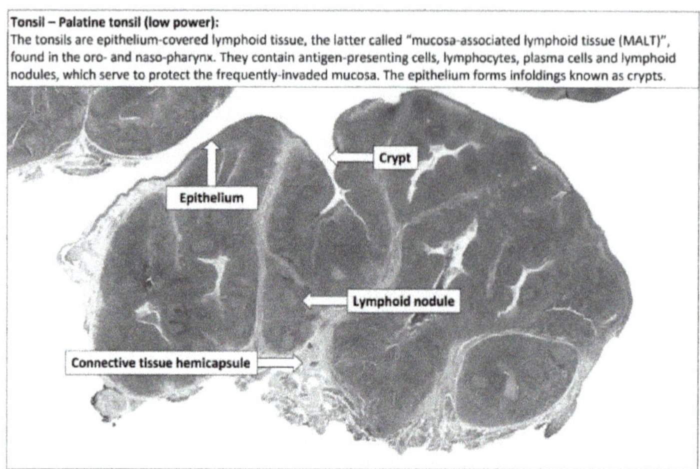
Tonsil – Palatine tonsil (low power):
The tonsils are epithelium-covered lymphoid tissue, the latter called "mucosa-associated lymphoid tissue (MALT)", found in the oro- and naso-pharynx. They contain antigen-presenting cells, lymphocytes, plasma cells and lymphoid nodules, which serve to protect the frequently-invaded mucosa. The epithelium forms infoldings known as crypts.

Why is it important to have lymphoid tissues concentrated along the digestive and respiratory tracts?

Histology of the ileum of the small intestine showing the peyer's patches.

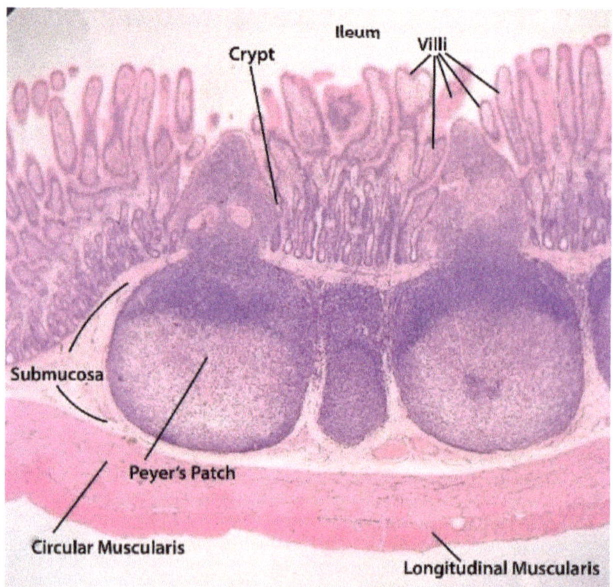

Chapter 6: Immune System and the Body's Defense

I. Overview of Diseases Caused by Infectious Agents

A variety of factors can cause disease; some have causes within our bodies (*e.g.*, genetic disorders and cancer), and some have external causes (*e.g.*, injury or infection). Infectious agents that cause harm to the body are often called **pathogens**. The chart below describes five categories of infectious agents : bacteria, viruses, fungi, protozoans, and multicellular parasites.

Type of pathogen	Description	Human diseases caused by this type of pathogen
BACTERIA	Single celled organisms without a nucleus	tuberculosis, diphtheria, typhoid, cholera, tetanus, dysentery and pneumonia
VIRUSES	Non-living particles that can only reproduce within a living cell	Common cold, herpes, measles, AIDS, chicken pox, small pox
FUNGI	Simple organisms including mushrooms and yeasts that grow as single cells or thread like filaments	Ringworm, athletes foot, candidiasis, histoplasmosis
PROTOZOA	Single celled organisms with a nucleus	Malaria, giardiasis, chagas disease, sleeping sickness, leishmaniasis
MULTI-CELLULAR PARASITES	Living organism, that can live inside intestinal tract or blood stream	Round worm infections, tape worm infections

II. Overview of the Immune System

Mechanisms your body employs to defend against pathogens, cancers, etc. make up your body's immune system. This system differs from other systems studied in the class in that it is a system composed primarily of individual cells spread throughout the body rather than a discrete system of organs. However, the cells of the immune system work co-operatively to provide a clear function for the body.

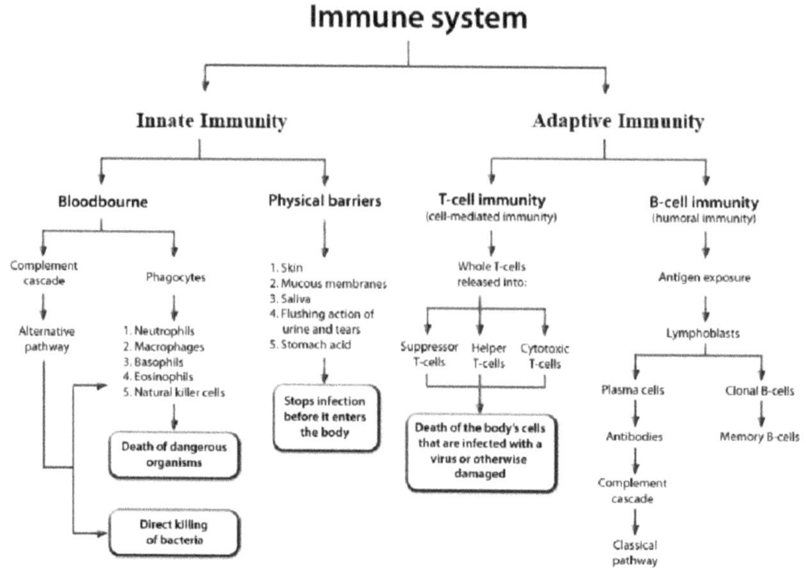

Immune cells and their locations

The primary cells of the immune system are the leukocytes:

- Granulocytes: neutrophils, eosinophils, and basophils.
- Monocytes: become macrophages when they leave the blood and enter other tissues.
- Lymphocytes: include T-cells, B-cells, and natural killer (NK) cells.

Although leukocytes circulate in the blood, most of the leukocytes in the body are found in other locations:

- Lymphatic tissue–macrophages and lymphocytes are housed in secondary lymphatic structures.
- Select organs–macrophages can be found in most organs of the body, including the lungs, brain, and others.
- Epithelial layers of the skin and mucous membranes–dendritic cells are phagocytes (related to monocytes) that patrol the skin and mucous membranes, looking to devour pathogens.

- Connective tissues–mast cells (similar to basophils) are located throughout the body's connective tissues, especially in the dermis of the skin and those tissues that line the respiratory, digestive, and urogenital tracts. You should recall their functions from A&P I.

Comparison of innate immunity and adaptive immunity

The immune system provides defense in two general ways: (1) The **innate** (or *non-specific*) **system** attempts to protect the body from all invaders. The skin and mucosae provide the first line of defense in the innate system, which act as physical barriers to invasion. The second line of defense includes phagocytic cells and antimicrobial proteins, which attempt to contain the spread of any invaders that get past the first line of defense. The innate system provides a generic but immediate defense against toxins and pathogens. (2) The **adaptive** (or *specific*) **system** can mount specific attacks against specific invaders. This system can provide a more effective defense against a specific invader, but it takes more time for the adaptive system to coordinate an attack. The two systems generally act together to fight an infection.

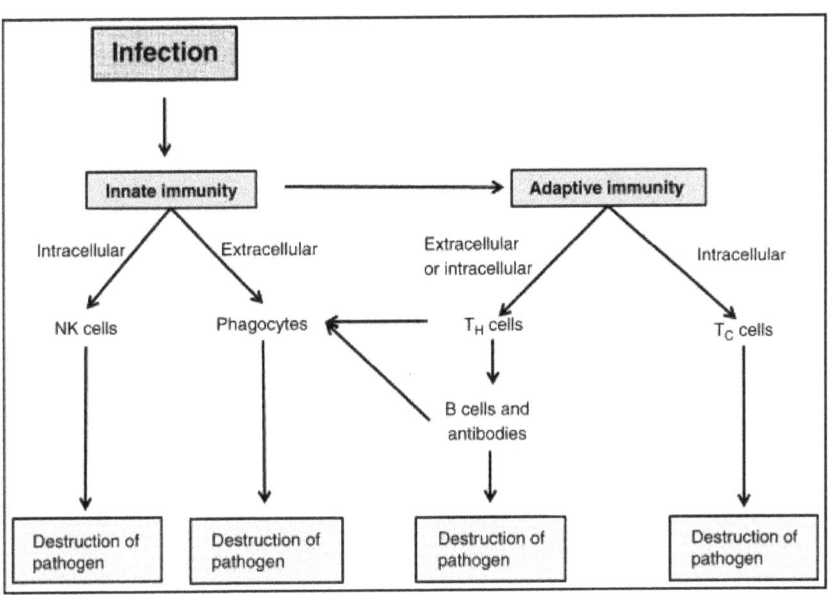

III. Innate Immunity

Preventing entry

The thick, keratinized stratum corneum provides a nearly impenetrable barrier to pathogens that might try to enter the body through intact skin. Mucous membranes also provide an effective barrier against most potential invaders. Various secretions also help prevent entry into the body:

1. Fluids secreted onto the skin (*e.g.,* sweat and sebum) contain chemicals that inhibit the growth of microorganisms.
2. The stomach secretes hydrochloric acid and proteases that kill microorganisms in the food we eat.
3. Saliva and tears contain **lysozyme** to kill bacteria residing in the mouth or on the surfaces of the eye.
4. Mucous membranes produce mucus to trap microorganisms. Cilia lining internal passages sweep mucus out of the body or into the digestive tract. Mucus also contains lysozyme.

Various harmless (and often beneficial) microorganisms live on your skin and within exposed body tracts. Many of these microorganisms inhibit the growth of potentially harmful microorganisms.

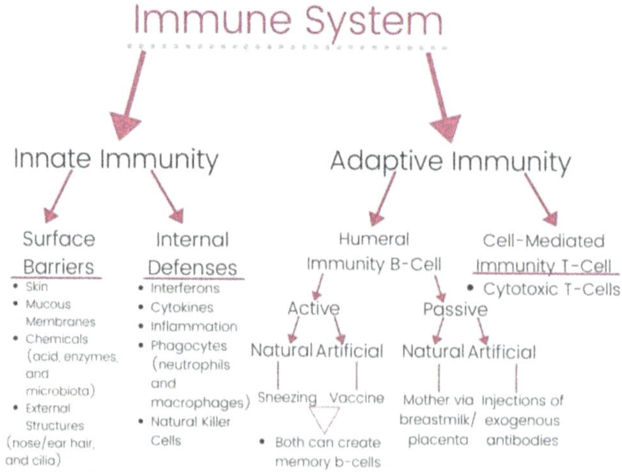

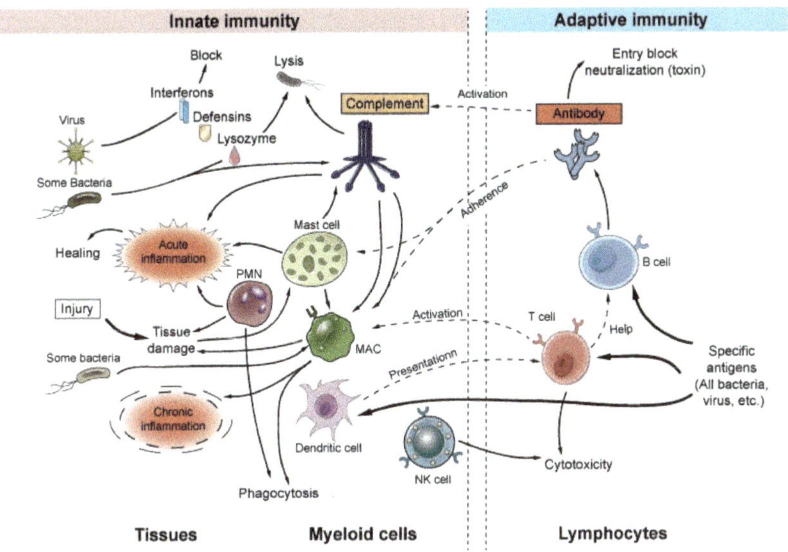

Cellular defenses

Phagocytic cells patrol the body's tissues in search of invaders that have breached the surface barriers. The chief phagocytes are *macrophages* and *neutrophils*. They engulf pathogens (as well as damaged cells or cellular debris) and chemically digest them inside structures called phagolysosomes.

In response to an infection, *basophils* and *mast cells* release chemicals that cause inflammation and serve as attractants for other cells in the immune system. You should already be familiar with the chemicals, histamine and heparin.

Natural killer (NK) cells are lymphocytes that specialize in detecting and killing cancer cells and cells infected by viruses. Whereas other lymphocytes act against specific targets, NK cells are less particular and can kill a broad range of targets. NK cells destroy cells by attacking the target cell's membrane. They produce chemicals called **perforins**, which create channels in the target cell membrane and lead to the destruction of the target cell.

Activation of NK cell

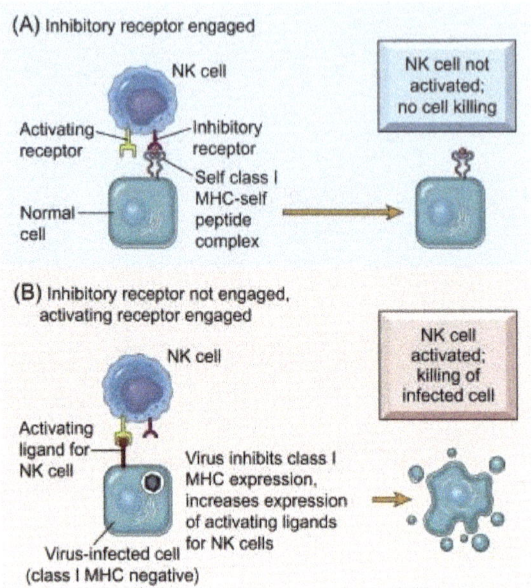

Eosinophils are most important in defending against invaders too large for phagocytosis, such as parasitic worms. They can also phagocytize antigen-antibody complexes.

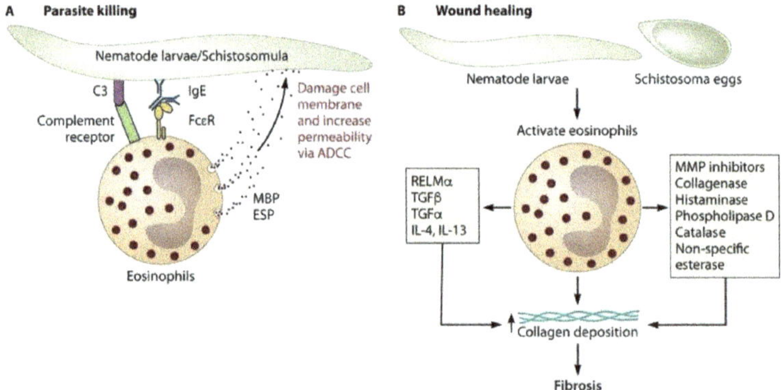

Antimicrobial proteins

Various antimicrobial proteins assist in defense by attacking pathogens directly or hindering their ability to reproduce. **Interferons (IFNs)** are released by some cells that viruses have infected. They can hinder virus reproduction and activate macrophages and NK cells to seek and destroy other virus-infected cells.

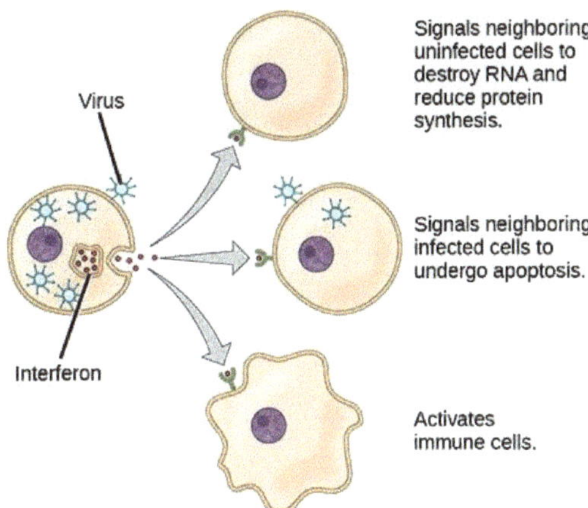

Complement is a system of at least thirty proteins that circulate in the blood in an inactive state. Activated complement amplifies the inflammatory response, and it directly kills cells. Activation of the complement system results in several effects that assist in the defense of the body:

- **Opsonization**–binding of complement proteins or antibodies to a pathogen makes it an easier target for phagocytosis.
- **Inflammation**–complement enhances the inflammatory response by activating mast cells and basophils.
- **Cytolysis**–several complement proteins work together to form a **membrane attack complex (MAC)**, which puts holes in the target cell plasma membrane, causing lysis of the target cell.

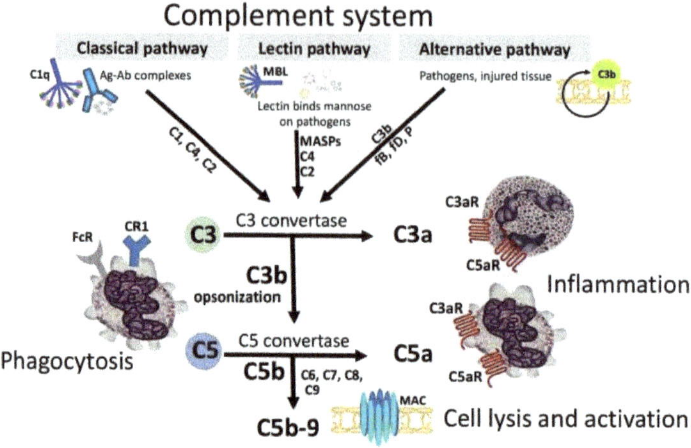

Inflammation

Infection, chemicals, heat, and physical trauma produce the **inflammatory response** in tissue. Inflammation involves *redness*, *heat*, *swelling*, and *pain* in the injured or infected area. The inflammatory response begins with the release of chemicals from injured tissue, phagocytes, lymphocytes, mast cells, basophils, and the blood. These chemicals include histamine, **leukotrienes**, **prostaglandins**, and others.

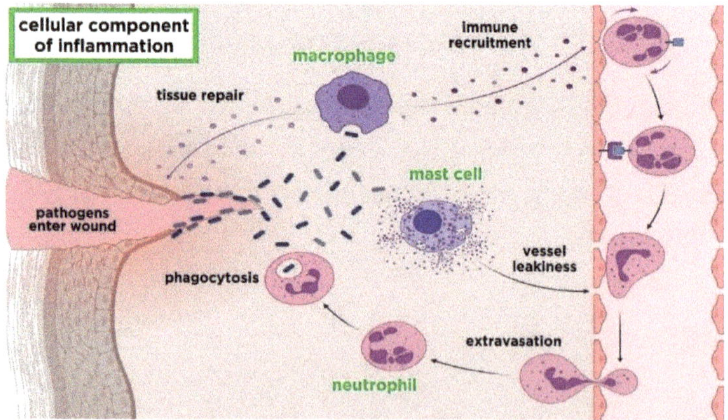

One effect of these chemicals is the dilation of small blood vessels and increased permeability of capillaries in the vicinity of the injury or infection. This causes both **hyperemia**, increased blood flow, and **edema**, swelling at the site of infection or injury.

Also, endothelial cells of the capillaries produce **cell-adhesion molecules** (CAMs) that enable leukocytes to stick to capillary walls. Leukocytes traveling through the blood can stick to the CAMs, exit the blood vessels, and migrate to where they are needed.

Inflammation is the result of a net movement of fluid from the blood to the infected/injured tissue. This fluid brings with it chemicals, nutrients, and cells that can fight infection and heal damaged tissue. Edema causes an increase of pressure in the interstitial fluid, which drives more fluid into the lymphatic system. This fluid is likely to contain pathogens that have entered the body, which can be detected by cells in the lymphatic system.

Eosinophils, found in large number in the respiratory system and GI tract, multiply in allergic and parasitic disorders, Although their phagocytic function isn't clearly understood, evidence suggests that they participate in host defense against parasites.

IV. Adaptive Immunity: An Introduction

Introducing a foreign substance into the body may initiate an adaptive immune response. The result is the multiplication (cloning) of lymphocytes to create an army of cells that can recognize and fight the specific invader. Whereas the innate defenses can act immediately (or at least very quickly), the adaptive system takes several days to develop a full response. T-lymphocytes are responsible for carrying out what is called the **cell-mediated response** of adaptive immunity, and B-lymphocytes carry out the **humoral response**.

Fever, defined as an abnormally high body temperature (99.0 degrees Farenheight or higher while normal body temperature is 98.6 degrees F), is a systemic response to infection. After exposure to foreign invaders, leukocytes may release chemicals called **pyrogens**, which raise the body temperature. Fever appears to cause the spleen to sequester zinc and iron, which bacteria require to multiply. Increased temperature also elevates the body's metabolic rate, which speeds up tissue repair.

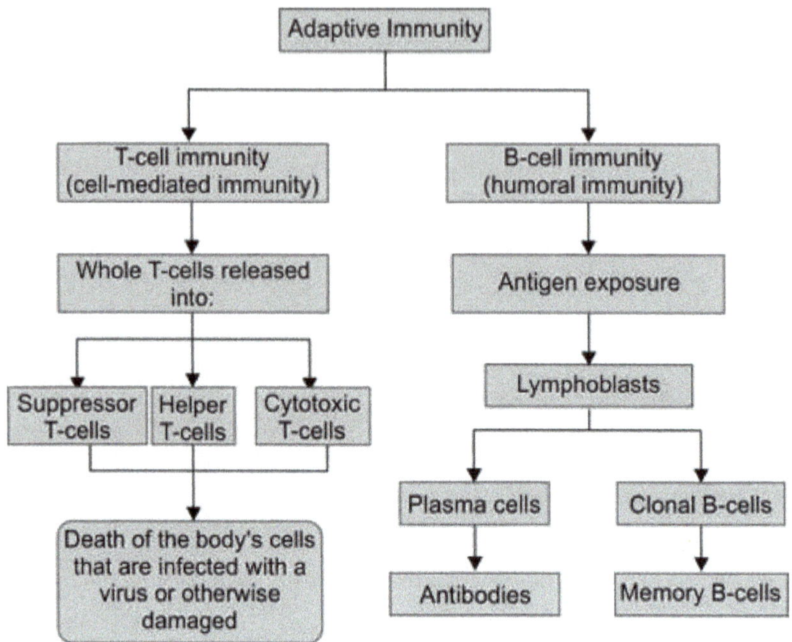

Overall activation of adaptive immunity

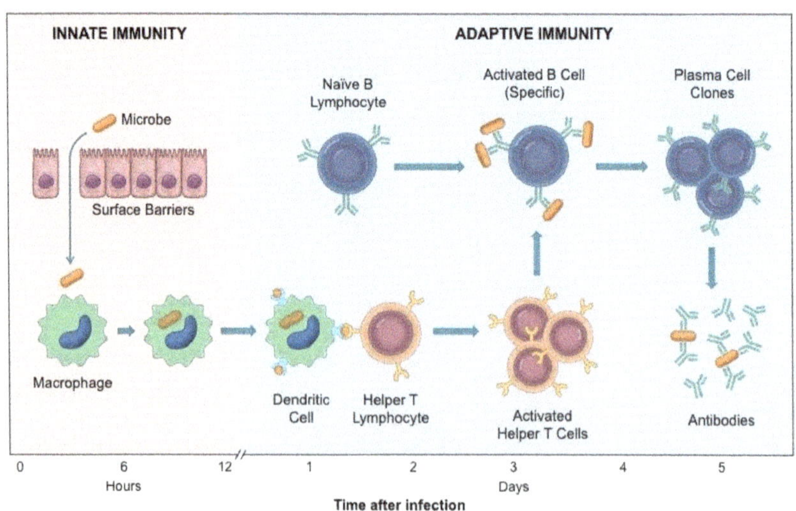

Antigens

An **antigen** is a substance that can provoke an adaptive immune response. Most antigens are proteins; some are polysaccharides. It makes sense that most antigens are proteins, as proteins are the organic molecules with the most diversity among organisms. Consider that sugars like glucose, fructose, and sucrose, as well as lipids such as cholesterol and fatty acids, are used by many (maybe most?) organisms. Thus, you would not want these molecules to trigger the immune system.

An antigen foreign to the body is, simply enough, called a **foreign antigen**. For example, if a streptococcus bacterium enters my body, my adaptive immune system cells would recognize various proteins on the bacterium's surface. These molecules would be considered foreign antigens.

The term **self-antigen** describes various proteins found on the surfaces of cells that are recognized as self by that person's immune system but would be seen as foreign if placed in another individual.

It is typical for the immune system to recognize only part of an antigen molecule as antigenic. This part of the antigen that the immune system recognizes is called the **antigenic determinant** (or **epitope**). Many antigens are large enough molecules that they have more than one epitope. Also, be aware that a particular pathogen may contain numerous different antigenic molecules. So, a bacterium may have multiple antigens, some with more than one epitope, all capable of triggering a host's immune system.

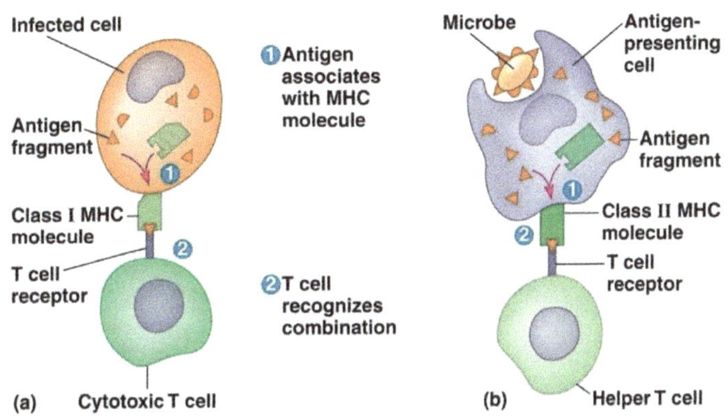

General structure of lymphocytes

Each T-cell and B-cell produces a rather unique receptor for a particular antigen. Each receptor is composed of several proteins that form a "receptor complex," and each lymphocyte may have about 100,000 copies of the receptor complex on its outer surface. The receptor complex is referred to as a **TCR** on a T-cell or a **BCR** on a B-cell.

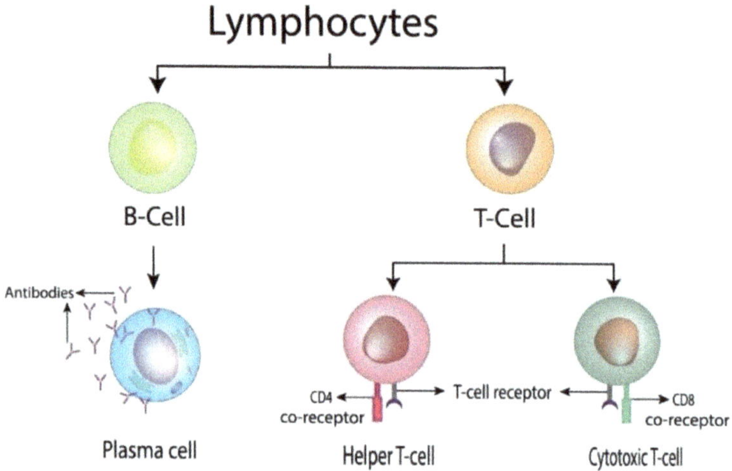

As noted above, a given lymphocyte makes receptors that bind to a particular antigen. This is not the result of any specific intent for the cell to recognize a particular antigen. In other words, your body does not intentionally make cells with receptors to match an antigen on the streptococcus bacterium. Rather, your body makes billions of lymphocytes, each with receptors that will recognize something. Because there are so many lymphocytes, if streptococcus (or any other pathogen) gets into your body, chances are pretty much 100% that some lymphocytes in the body will be able to recognize it.

A B-cell is able to bind directly to an antigen and begin its response. A T-cell requires that an antigen be presented to the T-cell and its receptors by another cell. Each T-cell has coreceptors that allow it to recognize this other cell:

- Cells known as **helper T-cells** have coreceptors called **CD4 receptors**. Each TCR on a helper T-cell is associated with a CD4 receptor.
- Cells known as **cytotoxic T-cells** have coreceptors called **CD8 receptors**. Each TCR on a cytotoxic T-cell is associated with a CD8 receptor (see diagram below).

Antigen presentation and MHC molecules

As mentioned in the previous section, in order for a T-cell to recognize an antigen, the antigen must be presented. There are certain cells of the immune system that have the specific function of presenting antigen to helper and cytotoxic T-cells. These cells are called **antigen-presenting cells (APCs)**, and they include dendritic cells, macrophages, and B-lymphocytes. However, you will soon learn that most cells of your body have the ability to present antigens to the immune system.

The presentation of antigen to T-cells requires that the antigen be attached to a special group of glycoproteins, called **MHC**, found on the surfaces of cells. These proteins are coded by a group of genes called the **major histocompatibility complex** (hence the abbreviation MHC).

Millions of different combinations of MHC genes can be found in the human population, so generally, only identical twins have the same MHC molecules. The combination of MHC molecules on cells in a person's body is the combination that is recognized as self. If cells with another combination of MHC enter the body (as may occur during an organ transplant), then they will be recognized as foreign.

Action of Antigen-presenting cells (APC) in the endogenous and exogenous pathways

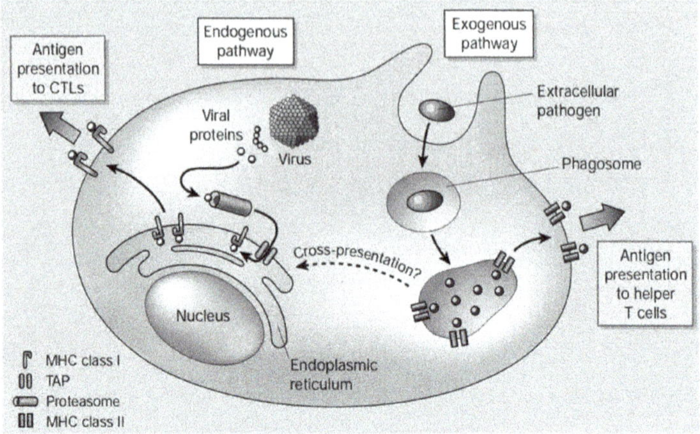

There are two general classes of MHC molecules found within a person:

A closer look of the APC-MHC I and II presentation pathways

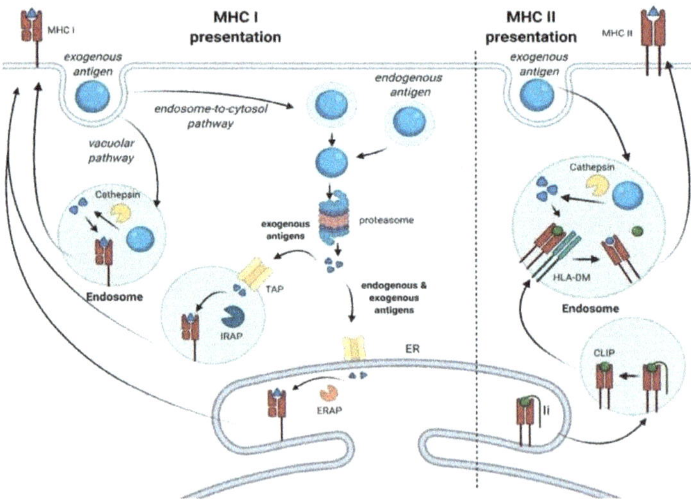

Class I MHC molecules are displayed by nearly all cells of the body. Class I MHC molecules are made in the rough ER, and they bind fragments of protein (peptides) that come from within the cell. These MHC molecules and associated peptides are then displayed on the cell's outer surface. Most of the time, these peptides are parts of normal cellular proteins, and the immune system recognizes them as self. However, if a cell has been infected or becomes cancerous, it will typically produce abnormal proteins. Fragments of these abnormal proteins are displayed on the cell's surface with the class I MHC molecules, where cytotoxic T-cells can recognize the abnormal particles as foreign antigens. The T-cell's CD8 receptors bind to the class I MHC molecules, and its TCR binds to the antigen.

This is essentially a way for an infected or cancerous cell to advertise its condition to the immune system and prepare for destruction by cytotoxic T-cells.

Class II MHC molecules are displayed on the surfaces of APCs (APCs also display class I MHC). Class II MHC molecules are made in the rough ER. They bind peptide fragments from foreign molecules engulfed by the APC. Because these antigens have been brought into the APC from the outside, they are called **exogenous antigens**. After the exogenous antigen is bound to the MHC molecule, the MHC molecule migrates to the cell's surface to display the exogenous antigen (which is foreign). Class II MHC molecules are recognized by helper T-cells, with CD4 receptors binding to the class II MHC molecules and TCR binding to the antigen.

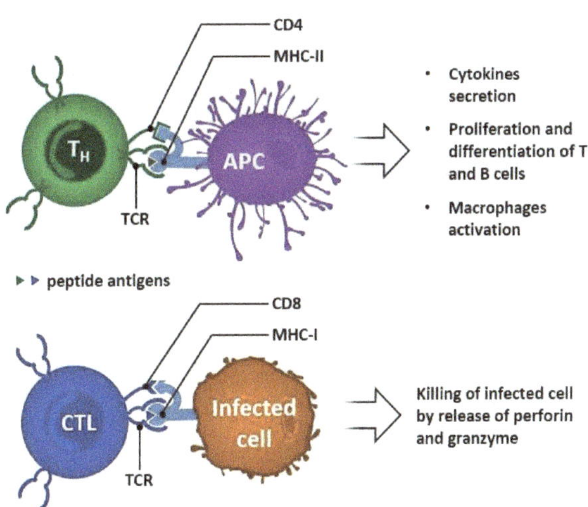

This alerts helper T-cells to the presence of an infection or other danger to the body that requires action by the immune system.

V. Activation and Clonal Selection of Lymphocytes

Recall the basics of lymphocyte formation from hemocytoblasts from Chapter 2. Immature lymphocytes are basically identical, but at some point, they become **immunocompetent**—able to recognize specific antigens. T-cells become immunocompetent in the thymus, and B-cells become immunocompetent in the bone marrow. Once a lymphocyte becomes immunocompetent, it displays its receptors (TCRs or BCRs) on its surface. These receptors are unique to the specific lymphocyte and will bind only to a specific antigen.

Until a lymphocyte encounters an antigen it can recognize, it is said to be **naïve**. Immunocompetent but naïve lymphocytes spread to the lymph nodes, spleen, and other lymphoid organs, where they await encounters with antigens. A blood-borne antigen may meet up with a lymphocyte in the spleen; an antigen that entered a break in the skin may be taken by a dendritic cell to a nearby lymph node; an antigen that enters the body through a mucous membrane may meet up with a lymphocyte in the tonsils or MALT. Once a lymphocyte encounters its antigen, it differentiates into an **activated** lymphocyte. The initial contact between a lymphocyte and an antigen is called an **antigen challenge**.

Be sure to understand the meanings of the words *immunocompetent*, *naïve*, and *activated* in relation to T and B-cell development.

Activation of T-lymphocytes

T-cell activation requires a two-step process:

Step 1—first stimulation. As described previously, helper T-cells bind to antigens attached to class II MHC displayed by APCs. Cytotoxic T-cells bind to antigens attached to class I MHC.

Step 2—second stimulation. A helper T-cell that has bound its antigen will release **interleukin 2** (IL-2), which stimulates the helper T-cell to divide and form a clone of helper T-cells that can recognize the particular antigen (they will all have the same TCR). A cytotoxic T-cell that has bound its antigen also requires stimulation by IL-2 released from helper T-cells. This stimulates the cytotoxic T-cell to divide and form a clone.

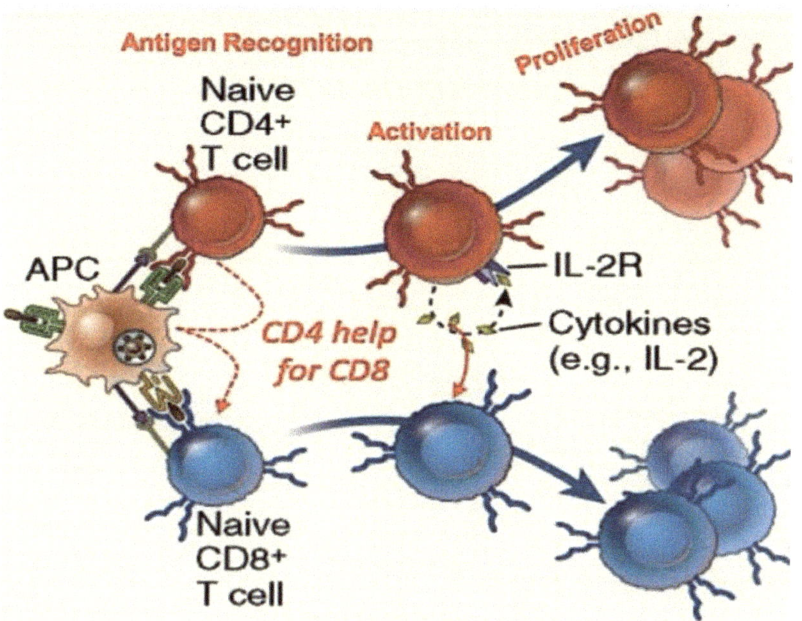

If helper T-cells are not proliferating during an infection, what is the effect on cytotoxic T-cells?

What is the purpose of forming a clone of lymphocytes?

When a T-cell multiplies to form a clone, most cells in the clone are activated T-cells. These cells fight the antigen and then die off. A clone of T-cells also contains **memory T-cells**, which remain in the body for an extended period of time (months or years). The memory T-cells enable the body to mount an immune response rapidly if the body is exposed to the antigen again.

Activation of B-lymphocytes

B-cell activation is also a two-step process (see illustration below):

Step 1–first stimulation. Whereas T-cells require antigen presentation by APCs, B-cells can bind directly to their antigens.

Step 2–first stimulation. **Interleukin 4** (IL-4) released from activated helper T-cells stimulates the B-cell to divide and form a clone.

When a B-cell multiplies to form a clone, most cells in the clone are **plasma cells**. Plasma cells produce antibody molecules against the antigen and then die off. A clone of B-cells also contains memory B-cells.

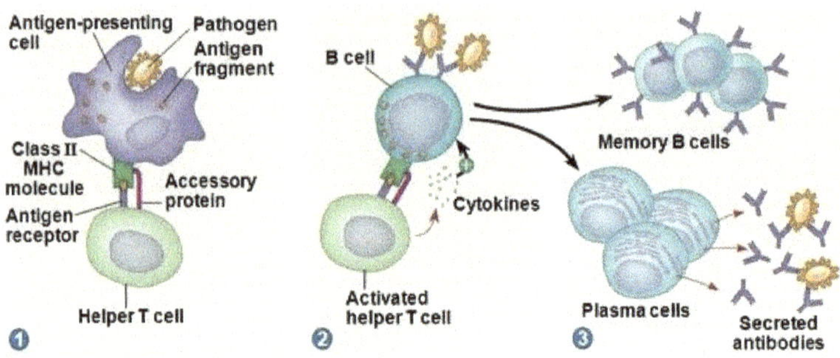

Lymphocyte recirculation

It is estimated that only 1 in 100,000 to 1,000,000 lymphocytes can bind to a particular antigen upon the first exposure. Therefore, a considerable period of time may pass from when an antigen enters the body to when it is detected by a lymphocyte that can recognize it. Immunocompetent lymphocytes regularly circulate throughout the body, traveling through the blood and lymph. This reduces the average time it takes for a lymphocyte to detect its antige, compared to what it would be if each lymphocyte stayed in one location.

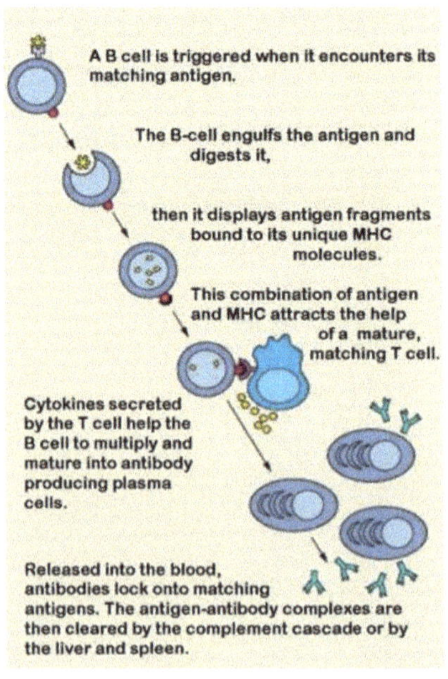

VI. Effector Response at Infection Site

Effector response of T-lymphocytes

Several days after exposure to an antigen, an activated helper T-cell leaves its secondary lymphatic structure and migrates to where it is needed to defend the body. Helper T-cells stimulate the proliferation of other T-cells and B-cells already bound to the antigen. In fact, without signals from helper T-cells, there is generally little or no specific immune response (think of the disease AIDS). Cytokines released by helper T-cells also enhance the functions of non-lymphocyte WBCs to assist in defense.

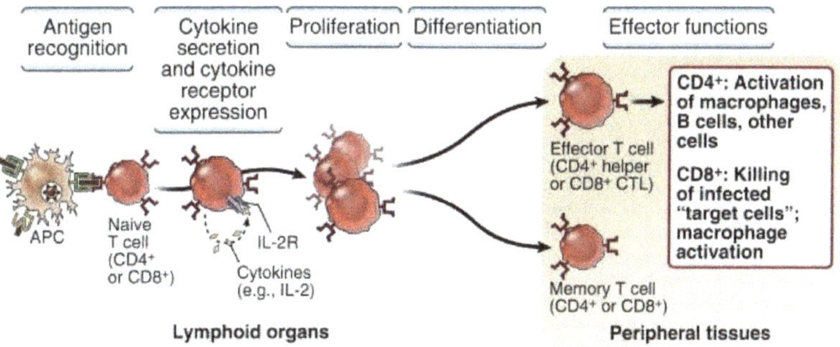

Cytotoxic T-cells are the only T-cells that can kill other cells. Hence, they are sometimes called "killer T cells." Activated cytotoxic T-cells leave the secondary lymphatic structures and roam the body looking for cells displaying the proper MHC-antigen combination. Once a cytotoxic T-cell finds such a target, it attaches to the target cell membrane and releases **perforin** molecules to lyse it. The T-cell is then able to move on and attack another target. Although cytotoxic T-cells have the same killing method as NK cells and similar-sounding names, they are different! NK cells are non-specific, and cytotoxic T-cells are specific.

The roles of T-cells in the immune system are often referred to as **cell-mediated immunity**.

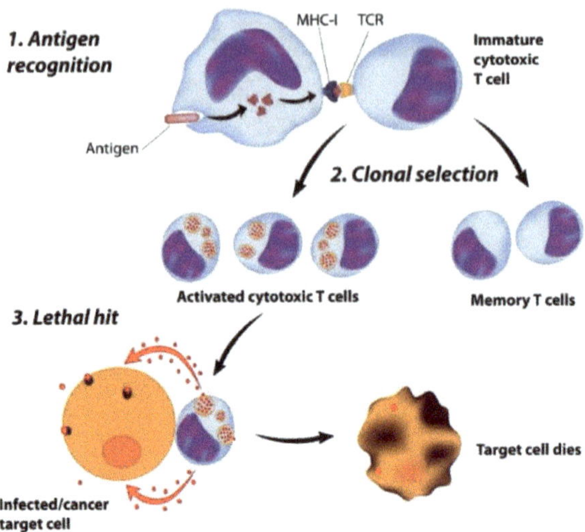

Effector response of B-lymphocytes

The primary role of a plasma cell is to produce antibody molecules. A plasma cell typically remains in its secondary lymphatic structure and releases antibodies circulating through the body via lymph and blood. A single plasma cell can produce hundreds of millions of antibody molecules over its life span of about five days. The structure and effects of antibody molecules are discussed next.

Because antibody molecules circulate in the body fluids (humors), the role of B-cells in the immune system is often referred to as **humoral immunity**.

VII. Immunoglobulins

The word "antibody" is a term for a type of protein, also known as an **immunoglobulin**. An antibody molecule does not destroy its target directly; instead it “tags” the target for destruction by other components of the immune system.

Structure of immunoglobulins

The body can produce a nearly limitless variety of immunoglobulin molecules against an almost infinite variety of antigens. Despite the tremendous variety in antibody molecules, they all share a common basic structure. Each antibody is made of four polypeptide chains (two "heavy chains" and two "light chains"). The four chains are bonded together to form a Y-shaped molecule called an **antibody monomer**.

The **variable regions** of the antibody form the tips of the Y. These are the parts that are different among all the antibodies, and these variable regions form the two parts of the antibody that can bind to antigen (the **antigen-binding sites**). Because each antibody has two binding sites, a single antibody can simultaneously bind two antigens, cross-linking them together in a process called **agglutination**. The **constant regions** of the antibody form the stem of the Y. The structure of the constant region determines which one of the five categories the antibody belongs to.

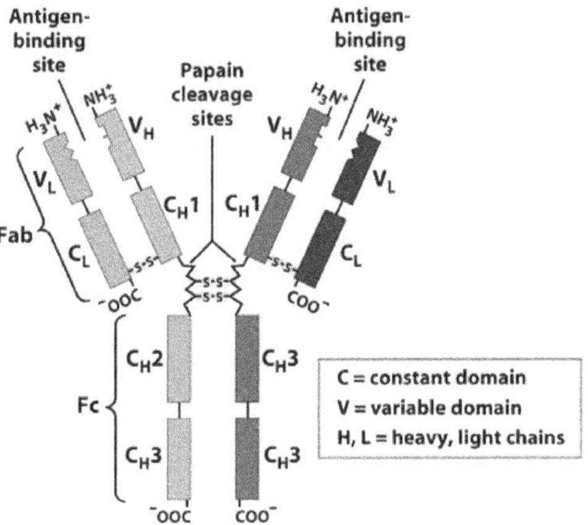

Actions of antibodies

As noted above, antibodies cannot by themselves destroy antigens. However, they are able to neutralize and assist in their destruction in the following ways:

- *Neutralization*—binding of antibodies to harmful chemicals or various sites on bacteria or viruses often renders the chemical or pathogen ineffective.
- *Agglutination*—agglutination of cellular targets inactivates them and makes them easy prey for phagocytes.
- *Complement fixation*—the constant region of some antibodies binds to complement and helps complement attack the target.
- *Opsonization*—when antibodies are attached to an antigen, they make it easier for phagocytic cells to recognize and destroy the antigen.
- *Activation of NK cells*–certain antibodies trigger the activity of NK cells.

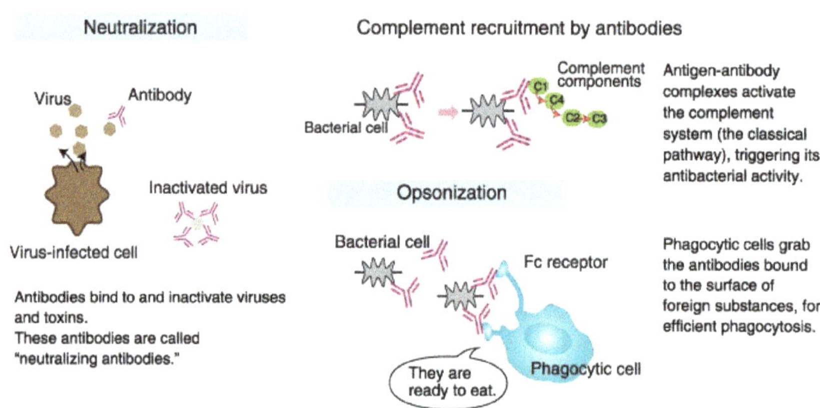

Classes of immunoglobulins

The five major immunoglobulins (Ig) classes are as follows (see Table below). The acronym GMADE might help you to remember.

IgG is the most abundant antibody, and it can be found in the blood, lymph, and other body fluids during an immune response. These antibodies can perform all of the functions described in the previous section. IgG can cross the placental boundary to confer immunity to a fetus.

IgM exists primarily as a pentamer (five molecules joined together). The monomeric form is found on the B-cell surface, where it acts as a receptor for antigens. IgM is generally the first type of antibody released by a plasma cell during a primary immune response (see below), and it circulates in the blood plasma. The pentameric form has ten antigen-binding sites and is, therefore, particularly good at agglutination. It also fixes and activates complement.

IgA exists primarily as a *dimer* and is found in secretions such as saliva, sweat, intestinal juice, and milk. It helps prevent pathogens from attaching to epithelial cell surfaces.

IgD is attached to the surface of B-cells (along with the monomeric form of IgM), where it acts as the receptor for antigens.

IgE is produced by plasma cells in the skin, mucosae of the digestive and respiratory tracts, and tonsils. Basophils and mast cells can bind to the stem portion of IgE, and binding causes the release of histamine and other chemicals that are important to inflammation and allergic responses.

Name	Properties	Structure
IgA	Found in mucous, saliva, tears, and breast milk. Protects against pathogens.	
IgD	Part of the B cell receptor. Activates basophils and mast cells.	
IgE	Protects against parasitic worms. Responsible for allergic reactions.	
IgG	Secreted by plasma cells in the blood. Able to cross the placenta into the fetus.	
IgM	May be attached to the surface of a B cell or secreted into the blood. Responsible for early stages of immunity.	

VIII. Immunologic Memory and Immunity

Immunologic memory

The first time the body is challenged with an antigen, the body produces a **primary immune response**. Because it takes time for the activated B-cells to respond to the challenge and proliferate, the peak of antibody production occurs about two weeks after the initial challenge. If the body is exposed to the same antigen at a later time, memory cells initiate a **secondary immune response** that is both more rapid (peaking in less than a week) and more intense (with much higher levels of antibody) than the primary immune response.

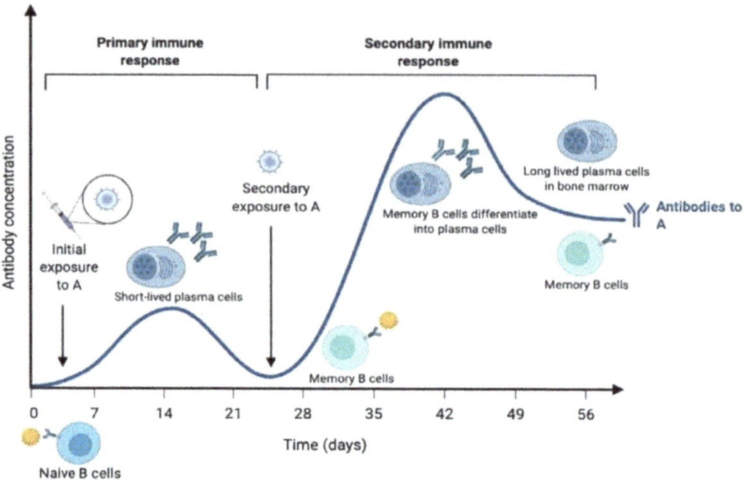

Active and passive immunity

Immunity is generally acquired *naturally*; a person encounters an antigen as a normal part of life, and the body reacts. Immunity can also be acquired *artificially* in the form of a **vaccine**. The enhanced speed and intensity of the secondary immune response is the basis for the success of vaccination.

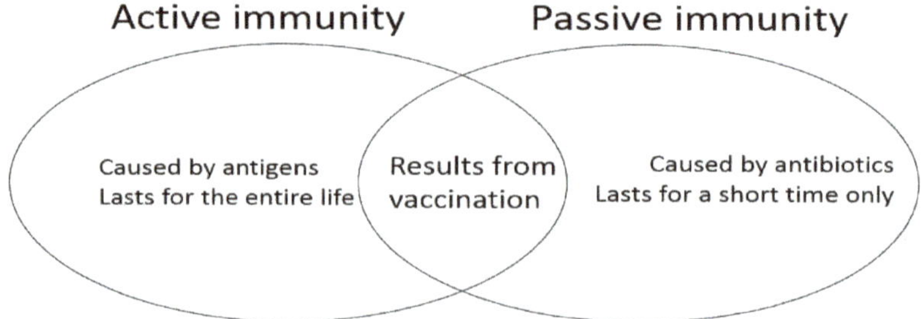

Whether it is acquired naturally or artificially, humoral immunity is said to be **active** if the person's own B-cells produce the antibodies. A person can also acquire antibodies from other sources, in which case the immunity is acquired **passively**. For example, a developing fetus gets antibody molecules from its mother's blood. These antibodies remain in the baby for several months after birth. A baby also obtains antibodies from the mother by drinking the mother's milk. Passive immunity may also be acquired by an injection of antibodies. For example, gamma globulin may be given to a person with hepatitis, and antivenoms for snake bites are antibody molecules against the particular venom. Injection of antibodies can provide rapid protection when the danger is too great to wait for the body to produce its own antibodies.

Passive Humoral Immunity

- Differs from active immunity in the antibody source and the degree of protection
  - B cells are not challenged by antigens
  - Immunological memory does not occur
  - Protection ends when antigens naturally degrade in the body
- Naturally acquired – from the mother to her fetus via the placenta
- Artificially acquired – from the injection of serum, such as gamma globulin

Chapter 7: Respiratory System

The word "respiration" generally means breathing, but there are some specific uses of the term of which you should be aware. **External respiration** refers to gas exchange between the air in the lungs and the blood; this is a function of the lungs. **Pulmonary ventilation** aids external respiration. It is the movement of the respiratory medium (in the case of humans, this is air) across the respiratory surface. **Internal respiration** refers to the exchange between systemic blood and tissues; this is a function of the circulatory system. **Cellular respiration** occurs in the mitochondria. Oxygen is used to oxidize organic molecules and provide ATP (energy). External respiration and pulmonary ventilation will be our main concerns in this chapter, although we will also consider some aspects of internal respiration. Cellular respiration is addressed in Chapter 4.

I. Introduction to the Respiratory System

General functions of the respiratory system

The respiratory system is involved in the following functions:

A. *Air passageway.* Many of the structures of the respiratory system serve primarily as passageways for the movement of air between the alveoli of the lungs and the outside atmosphere.

B. *Exchange of oxygen and carbon dioxide.* This is the process of external respiration referred to above. It takes place in the alveoli of the lungs.

C. *Detection of odors.* Sensory receptors in the nasal cavity allow us to detect odors in the air we breathe.

D. *Sound production.* The vocal cords of the larynx allow us to speak and produce other sounds.

General organization of the respiratory system

The respiratory system can be divided into two structural regions:

A. The **upper respiratory tract** includes the nose, nasal cavity, and pharynx.

B. The **lower respiratory tract** includes the larynx, trachea, bronchi, bronchioles, alveolar ducts, and alveoli.

These terms are used when someone refers to an upper or lower respiratory tract infection.

Similarly, the respiratory system can be divided into two functional zones (see illustration below):

A. The **conducting zone** consists of structures extending from the nose to the terminal bronchioles. The conducting zone conducts air to the respiratory surfaces. It also warms and filters incoming air.

B. The **respiratory zone** consists of the respiratory bronchioles, alveolar ducts, and alveoli. It is in this portion of the respiratory system that external respiration takes place.

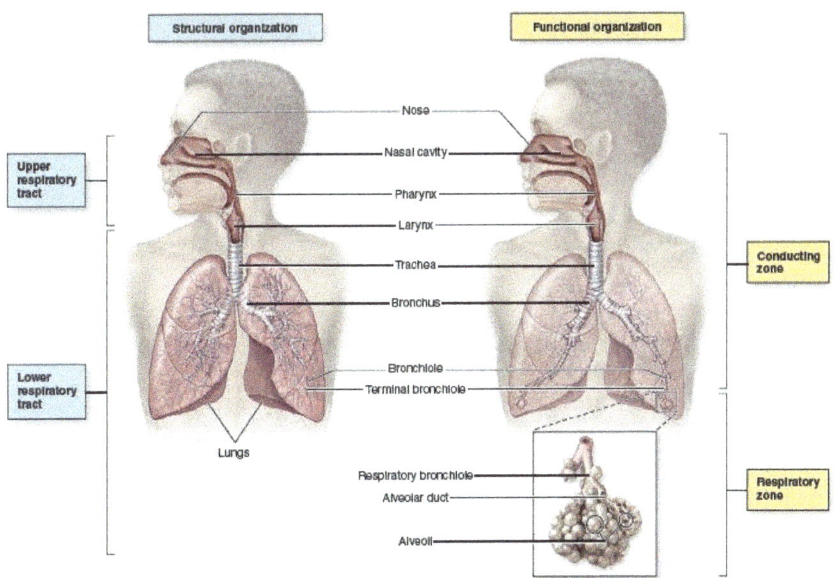

Mucosal lining

The **respiratory mucosa** lines the conducting portion of the respiratory system. Most of the conducting zone is lined with ciliated epithelium. This tissue generally gets thinner as you go deeper into the respiratory tract, beginning with pseudostratified ciliated columnar epithelium in the nasal cavity, trachea, and larger bronchi, and changing to simple ciliated columnar and simple ciliated cuboidal epithelium in the smaller bronchi and bronchioles. Portions of the respiratory tract that are regularly exposed to abrasion (*e.g.*, by food) are covered in stratified squamous epithelium. These areas include parts of the pharynx and larynx.

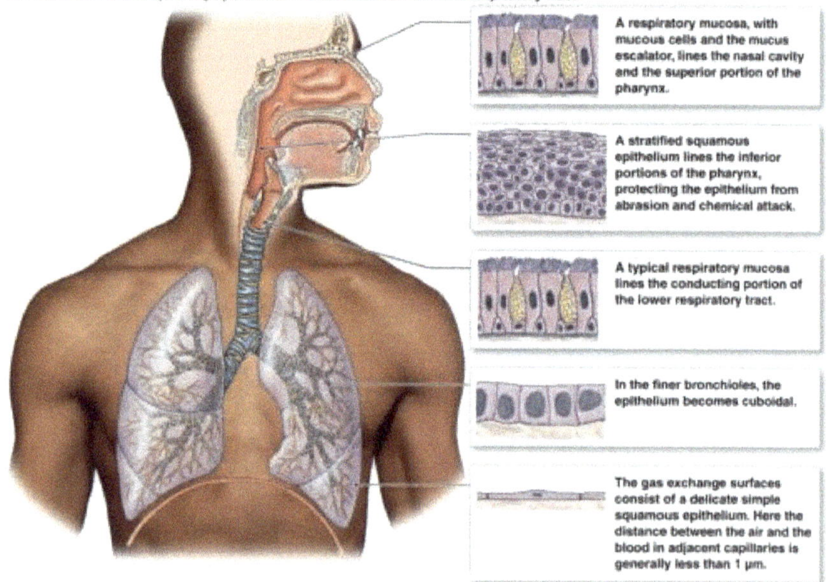

II. Upper Respiratory Tract

Nose and nasal cavity

The nose is the entry/exit of the respiratory system. It opens externally through paired **nostrils**. The **nasal vestibule** is the space contained within the flexible outer portion of the nose.

The **nasal cavity** is located behind the nose. The roof of the nasal cavity is formed primarily by the *ethmoid bone*, and the floor of the nasal cavity is formed by the **hard palate** (*palatine bones* and *maxillae*). The nasal septum partitions the nasal cavity into left and right sides. The nasal cavity contains passages called the **superior, middle, and inferior meatuses**. These passages are formed by convolutions of the *ethmoid bone* and *inferior nasal conchae*. The nasal cavity opens posteriorly to the pharynx through two holes called the **posterior nasal apertures**.

The passages of the nasal cavity are variously covered with hairs, cilia, and/or mucus, which filter particles from the air. The convolutions in the nasal cavity create turbulence in the air, which helps to force small airborne particles into contact with the mucous that lines the cavity. The superior region of the nasal cavity is also lined with **olfactory receptors** for smell.

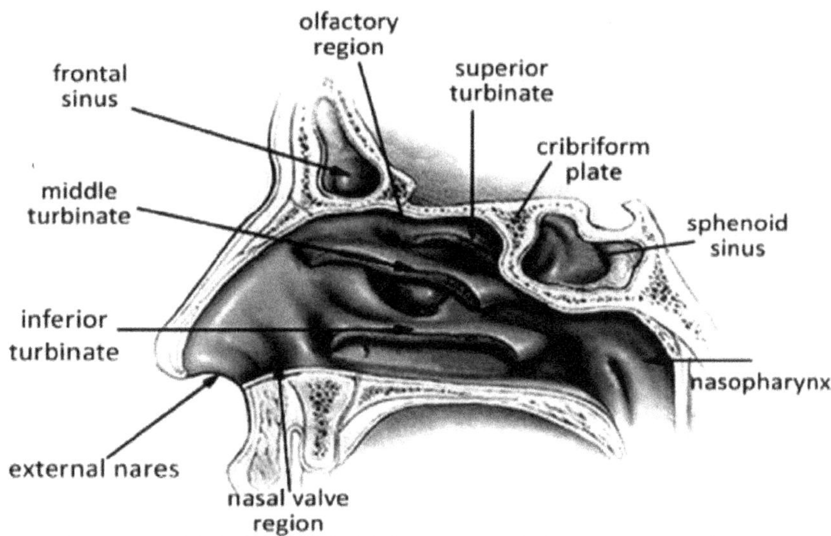

Paranasal sinuses

The nasal cavity is surrounded by **paranasal sinuses** (see illustration below), which are cavities inside various bones of the skull (specifically the *frontal*, *sphenoid*, *ethmoid*, and *maxillary* bones). The mucus produced in the sinuses drains into the nasal cavity where it can trap particles in inhaled air.

Lateral Anterior

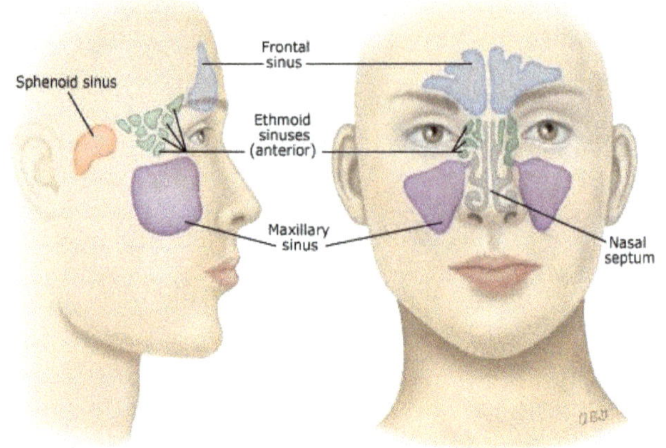

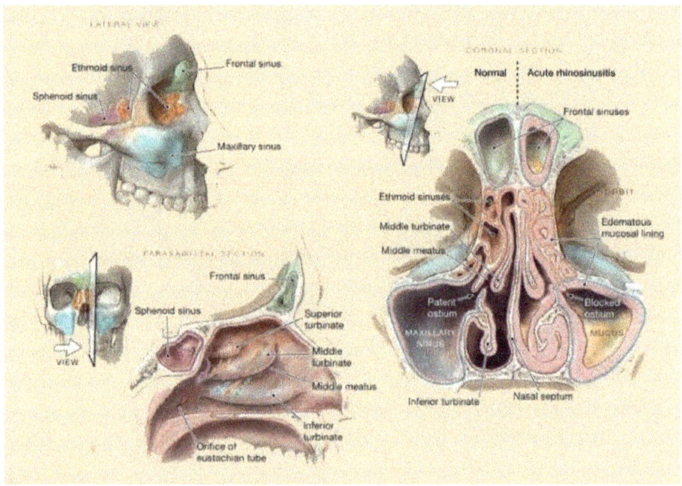

Pharynx

The **pharynx** is the passage that runs from the nasal cavity to the esophagus and the larynx. It is located dorsal to the nasal cavity, oral cavity, and larynx and ventrally to the spinal column. The pharynx is lined with skeletal muscle and respiratory mucosa.

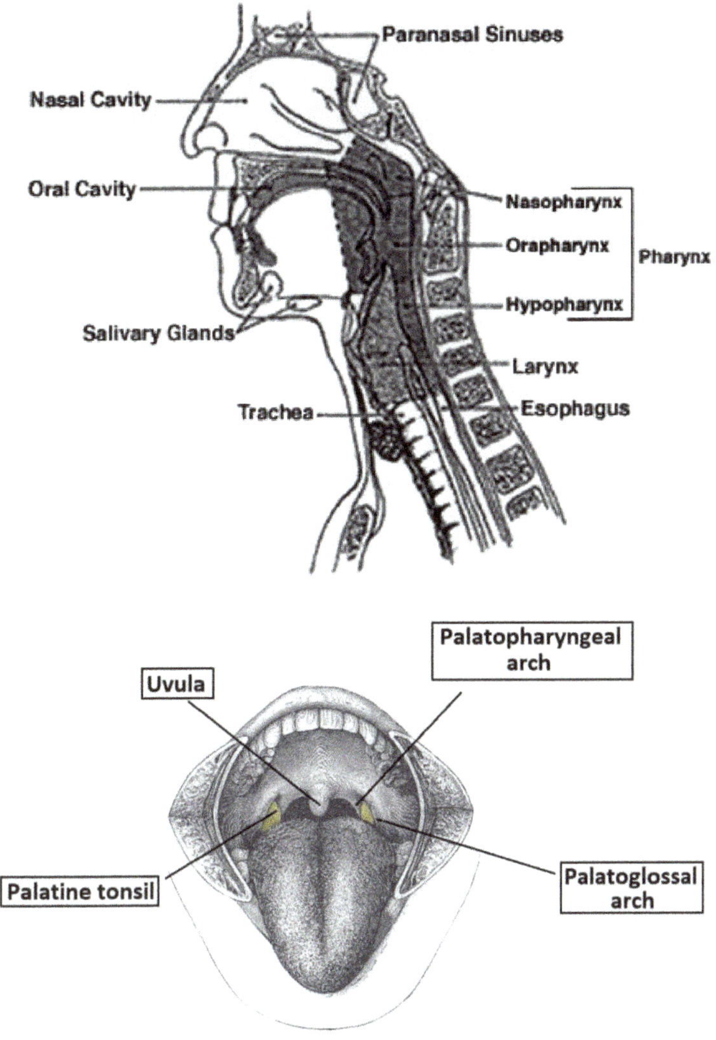

The pharynx can be divided into three regions:

A. The **nasopharynx** is dorsal to the nasal cavity, extending from the posterior nasal apertures to the soft palate. It is normally a passageway for air only. **Auditory tubes** open into either side of the nasopharynx.

B. The **oropharynx** is dorsal to the oral cavity, extending from the soft palate to the hyoid bone. It is a passageway for both air and food.

C. The **laryngopharynx** is dorsal to the larynx, extending from the epiglottis to the cricoid cartilage. It is a passageway for both air and food, but it is also where the paths for air and food diverge: air flows to and from the larynx, and food moves to the esophagus.

The nasopharynx is lined with pseudostratified ciliated columnar epithelium, but in the oropharynx, the lining changes to stratified squamous epithelium. Why is this important? (see epithelial tisse

III. Lower Respiratory Tract

Larynx

The **larynx** (or "voice box") connects the pharynx to the trachea. It lies ventral to the pharynx. Inhaled air enters the larynx through the **glottis**, the superior opening of the trachea.

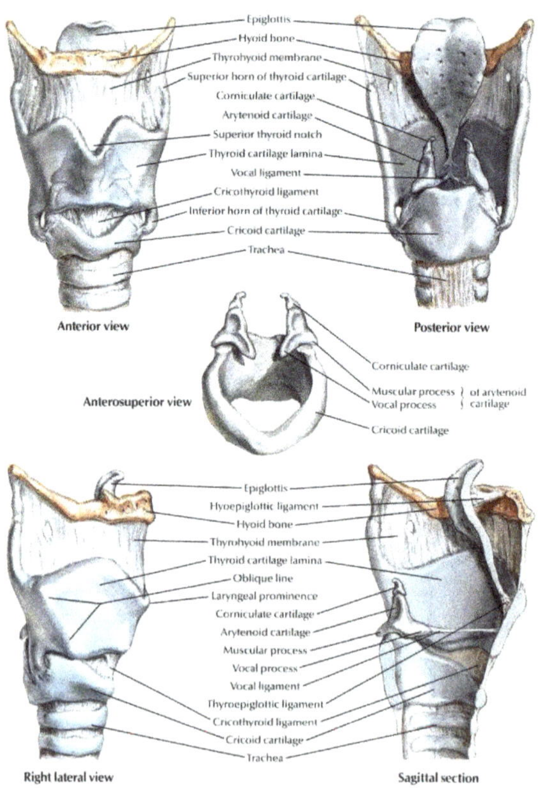

Superior view of larynx Midsagittal section of larynx

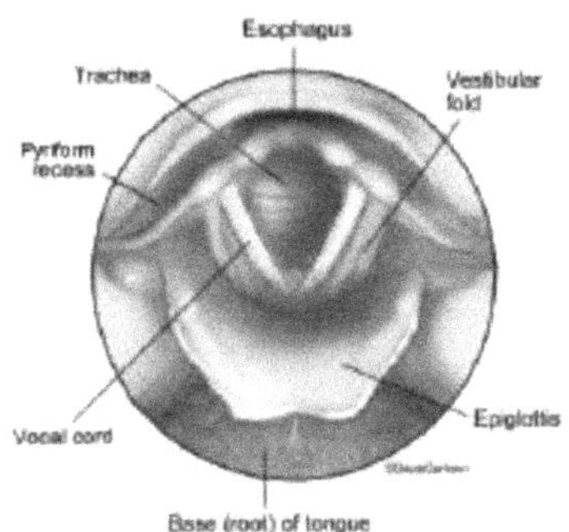

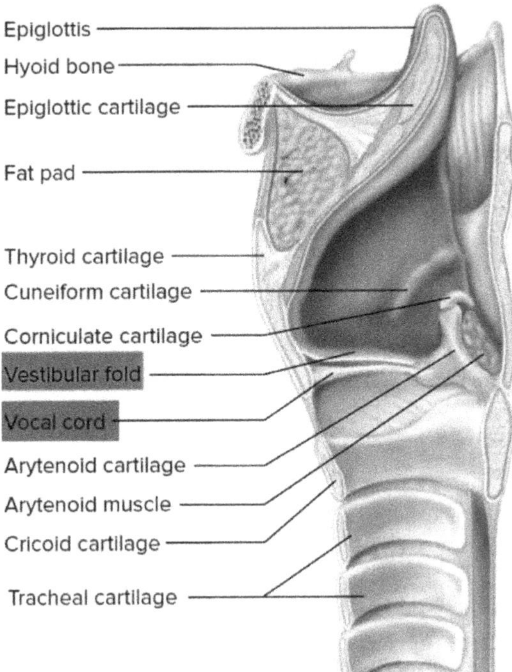

Trachea

The **trachea** (or "windpipe") leads from the larynx to the primary bronchi (se below). It is a tube that is lined with 15 to 20 rings of **tracheal cartilage**. These "rings" are actually C-shaped and provide protection to the airway and prevent the trachea from collapsing as air is inspired. The tracheal cartilages have an open end facing dorsally toward the esophagus. A band of smooth muscle called the

trachealis muscle connects the ends of each tracheal cartilage. The trachea is lined with cilia and mucus to help trap and remove particles from the air (recall pseudostratified ciliated columnar epithelium from the body tissues section).

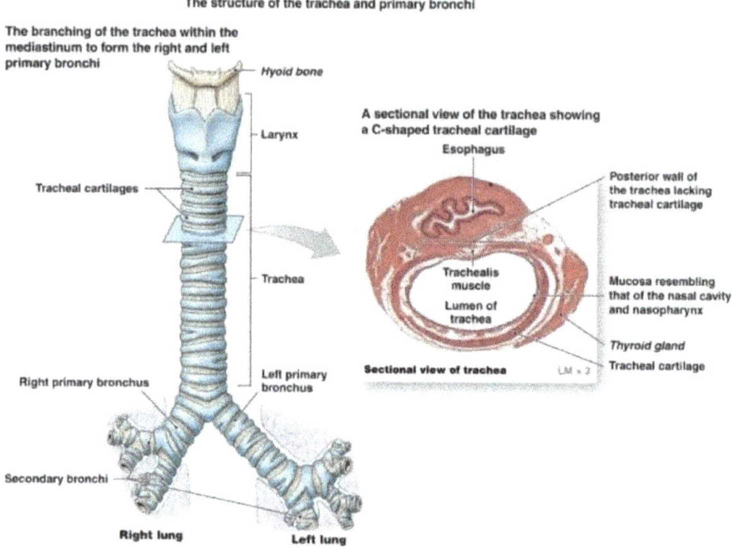

Bronchial tree

Inferiorly, the trachea branches into the left and right **main bronchi**. Each main bronchus enters its lung at a groove called the **hilus**, which lies on the medial surface of the lung. (The hilus also provides entry or exit for the pulmonary vessels and nerves.) After the main bronchi enter the lungs, they branch into **lobar bronchi**. There are three lobar bronchi in the right lung and two in the left lung–one for each lobe. Each lobar bronchus divides into numerous **segmental bronchi**, and they branch into smaller tubes that we will just call "bronchi." The branching pattern formed by the bronchi and bronchioles is often called the **bronchial tree**.

Bronchi, like the trachea, are lined with pieces of cartilage to provide support. They also are lined with cilia and mucus. As the bronchi become smaller, the amount of cartilage lining each bronchus decreases.

Bronchi lead to still smaller passages called **bronchioles**, which lack cartilage completely and are lined with smooth muscle. **Terminal bronchioles** form the end of the conducting zone.

The diameter of each bronchiole is controlled by the CNS to regulate airflow into the alveoli. The sympathetic nervous system causes **bronchodilation**, and the parasympathetic causes **bronchoconstriction**. Histamine also stimulates bronchoconstriction.

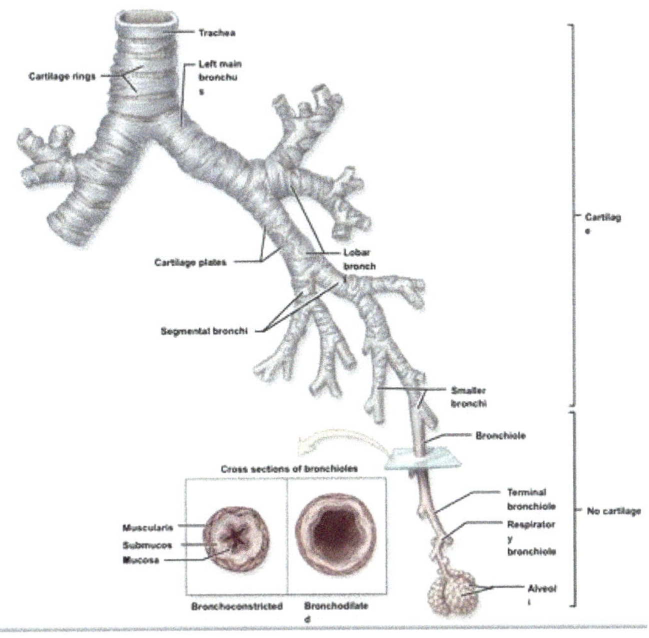

Respiratory zone: respiratory bronchioles, alveolar ducts, and alveoli

Terminal bronchioles lead into passageways called **respiratory bronchioles**, which in turn lead to **alveolar ducts**. Alveolar ducts open into **alveolar sacs**, each of which contains numerous **alveoli**.

Each lung contains approximately 300 to 400 million alveoli. Alveoli are surrounded by capillaries for gas exchange, and networks of elastic tissue help the alveoli retain their shape. The alveolar epithelium consists primarily of **type I cells**, which make up a layer of simple squamous epithelial tissue. The alveolar epithelium also contains scattered **type II cells**, which produce an oily secretion called **surfactant**, which coats the alveolar walls. Surfactant prevents water on the alveolar walls from generating enough surface tension to collapse the alveoli.

Alveolar macrophages (also called "dust cells") patrol the alveoli to remove foreign particles that get past the cilia and mucus.

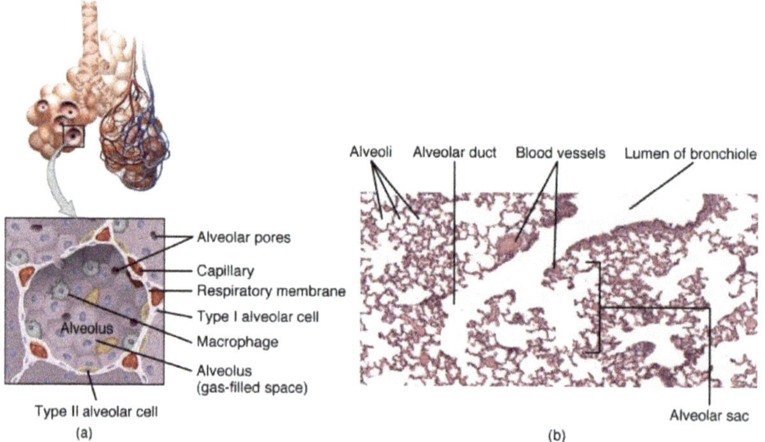

Respiratory membrane

Together, the type I cells and capillary walls form the **respiratory membrane**, across which gas exchange occurs.

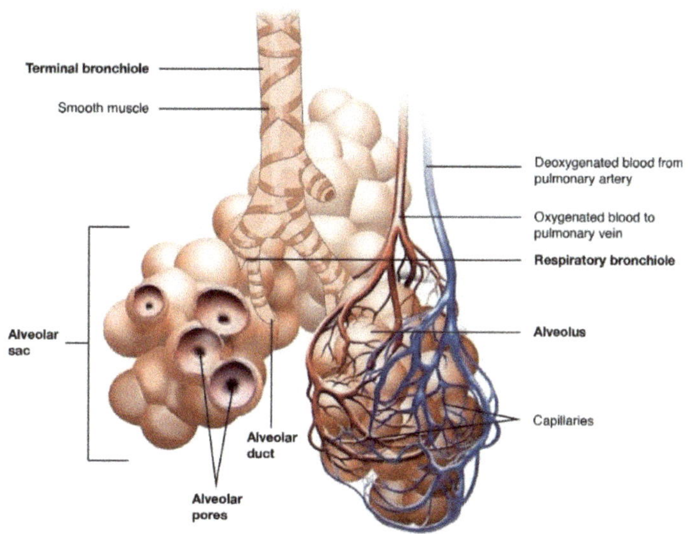
Terminal bronchiole
Smooth muscle
Deoxygenated blood from pulmonary artery
Oxygenated blood to pulmonary vein
Respiratory bronchiole
Alveolar sac
Alveolus
Capillaries
Alveolar duct
Alveolar pores

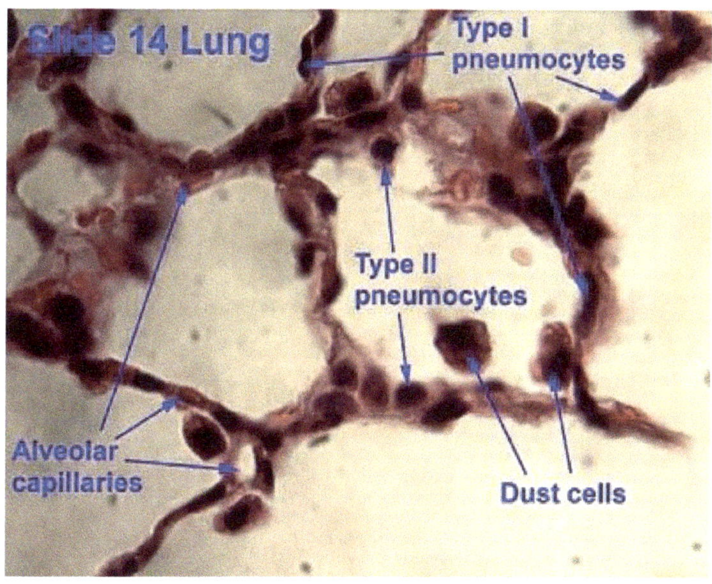
Slide 14 Lung
Type I pneumocytes
Type II pneumocytes
Alveolar capillaries
Dust cells

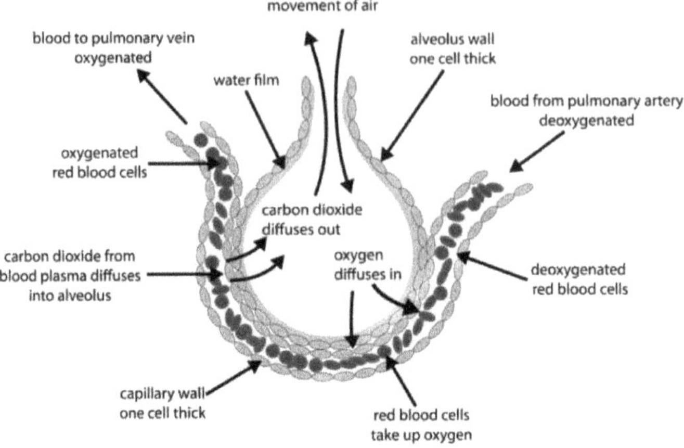

IV. Lungs

Gross anatomy of the lung

Each **lung** has a conical appearance, with a superior **apex** and an inferior **base**. Each lung is divided into distinct lobes: the right lung has three **lobes** (*superior, middle,* and *inferior*), and the left lung has two lobes (*superior* and *inferior*).

You should already be familiar with the meanings of the following terms: **mediastinum, thoracic cavity, pleural cavity, visceral pleura, parietal pleura, pleural fluid, intercostal muscles**, and **diaphragm**.

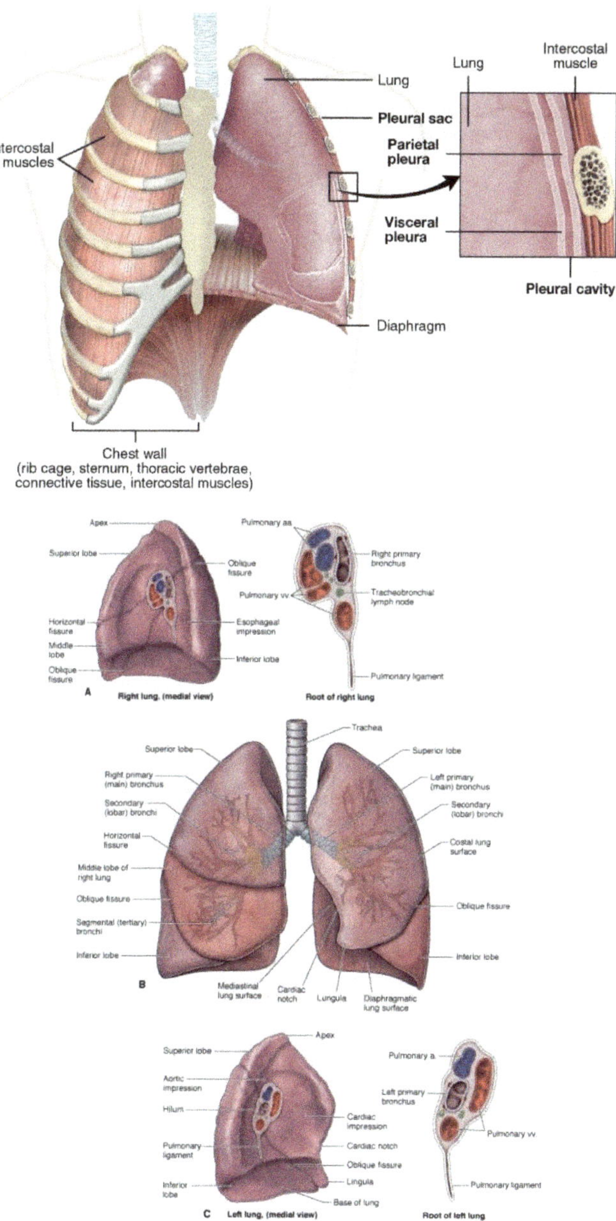

V. Respiration: Pulmonary Ventilation

As noted in the first paragraph of these notes, pulmonary ventilation is the movement of air over the respiratory surface. In humans, this means moving air back and forth from the atmosphere to the alveoli.

Introduction to pulmonary ventilation

Define the following terms:

inspiration–

expiration–

quiet breathing–

forced breathing–

Mechanics of breathing

List the muscles used for the following actions:

quiet breathing–

forced inspiration–

forced expiration–

Breathing Mechanism--Expiration

Quiet/Restful Breathing

- ELASTIC RECOIL
  - Ribcage
  - Diaphragm
  - Abdominal contents
  - Lung tissue

Forced Expiration

- ALL SECONDARY
  - Abdominals
  - Obliques
  - Internal Intercostals

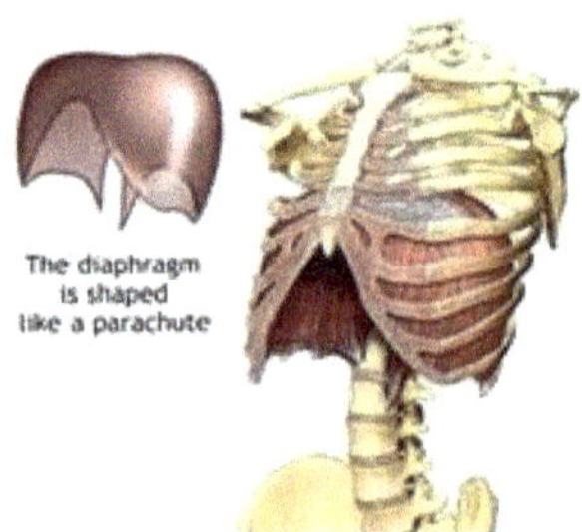

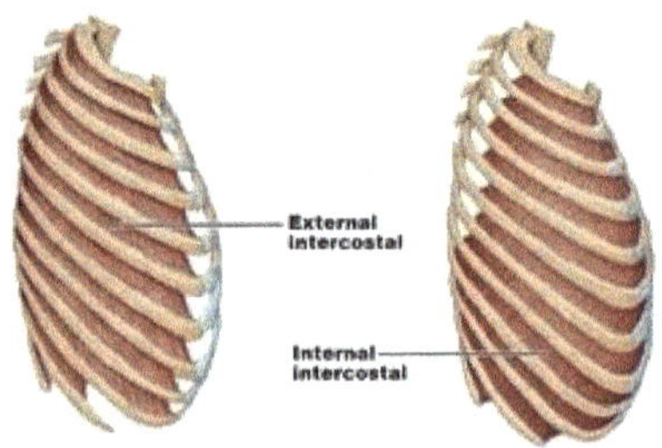

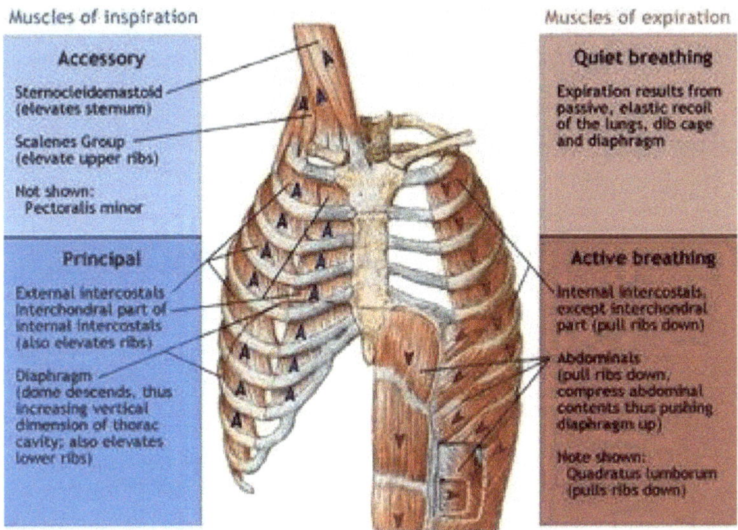

Define the following terms:

atmospheric pressure–

intrapulmonary pressure–

intrapleural pressure–

Just as blood flows through vessels from high pressure to low pressure, air also moves from areas of high pressure to low pressure. This fact and the pressure-volume relationship are key to understanding ventilation of the lungs.

Boyle's law states that pressure is inversely proportional to volume, or $P \alpha 1/V$. Therefore, an increase in the volume of the lungs will result in decreased pressure within the lungs, and a decrease in the volume of the lungs will result in increased pressure within the lungs.

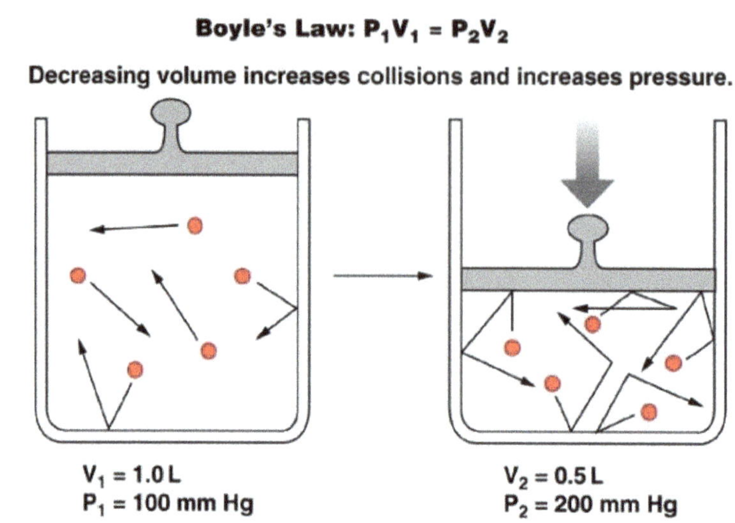

Another key to understanding pulmonary ventilation is that changes in the volume of the lungs are *not* accomplished *directly* by the action of muscles. Muscles act directly to change the volume of the thoracic cavity, and changes in the volume of the lungs follow changes in the volume of the thoracic cavity.

Other factors that influence ventilation are:

Elastic recoil of the lungs. In the absence of other forces acting on the lungs, the lungs tend to contract to their smallest size.

The surface tension of the alveoli also tends to draw the alveoli closed. What substance acts to reduce surface tension in the alveoli?

***Inspiration**:* The illustration below shows that contraction of the diaphragm or elevation of the ribs and sternum expands the volume of the thoracic cavity. As the thoracic cavity expands, negative intrapleural pressure causes the lungs to expand with the walls of the thoracic cavity. Intrapulmonary pressure falls, and air moves from the atmosphere into the lungs.

***Expiration**:* When the diaphragm and external intercostals relax, the volume of the thoracic cavity decreases. This leads to increasing intrapulmonary pressure and the movement of air out of the lungs.

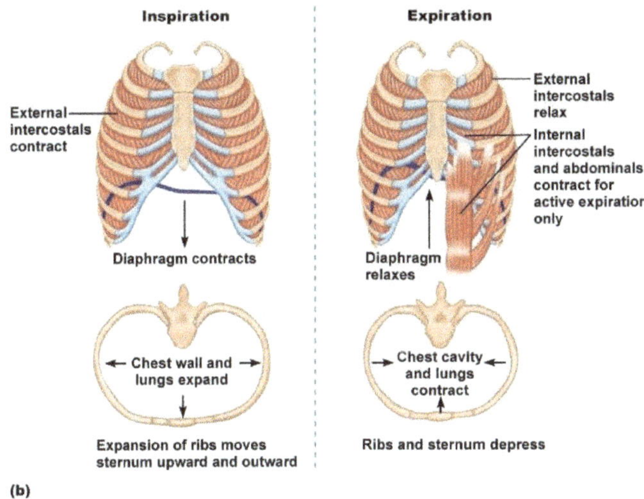

Nervous control of breathing

Breathing is primarily under the control of the medulla and the pons (see diagram below). Two particular areas of the medulla are especially critical to breathing:

A. The **ventral respiratory group (VRG)** appears to be the pacemaker for breathing. Certain neurons in the VRG (the "inspiratory" neurons) send signals along the phrenic and intercostal nerves to stimulate the diaphragm and external intercostal nerves. This stimulation causes inspiration. When another group of neurons in the VRG fires (these are the "expiratory" neurons), signals to the phrenic and intercostal nerves cease. Passive expiration follows. The normal rate of stimulation by the VRG is about 12-15 breaths per minute. This normal respiratory rhythm is called **eupnea**.

B. The **dorsal respiratory group (DRG)** appears to integrate information from stretch and chemoreceptors. This information is sent to the VRG, but the exact role of the DRG is unclear.

Respiratory centers in the pons, the **pontine respiratory centers**, play roles in fine-tuning the rate and depth of breathing.

Remember that the muscles used for breathing are all skeletal muscles, and they are connected to the CNS by somatic motor neurons. Thus, you can consciously control the rate and depth of breathing. However, we rarely think about breathing, and the rate and depth of breathing are generally controlled by reflexes.

The reflexes that normally control breathing begin with chemoreceptors, which monitor the pH of cerebrospinal fluid and blood. In brief, a drop in pH likely signals an increase in CO_2, which triggers a breath.

One might think that breathing should be adjusted in response to the levels of oxygen in the blood, but the concentration of oxygen in the blood is typically not a factor. However, if arterial P_{O2} drops below 60 mm Hg, then chemoreceptors stimulate the respiratory centers to increase ventilation.

Stretch receptors in the lungs are stimulated as the lungs expand. These receptors send inhibitory signals to the medullary respiratory centers that may end inspiration. Once the lungs recoil, the stretch receptors stop sending signals, and this inhibition of the medulla ends. This reflex is called the **Hering-Breuer reflex**. Although it may be important in protecting the lungs from over-stretching, this reflex does not appear necessary for normal breathing.

Sneezing and coughing are reflexes designed to clear the respiratory passages. They involve taking a deep breath followed by closure of the larynx and contraction of the abdominal muscles to build pressure. Opening the larynx results in an "explosive" exhalation.

The hypothalamus can adjust breathing in response to strong emotions and pain (consider a gasp of fear). It causes an increase in respiratory rate in response to rising body temperature and a decrease in rate in response to cold.

The cortex can exert conscious control over breathing, altering both rate and depth. However, the cortex has its limits—you cannot hold your breath long enough to kill yourself!

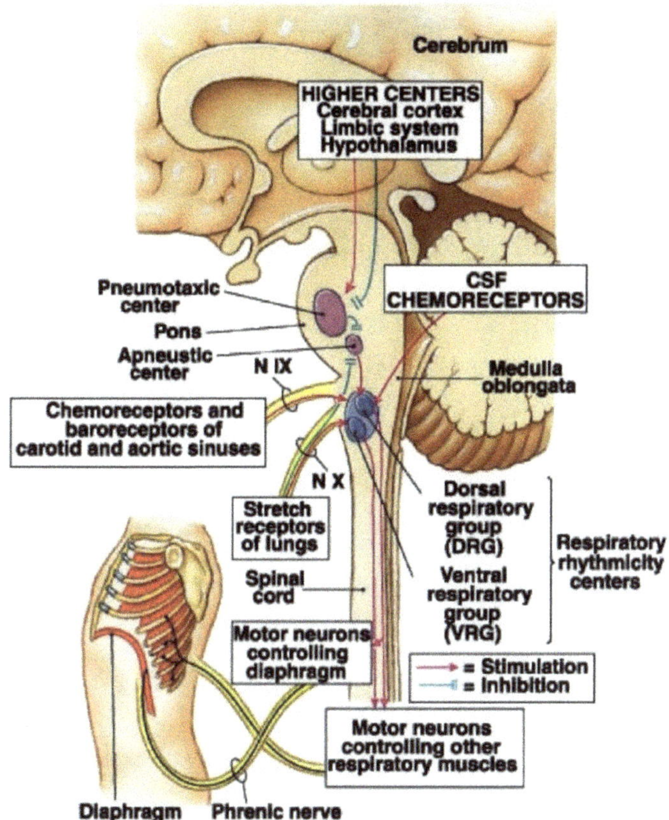

VI. Respiration: Alveolar and Systemic Gas Exchange

Chemical principles of gas exchange

The air we breathe is a mixture of gases, primarily nitrogen (78.6%) and oxygen (20.9%). Carbon dioxide accounts for only about 0.04% of the air. At sea level, the pressure exerted by all the components of the air equals 760 mm Hg (equivalent to 1 atm, 760 torr, or 101 kPa). **Dalton's law** states that the total pressure of a gas is equal to the sum of the pressures of the individual gases in the mixture. The individual pressure, or **partial pressure (P)**, can be defined as follows:

$$P = \text{fractional content x total pressure}$$

For example, the partial pressure of oxygen in air at sea level is given by the following equation:

$$P_{O2} = 0.21 \times 760 \text{ mm Hg} = 159 \text{ mm Hg}$$

The partial pressure of a gas provides an estimate of the concentration of that particular gas in the air.

When a gas is in contact with a liquid (as is the case with air in your lungs and the blood in the capillaries), molecules of the gas will move into and out of the liquid. According to **Henry's law**, the amount of a particular gas in the liquid is directly proportional to the partial pressure of the gas in the air.

The **solubility** of a gas in a liquid is the amount of the gas that can dissolve into the liquid. The solubility varies from gas to gas. For example, CO_2 is over twenty times more soluble in water than O_2. Temperature also affects solubility: the warmer the water, the lower the solubility.

By now, it should be familiar to all the idea that things move from high to low pressures. Gases as well, move from higher partial pressures to lower partial pressures. Review the previous illustrations to understand the movement of gases described below.

Alveolar partial pressures

During external respiration, oxygen moves from the alveolar air into the blood. The P_{O2} in the alveoli is about 104 mm Hg, and the P_{O2} in the venous blood entering the lungs is about 40 mm Hg. The difference in oxygen concentrations drives the diffusion of oxygen into the blood. This diffusion occurs so rapidly that the P_{O2} in the blood rises to the P_{O2} in the air (104 mm Hg) within a fraction of the time that the blood spends in the lungs.

What keeps the P_{O2} in the lungs nearly constant at 104 mm Hg even as oxygen leaves the lungs to enter the blood?

During external respiration, carbon dioxide moves from the blood into the alveolar air. The P_{CO2} in the venous blood is about 45 mm Hg, and the P_{CO2} in the alveoli is about 40 mm Hg. The difference in concentrations of carbon dioxide drives the diffusion of carbon dioxide into the alveoli. Although the concentration gradient is not as steep as the one for oxygen, a large quantity of carbon dioxide leaves the blood at the lungs because of its high solubility.

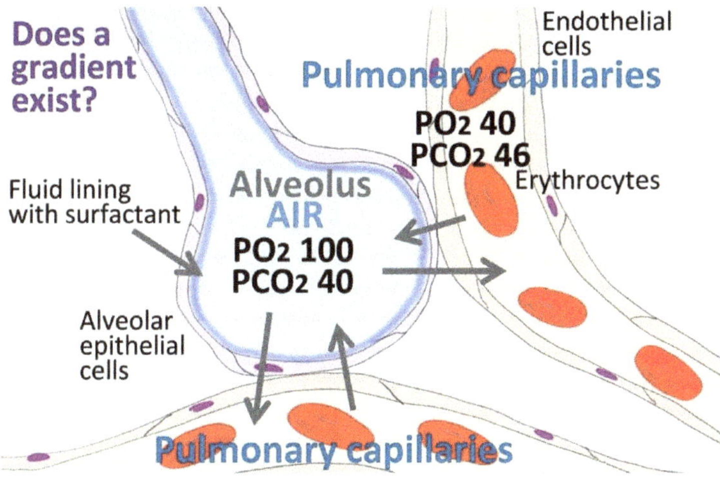

Partial pressures of systemic cells

During internal respiration, oxygen moves from the blood into the tissues. The P_{O2} in the blood entering the capillaries is about 100 mmHg, and the P_{O2} in the tissues is typically less than 40 mm Hg. The difference in concentrations of oxygen drives the diffusion of oxygen into the tissues.

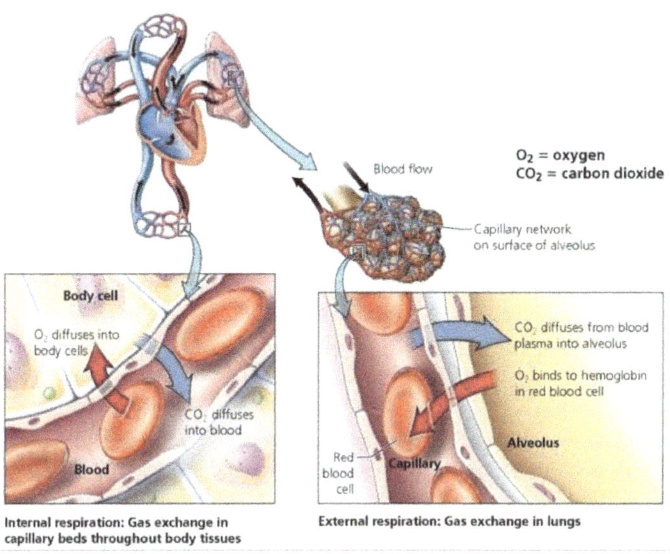

Internal respiration: Gas exchange in capillary beds throughout body tissues

External respiration: Gas exchange in lungs

During internal respiration, carbon dioxide moves from the tissues into the blood. The P_{CO2} in the tissues is greater than 45 mm Hg, and the P_{CO2} in the blood entering the capillaries is about 40 mm Hg. The difference in concentrations drives the diffusion of carbon dioxide into the blood.

What keeps the P_{O2} in the tissues low and the P_{CO2} in the tissues high?

VII. Respiration: Gas Transport

Oxygen transport

As you already know, one of the main functions of blood is to carry oxygen. Oxygen is transported in the blood in two ways: (1) Some oxygen is dissolved in the plasma. However, this accounts for only about 1.5% of the oxygen in the blood. (2) The majority of oxygen, about 98.5%, is carried by hemoglobin in the erythrocytes.

Each molecule of **hemoglobin (Hb)** may bind up to four molecules of oxygen according to the following reversible equation:

$$Hb + nO_2 \rightarrow Hb(O_2)_n$$

where n ranges from 1 to 4. "Hb" is referred to as the *deoxygenated* form of hemoglobin, or **deoxyhemoglobin**. "HbO_2" is referred to as the *oxygenated* form of hemoglobin, or **oxyhemoglobin**.

Hemoglobin may bind up to four molecules of oxygen because each hemoglobin molecule is made of four polypeptide chains, which I will refer to as "subunits." You should remember that each hemoglobin subunit contains a heme group, which can bind to an oxygen molecule.

Any increase in oxygen concentration shifts the equilibrium of the equation to the right, causing more oxygen to associate with hemoglobin. This is referred to as **oxygen loading**. Where in the body does the blood encounter an increase in P_{O2}? Where does oxygen loading occur?

Any decrease in the concentration of oxygen shifts the equilibrium to the left, causing more oxygen to dissociate from hemoglobin. This is referred to as **oxygen unloading**. Where in the body does the blood encounter decreasing P_{O2}? Where does oxygen unloading occur?

Carbon monoxide (CO) can also bind to the iron atoms of heme groups. However, CO binds much more tightly than oxygen. How does CO affect the body's ability to transport oxygen?

Carbon dioxide transport

Carbon dioxide is transported in the blood in three ways:

1. Transport of CO_2 dissolved in the plasma accounts for about 7-10% of total CO_2 transport. Recall that CO_2 has a higher solubility than oxygen.
2. About 20% of CO_2 transport is accomplished by hemoglobin. CO_2 can bind directly to the protein portion of the hemoglobin to form **carbaminohemoglobin** according to the following reaction. Note that CO_2 does not bind to the heme, so it does not interfere with O_2 binding.

$$CO_2 + Hb \rightleftarrows HbCO_2$$

3. The majority (about 70%) of CO_2 is transported in the blood as **bicarbonate** (HCO_3^-). As CO_2 enters the plasma, it may react with water to form carbonic acid. Carbonic acid then quickly dissociates into bicarbonate and protons. These two chemical reactions are shown by the following equation:

$$CO_2 + H_2O \rightarrow H_2CO_3 \rightleftarrows HCO_3^- + H^+$$

The concentration of carbon dioxide has a considerable effect on blood pH. As the level of CO_2 rises, the amount of carbonic acid increases, and blood pH falls. As the level of CO_2 falls, the amount of carbonic acid decreases, and blood pH rises.

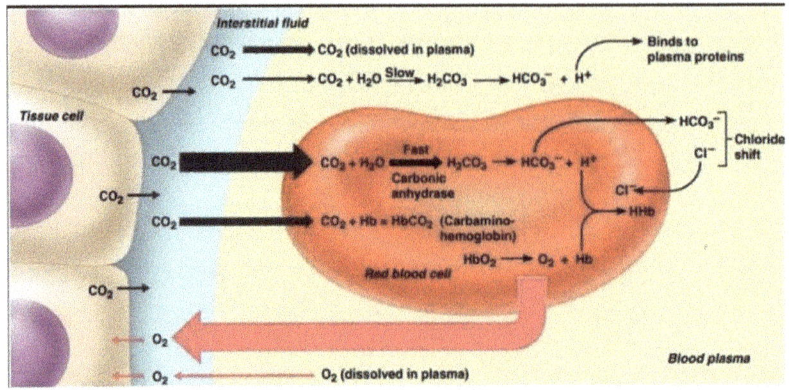

How does the rate of pulmonary ventilation affect blood pH?

Define the following terms:

Respiratory acidosis

Respiratory alkalosis

In the space below, distinguish respiratory acidosis and alkalosis from metabolic acidosis and alkalosis. Include some specific examples in your explanation.

Respiratory physiology chart

Pulmonary function test	Instrument	Measures	Function
Spirometry	Spirometer	Forced vital capacity (FVC)	Volume of air that is exhaled after maximum inhalation
		Forced expiratory volume (FEV)	Volume of air exhaled during one forced breath
		Forced expiratory flow, 25–75 percent	Air flow in the middle of exhalation
		Peak expiratory flow (PEF)	Rate of exhalation
		Maximum voluntary ventilation (MVV)	Volume of air that can be inspired and expired in 1 minute
		Slow vital capacity (SVC)	Volume of air that can be slowly exhaled after inhaling past the tidal volume
		Total lung capacity (TLC)	Volume of air in the lungs after maximum inhalation
		Functional residual capacity (FRC)	Volume of air left in the lungs after normal expiration
		Residual volume (RV)	Volume of air in the lungs after maximum exhalation
		Total lung capacity (TLC)	Maximum volume of air that the lungs can hold
		Expiratory reserve volume (ERV)	The volume of air that can be exhaled beyond normal exhalation
Gas diffusion	Blood gas analyzer	Arterial blood gases	Concentration of oxygen and carbon dioxide in the blood

The following formulas will be useful to calculate volumes of exhaled air. The instrument is normally used bedside for respiratory exercise of post surgical patients in a clinical setting. This may be part of pulmonary function test (**PFT**).

The graph below shows the total lung capacity (**TLC**) of a healthy adult to be approximately 6 liters or 6,000 mL (1L = 1,000 mL), and the tidal volume (**TV**) is normally around 500mL in a healthy adult as well. Keep in mind that the TV is the amount of that is inhaled and exhaled during a normal respiratory cycle (quiet breathing). Inspiratory reserve volume (**IRV**) is the volume of air that can be inhaled in addition to the tidal inspiration. Residual volume (described in the chart above) tends to increase with increasing age, whereas vital capacity decreases as we get older.. The following formulas are useful in solving the respiratory physiology calculations. These formulas can be created by simply using the graph below:

VC = TV + ERV + IRV

TLC = VC + RV

FRC = ERV + RV

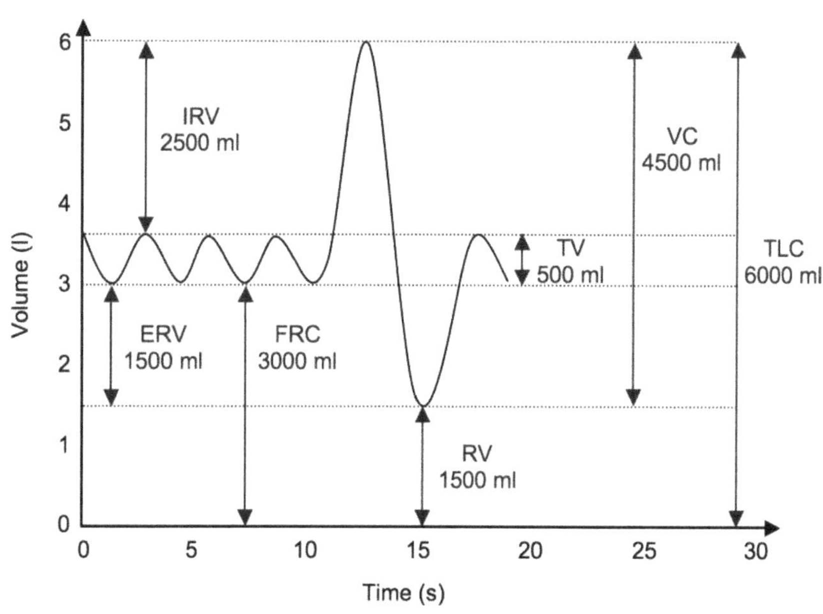

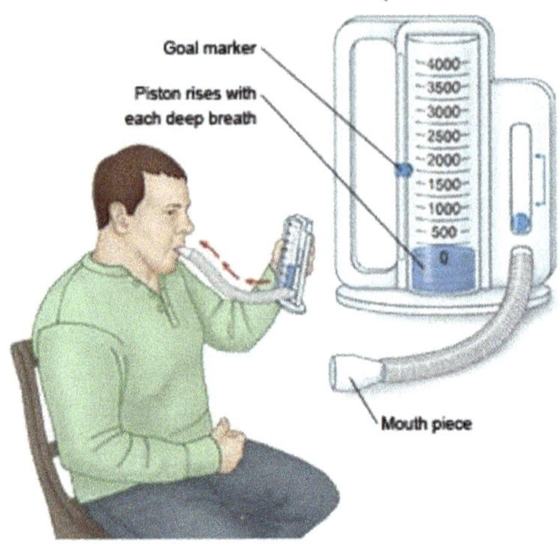

Now, let's try a couple of calculation exercises:

1. What is the tidal volume of the patient above if his VC = 4800 mL, IRV = 3100 mL, and his ERV = 1200 mL?
2. How would COPD (chronic obstructive pulmonary disease) affect vital capacity?
3. Calculate expiratory reserve volume (ERV) if TV = 450,000 µL, VC = 6.2 L, and IRV = 1700 mL. [1µL = 0.001 mL]
4. How would COPD affect residual volume? (See answers on page 417)

Chapter 8: Digestive System

Your body needs food for two primary purposes: growth and maintenance. Molecules and atoms in your food are generally used to either build new molecules in your body or provide the energy you need for metabolism. The overall functions of the digestive system are to digest food and absorb it into your body.

I. Introduction to the Digestive System

General functions of the digestive system

The digestive system functions in a series of integrated steps:

A. *Ingestion* is the active process of getting food, water, etc., into the system.

B. *Motility* is the movement of food through the alimentary canal. This includes swallowing and **peristalsis**.

C. *Secretion* of various fluids that contain digestive enzymes, acids, bile, mucus, etc.

D. *Digestion* is the process of breaking foodstuffs down into smaller and smaller particles until they can be absorbed into the tissues. **Mechanical digestion** is chewing or other physical manipulation of ingested materials. Mechanical processing aids in the passage of materials into the digestive tract and increases the surface area for chemical attack of the food. **Chemical digestion** is the breakdown of food into small molecules that the digestive epithelium can absorb. Chemical digestion is accomplished by enzymes and acids.

E. *Absorption* is the movement of organic molecules, electrolytes, vitamins, etc. into the body.

F. *Elimination* is the removal of waste products (generally either undigested food or wastes from the body itself) in the form of feces.

Organization of the digestive system

The adult digestive system can be divided into two major components (see illustration below):

A. The **gastrointestinal tract** (*i.e., alimentary canal*) is a hollow tube that runs from the mouth to the anus. What specific organs make up the digestive tract?

B. Working with the digestive tract is a variety of **accessory digestive organs**. What specific organs are included?

Diagram showing 2 components of digestive system: alimentary tract and accessory organs.

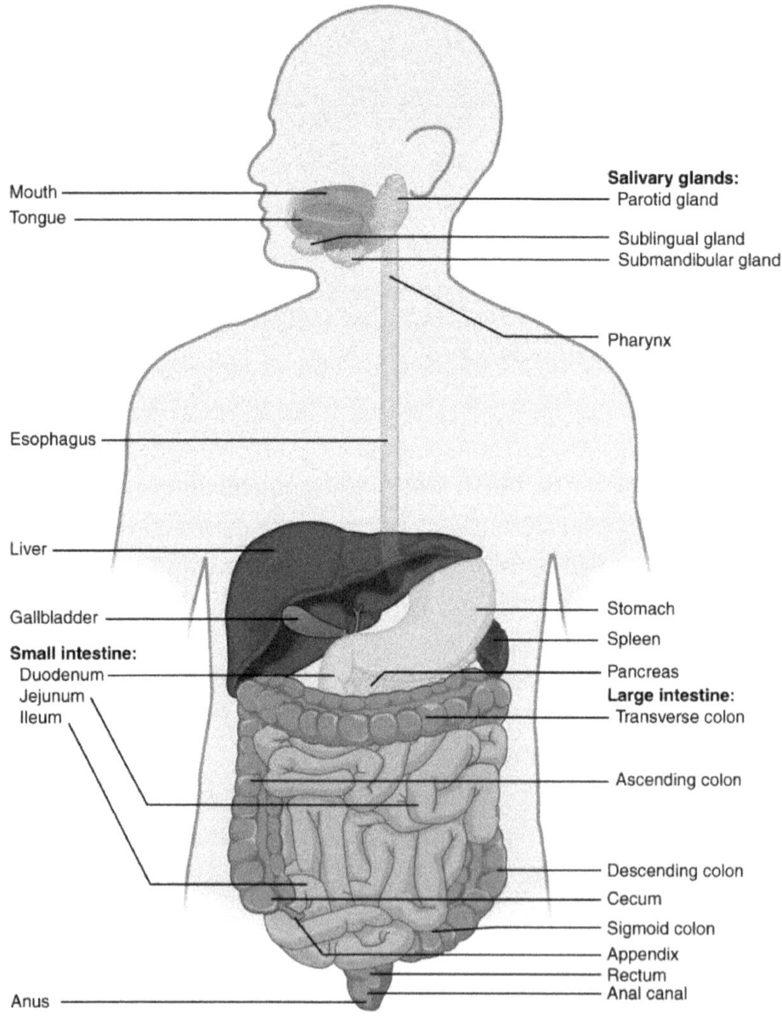

<u>Gastrointestinal tract wall</u>

The walls of the digestive tract consist of four tissue layers. Some details may vary from one part of the tract to another, but the basics of the four layers apply to the entire tract.

A. The **mucosa** is the layer exposed to the lumen. Like all mucous membranes, it is composed of an epithelial layer and an underlying layer of areolar tissue, the **lamina propria**. The epithelium is stratified in areas that receive the most mechanical stress (*e.g.,* the oral cavity, pharynx, and esophagus), and it is simple epithelium in the stomach, small intestine, and large intestine. The lamina propria contains blood vessels, mucous glands, and sensory nerve endings. In most areas of the digestive tract, the lamina propria also contains smooth muscle cells, which make up a layer of tissue called the **muscularis mucosae**. This muscle produces local movements of the mucosa and does not really move food through the tract. In the small intestine, this muscle makes folds in the mucosa to increase surface area.

B. The **submucosa** is a layer of moderately dense, irregular connective tissue. It contains blood vessels and some exocrine glands that secrete buffers and enzymes into the digestive tract's lumen.

C. The **muscularis externa** is a region dominated by smooth muscle cells, which are arranged in an inner, *circular*, layer and an outer, *longitudinal* layer. It is these layers of muscle that are responsible for peristalsis. In several places along the tract, particularly thick areas of circular muscle form **sphincters**.

D. The **serosa** is a serous membrane that lines most of the digestive tract inside the peritoneal cavity. The serosa is not present in the oral cavity, pharynx, esophagus, or rectum. In the pharynx, esophagus, and rectum the muscularis externa is lined by a network of collagen fibers called the **adventitia**.

Cross-sectional view of the esophagus

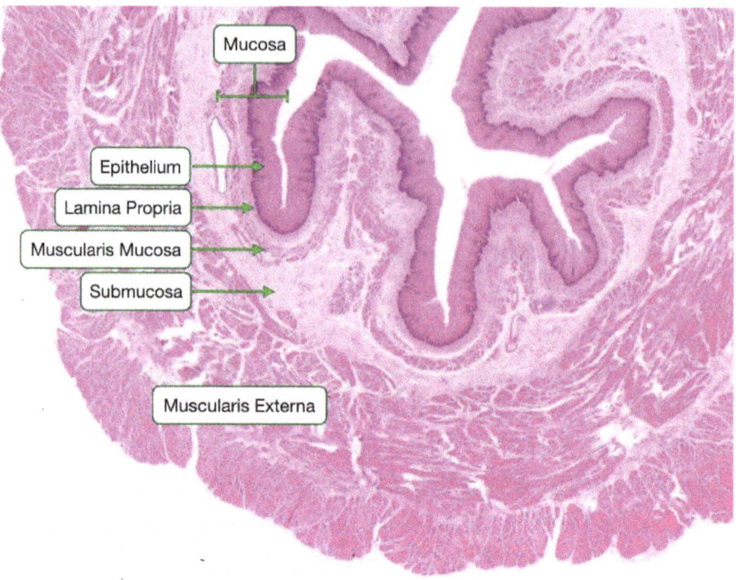

Serous membranes of the abdominal cavity

Portions of the digestive tract are suspended within the **peritoneal cavity** by **mesenteries**. Mesenteries are sheets of serous membrane that connect the **parietal peritoneum** to the **visceral peritoneum**. Serous fluid can be found in the space between the parietal and visceral peritoneum. Mesenteries provide passage for blood vessels, nerves, and lymphatic vessels, and they support the positions of digestive organs within the peritoneal cavity.

Portions of some digestive organs, including the pancreas and large intestine, lose their mesentery and lie posterior to the peritoneum. These structures are said to be **retroperitoneal**.

II. Upper Gastrointestinal Tract

Oral cavity and salivary glands

The **oral** (or *buccal*) **cavity** is the mouth, which contains the teeth and the tongue (see diagrams below). Functions of the oral cavity include (1) ingestion, (2) analysis of food before swallowing, (3) mechanical digestion, (4) lubrication with mucus and saliva, and (5) limited chemical digestion.

The oral cavity is bordered by the **palate, lips, tongue**, and **cheeks**. The palate enables you to chew food and breathe at the same time. Chewing food before swallowing allows quicker digestion and faster processing of energy. The **vestibule** is the space between the cheeks or lips and the teeth. The area within the teeth and gums is the **oral cavity proper**. The **gingivae** are the gums. The **uvula** is a dangling extension of the soft palate, which helps prevent food from entering the nasopharynx.

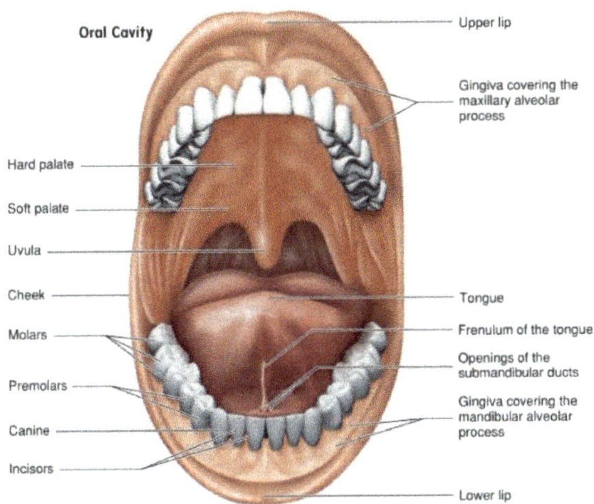

The tongue

The **tongue** occupies the lower portion of the oral cavity. It is made of skeletal muscle (voluntary control) covered in a layer of mucous membrane. The anterior portion of the tongue is referred to as the **body**, and the posterior portion is the **root**. The tongue moves food around the mouth during chewing, helps mix food with saliva, and forms food into a **bolus** for swallowing.

The superior surface of the tongue is covered with projections called the **lingual papillae**. The papillae contain taste buds and make the surface of the tongue rough. The front of the tongue is attached to the floor of the mouth by the **lingual frenulum**.

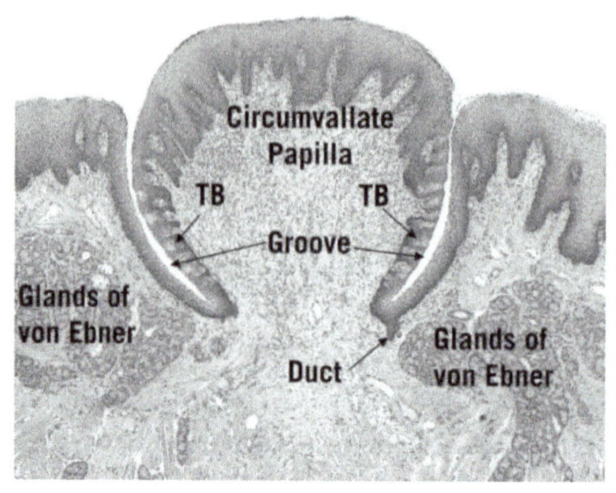

Papillae of Tongue

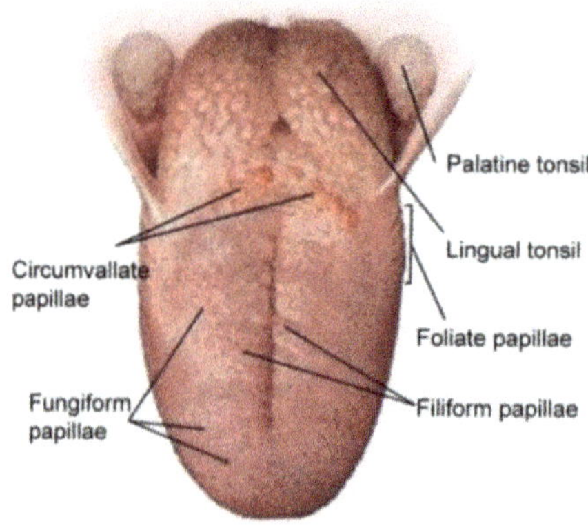

Microscopic view of taste buds in groove of tongue papilla below

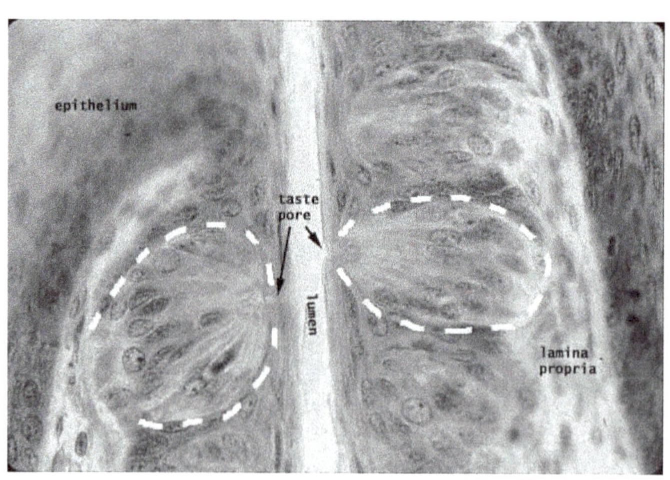

Filiform papillae

Filiform

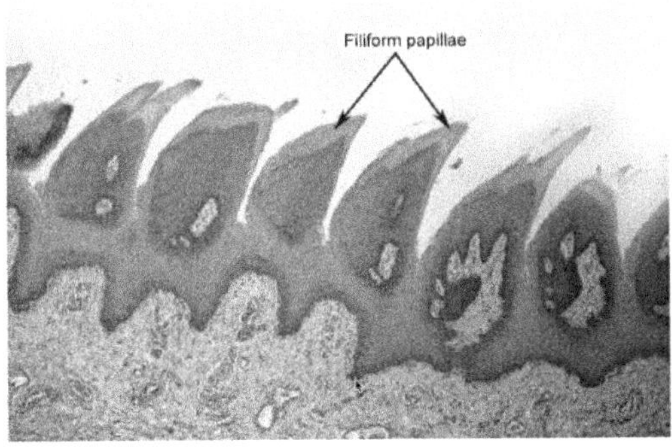

Salivary glands

Salivary glands secrete saliva into the oral cavity. The major functions of salivary glands are as follows:

1. Lubricate food. Saliva is 97 to 99.5% water, and it contains a protein called **mucin** (the main ingredient of mucus) that thickens saliva and lubricates ingested food.
2. Begin digestion. Saliva contains the enzyme amylase, which helps to break down carbohydrates.

Humans have three pairs of **extrinsic** salivary glands:

A. **Parotid glands** contain many **serous cells**, which produce thin, watery saliva that contains salivary amylase. Secretions of the parotid glands travel to the oral cavity via the **parotid duct**, which empties into the vestibule by the second upper molar.

B. **Sublingual glands** contain serous and **mucous cells**, which produce thicker saliva with higher amounts of mucin. The glands are located below the tongue and empty on either side of the **lingual frenulum**.

C. **Submandibular glands** lie on the posterior floor of the mouth, just inside the mandible. They contain a mix of serous and mucous cells and empty into the mouth on either side of the lingual frenulum.

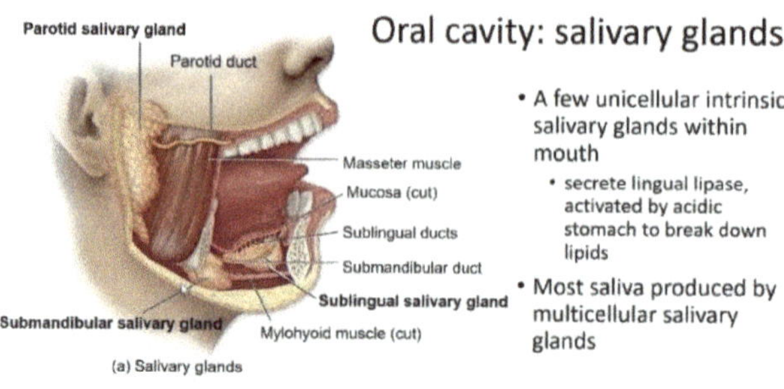

Briefly describe the **intrinsic** salivary glands:

Briefly describe the effects of the parasympathetic and sympathetic systems on salivary glands:

Pharynx and esophagus

We already covered the pharynx when we discussed the respiratory system.

The **esophagus** connects the pharynx to the stomach. The ANS moves food down the esophagus via peristaltic waves of smooth muscle contraction. The esophagus passes through an opening in the diaphragm called the **esophageal hiatus**. Compression of the abdomen may force parts of the lower digestive tract through this opening and into the thoracic cavity, a condition known as a **hiatal hernia**.

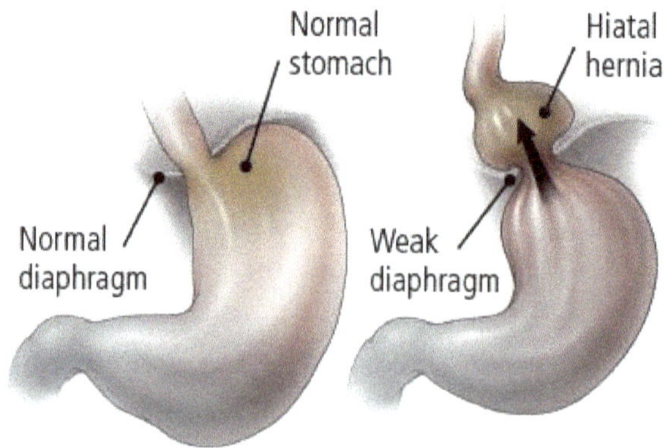

Muscles at the inferior end of the esophagus form the **gastroesophageal sphincter**, which is normally in a state of contraction to prevent the backflow of materials from the stomach. The opening of the esophagus into the stomach is referred to as the **cardiac orifice**. **Heartburn** is the condition that results when gastric juice is regurgitated through the cardiac orifice and into the esophagus.

Stomach

The esophagus connects to the **stomach**. The stomach performs four major functions:

1. Storage of ingested food.
2. Mechanical digestion.
3. Chemical digestion.
4. Production of **intrinsic factor**, which is required for absorption of vitamin B_{12}.

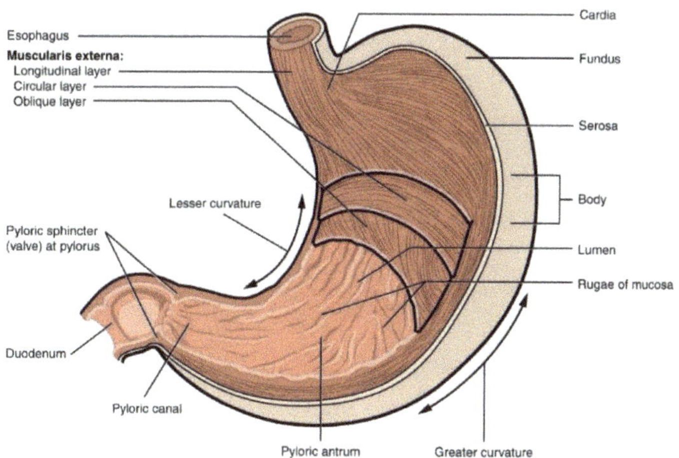

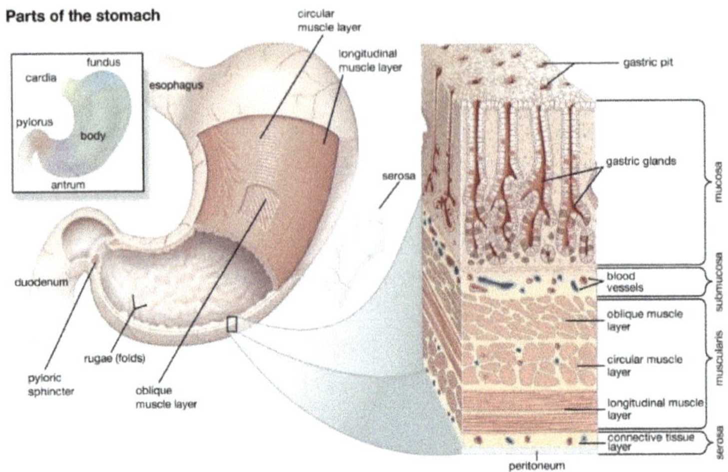

Histology of the stomach mucosa

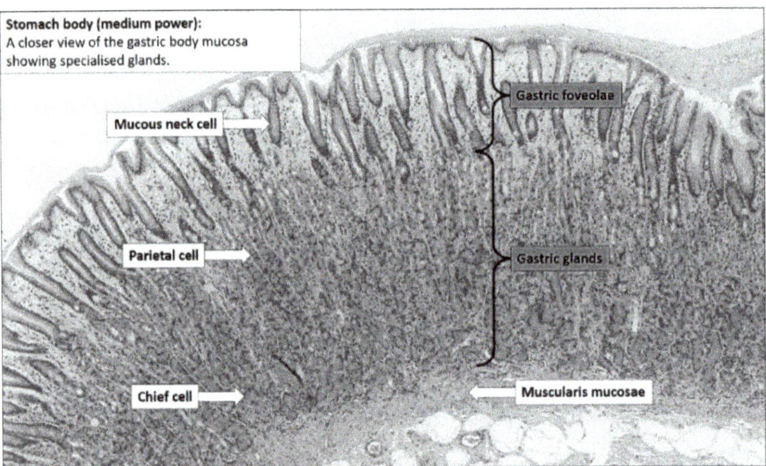

The anatomy of the stomach is shown in above. The **lesser curvature** of the stomach forms the medial edge of the stomach, and the **greater curvature** forms the lateral edge. There are four major anatomical regions of the stomach in humans:

A. The **cardia** (or *cardiac region*) is the superior end of the stomach within about 3 cm of the esophagus. It contains many mucous glands whose secretions help protect the esophagus from acidic secretions inside the stomach.

B. The **fundus** is the superior region of the stomach. Gastric glands in the fundus and body secrete most of the acids and enzymes involved in gastric digestion.

C. The **body** is the largest portion of the stomach. Much mechanical mixing of food occurs here.

D. The **pylorus** is the inferior part of the stomach. Much mechanical mixing of food occurs here. The **pyloric sphincter** regulates the passage of food into the small intestine.

The human stomach has folds called **rugae** within it. These folds allow the stomach to expand to accommodate a large meal. In addition to circular and longitudinal muscles, the muscularis externa of the stomach has **oblique muscles**, which help the stomach grind food.

Chemical breakdown is done primarily by secretions of **gastric glands**. These glands produce secretions that are collectively called **gastric juice**, and the gastric juice is released into the stomach via **gastric pits**. You should be familiar with the structures of gastric glands and gastric pits from your activities in the lab.

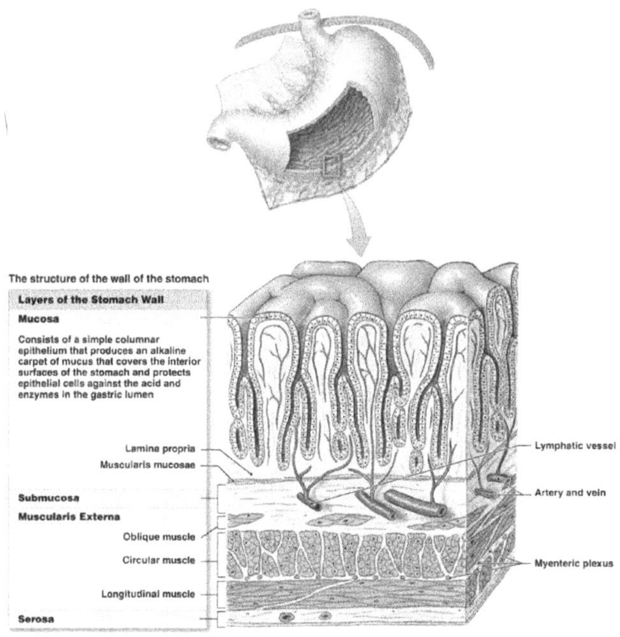

The gastric glands contain two types of secretory cells: (1) **Parietal cells** secrete the aforementioned intrinsic factor, and they secrete hydrochloric acid (HCl). This acid kills most microorganisms that enter the stomach, breaks down proteins and plant cell walls, and activates the enzyme **pepsin**. (2) **Chief cells** secrete the protein **pepsinogen**, which is converted to **pepsin** in the presence of stomach acid.

Pepsin digests proteins

Various hormones and the nervous system regulate the digestive processes of the stomach. These processes can be divided into three phases:

The **cephalic phase** of stomach activity is initiated by the thought, smell, sight, and/or taste of food. Conscious thoughts of food trigger the hypothalamus to send signals to the medulla oblongata. The medulla stimulates the stomach via the vagus nerve, causing an increase in motility of the stomach muscle and increased activity of gastric glands.

The **gastric phase** of stomach activity is when the stomach is actively digesting food. This is triggered by a stretch of the stomach by food, which further stimulates the motility of the stomach muscle and the activity of gastric glands. The presence of food in the stomach also stimulates the release of the hormone **gastrin**, which also contributes to the activity of stomach muscle and particularly the release of HCl into the stomach. Gastrin also stimulates contraction of the pyloric sphincter, to keep food in the stomach.

The **intestinal phase** of digestion involves the movement of chyme into the duodenum. Cells of the duodenum release **secretin** and **cholecystokinin (CCK)**. These chemicals inhibit secretion and motility of the stomach, stimulate the production of pancreatic juice and bile, stimulate contractions of the gall bladder, and cause relaxation of the hepatopancreatic sphincter.

III. Lower Gastrointestinal Tract

Small intestine

The **small intestine** is where most digestion and absorption of food occurs. The human intestine is approximately 20 feet long, and it can be divided into three major regions:

A. The **duodenum** begins at the pyloric sphincter and accounts for about the first 10" of the small intestine. It contains **duodenal glands** (or *Brunner's glands*), which secrete alkaline mucus to neutralize stomach acid.

B. The **jejunum** is the middle segment, about 8' long.

C. The **ileum** is the final segment, about 11' long. The **ileocecal valve** regulates the passage of material from the ileum into the large intestine.

The intestinal lining forms folds within the intestine called **circular folds** (or *plicae circulares*), which significantly increase the surface area for digestion. Unlike the rugae of the stomach, the plicae do not stretch. Another characteristic feature of the small intestine is the presence of **villi**. These are finger-like evaginations of the intestine that project into the lumen and further increase the surface area for absorption. The villi are, in turn, covered with cells with smaller projections (**microvilli**) that further increase the intestine's surface area. The microvilli region is often called the **brush border** of the intestine.

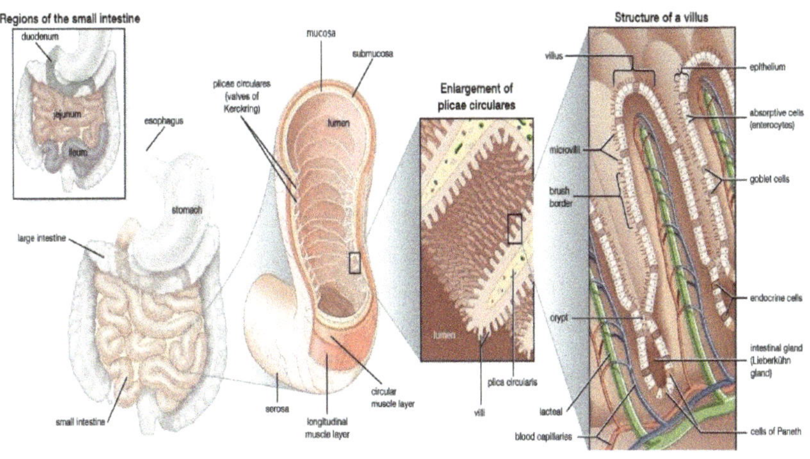

Accessory organs and ducts

Liver

The **liver** is the second-largest organ in the human body (the skin is the largest). Its tissue is divided into four **lobes**): the **right lobe**, the **left lobe**, the **caudate lobe**, and the **quadrate lobe**.

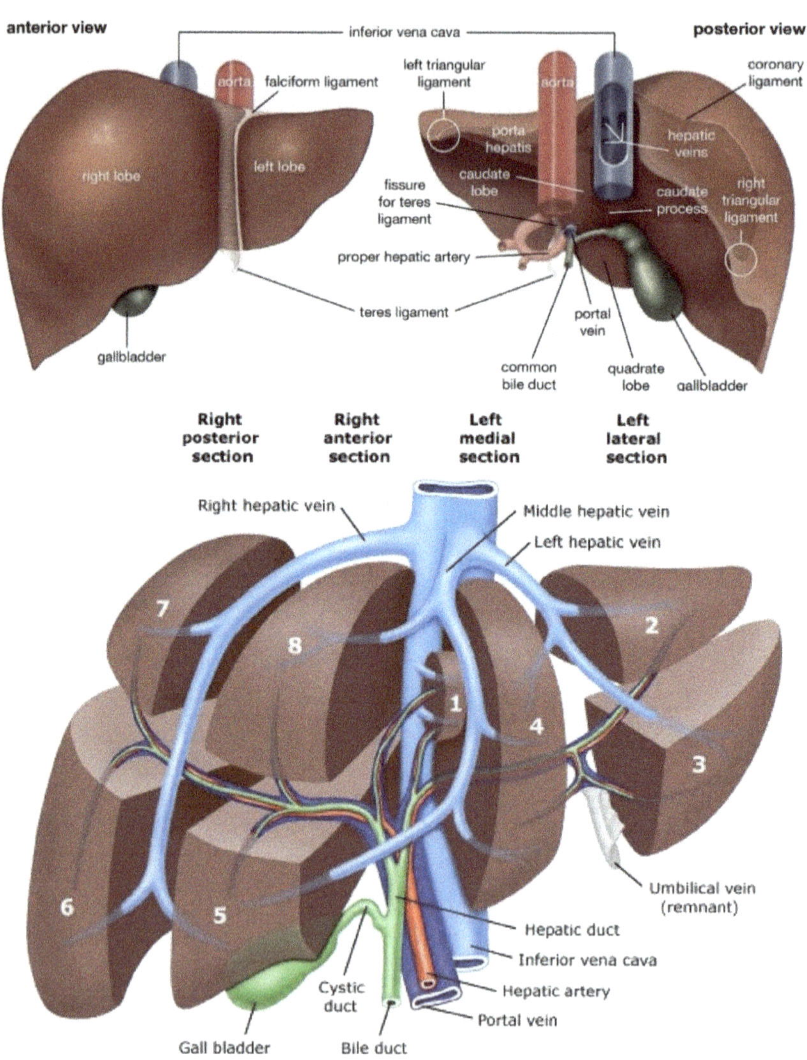

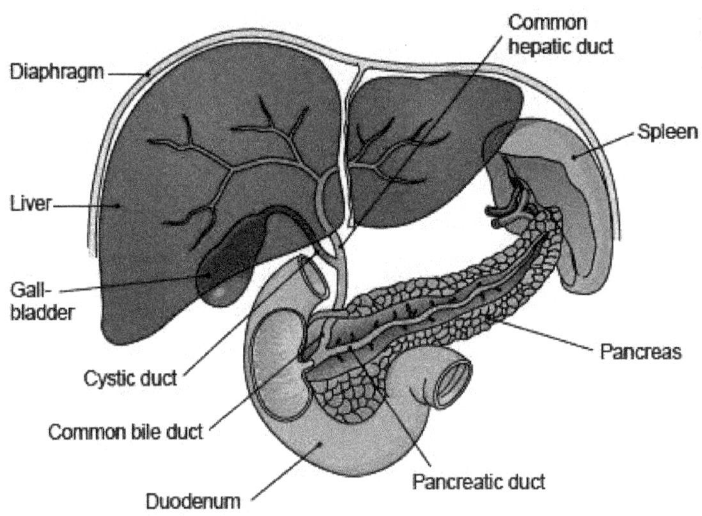

The primary functional cells of the liver are called **hepatocytes**. Hepatocytes are arranged into functional units called **hepatic lobules**. (Try to avoid confusing the words “lobe” and “lobule.”) Cells in a lobule are arranged around a **central vein**. Around the edges of the lobules are arterioles from the *hepatic artery* and venules from the *hepatic portal vein*. Blood from the hepatic artery and hepatic portal vein enters the lobules, nutrients are absorbed into the hepatocytes or released from the hepatocytes as necessary, and then blood passes into the central vein. The central vein eventually leads to the *hepatic veins*.

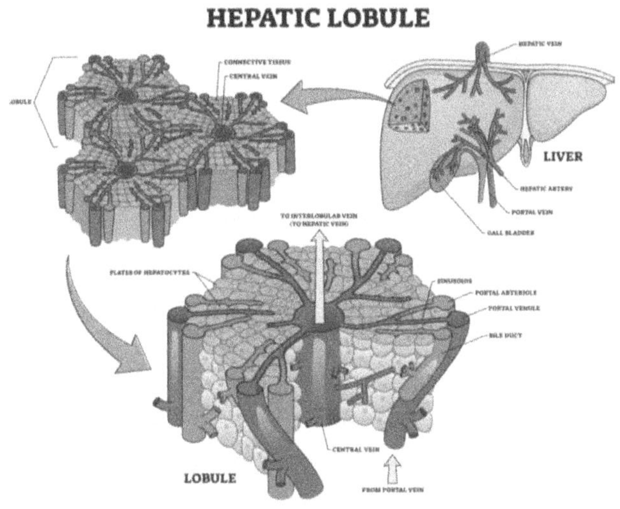

Portal triad

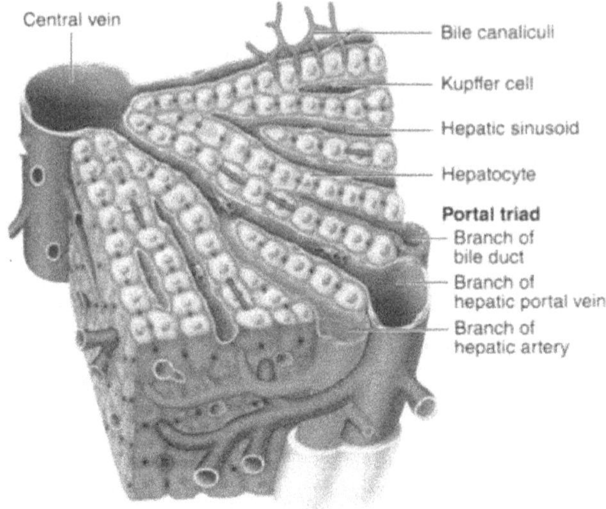

Hexagonal liver lobule with its structures

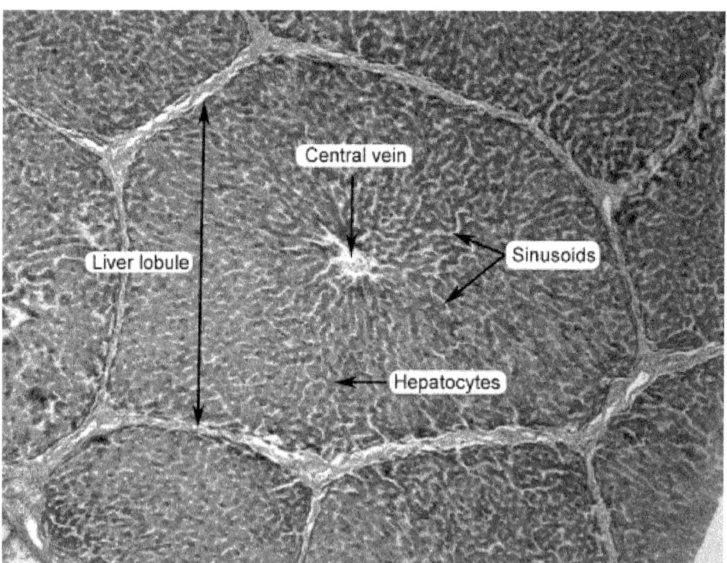

From which organs does most of the blood entering the liver come? Recall the flow of blood into and out of the liver and understand the functions of the hepatic artery, hepatic portal vein, and hepatic veins.

The primary glandular function of the liver is the secretion of **bile**. Bile secreted by the hepatocytes flows through tiny canals called the **bile canaliculi**. Bile from the canaliculi enters one of many **bile ducts** in the liver, and the bile ducts eventually lead to either the **right** or **left hepatic duct**. Bile is released from the liver via the **common hepatic duct**, which leads to (1) the **cystic duct**, which carries bile to the gallbladder, and (2) the **common bile duct**, which carries bile to the duodenum.

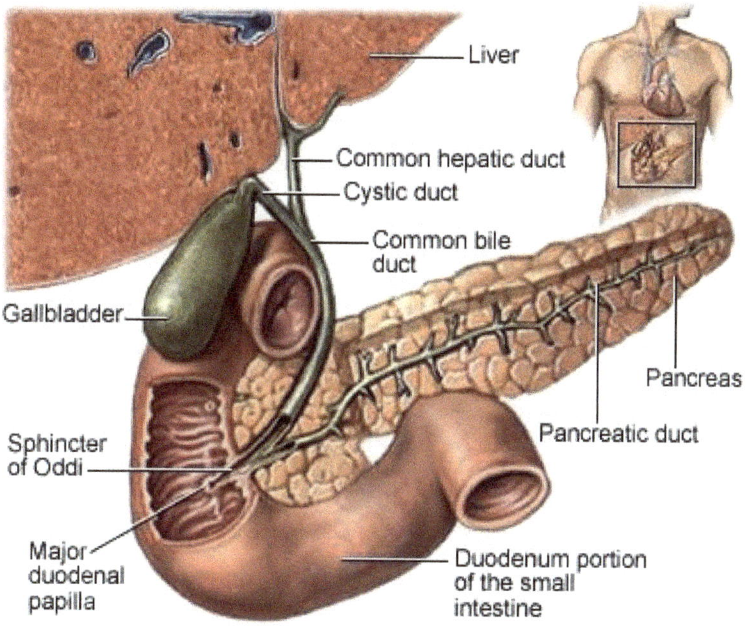

Under what conditions does bile flow from the liver to the gallbladder?

Under what conditions does bile flow from the gallbladder to the duodenum?

Reticuloendothelial cells (or *Kupffer cells*) are macrophages in the liver responsible for engulfing pathogens, cellular debris, and damaged blood cells.

The liver performs many functions for the body:

A. The liver's primary digestive function is producing bile. Bile is a mixture of water, ions, cholesterol, and a collection of lipids known as **bile salts**. Because lipids are insoluble in water, they tend to clump together in the digestive tract. Bile salts **emulsify** (break apart) these clumps of lipids so that digestive enzymes can digest the individual lipid molecules.

B. The liver works with the pancreas to regulate blood glucose. It stores glucose and releases it into the blood in response to glucagon.

C. The liver absorbs lactic acid from the blood and converts it to glucose.

D. The liver regulates the levels of triglycerides, fatty acids, and cholesterol in the blood.

E. The liver removes excess amino acids, hormones, and antibodies from the blood; the liver also removes various toxins from the blood.

F. The liver stores a variety of vitamins and minerals.

G. The liver produces the major plasma proteins.

Gallbladder

The **gallbladder** is located in a recess on the lower portion of the posterior side of the liver's right lobe. Bile enters the duodenum only when **chyme** is present, yet the liver is always producing bile. When bile is not released into the duodenum, it is stored in the gallbladder via the cystic duct.

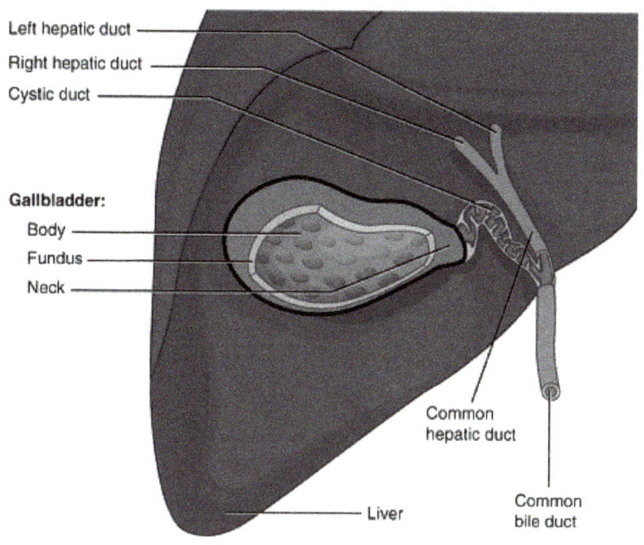
Left hepatic duct
Right hepatic duct
Cystic duct
Gallbladder:
Body
Fundus
Neck
Common hepatic duct
Liver
Common bile duct

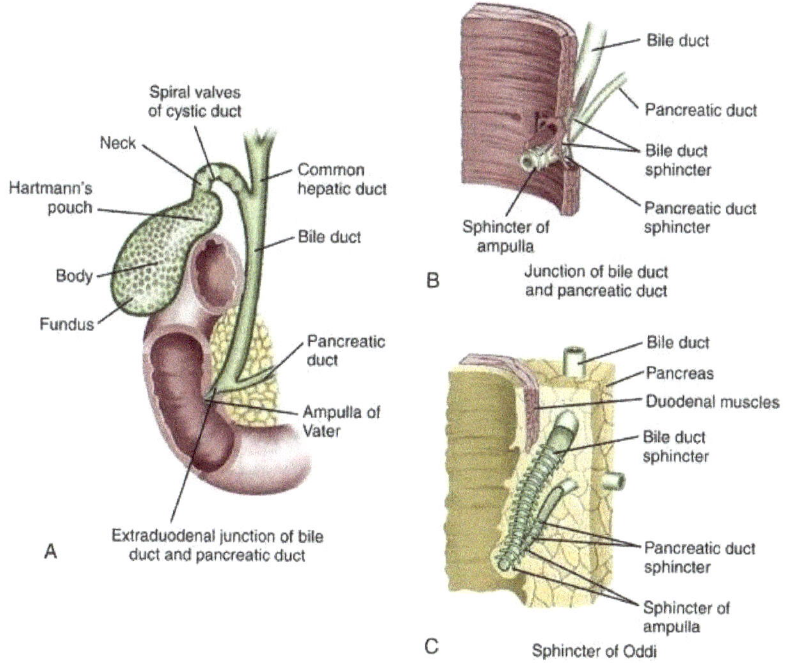
Spiral valves of cystic duct
Neck
Hartmann's pouch
Body
Fundus
Common hepatic duct
Bile duct
Pancreatic duct
Ampulla of Vater
A
Extraduodenal junction of bile duct and pancreatic duct
Bile duct
Pancreatic duct
Bile duct sphincter
Pancreatic duct sphincter
Sphincter of ampulla
B
Junction of bile duct and pancreatic duct
Bile duct
Pancreas
Duodenal muscles
Bile duct sphincter
Pancreatic duct sphincter
Sphincter of ampulla
C
Sphincter of Oddi

Biliary Tree / Tract

- Intrahepatic ducts
- Right & Left hepatic ducts
- Common hepatic duct
- Cystic duct
- (Common) bile duct
- Hepatopancreatic ampulla
  - Main pancreatic duct
- Major duodenal papilla
  - 2nd part of duodenum
  - Foregut → midgut

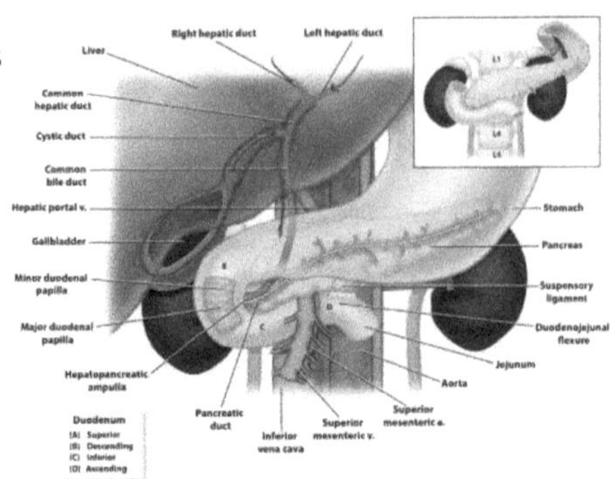

Pancreas

The **pancreas** lies posterior to the stomach and extends toward the duodenum. The **head** of the pancreas lies within a loop formed as the duodenum leaves the pylorus, the **body** extends along the stomach, and there is a short, rounded **tail**. Cells of the pancreas are divided into **lobules** by partitions of connective tissue. Each lobule contains cell clusters, forming pockets called the **pancreatic acini**. Small ducts lead from the acini to larger ducts, eventually leading to the **pancreatic duct**. The pancreatic duct carries exocrine secretions of the pancreas to the duodenum.

The acinar cells and epithelial cells that line the ducts have exocrine functions and secrete **pancreatic juice** into the ducts. Spread among the lobules are clusters of cells called **pancreatic islets** (or *Islets of Langerhans*). These cells have endocrine functions and account for only about 1% of the pancreatic cells.

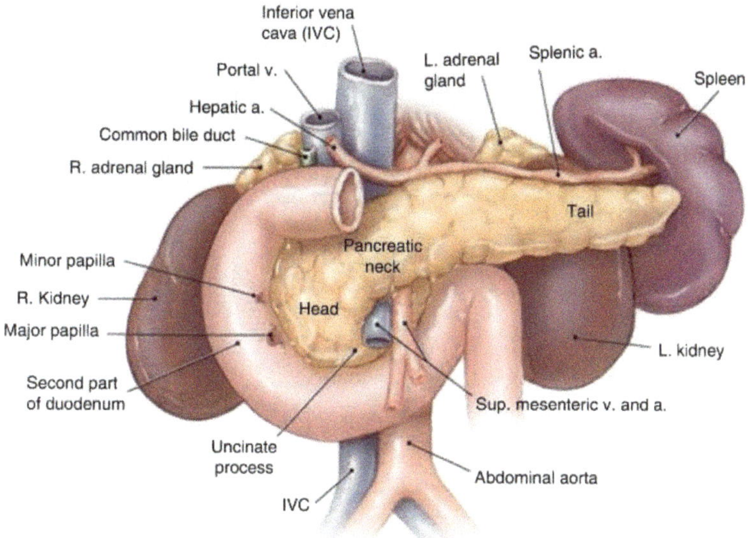

Sectioned pancreas showing pancreatic endocrine and exocrine characteristics and its various structures and histology.

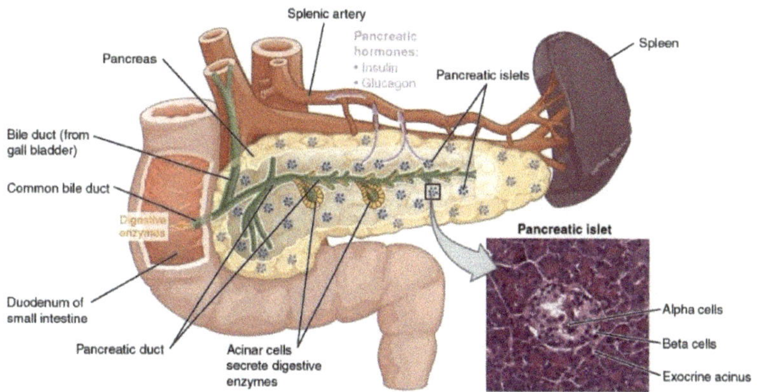

Exocrine functions of the pancreas involve the secretion of pancreatic juice into the intestine. This fluid is alkaline (why is this important?), and it contains water, enzymes, and various ions. Enzymes secreted by the pancreas include **amylase** to digest carbohydrates, **pancreatic lipase** to digest lipids, **ribonuclease**, and **deoxyribonuclease** to digest nucleic acids, and **proteases** (including **trypsin** and **chymotrypsin**) to digest proteins.

Endocrine functions involve the production of **insulin** and **glucagon**, which regulate blood glucose levels. **Beta cells** produce insulin, and

alpha cells produce glucagon. Both of these cell types are located in the pancreatic islets.

Understand the functions of both insulin and glucagon, and understand the conditions under which each is secreted.

Large intestine

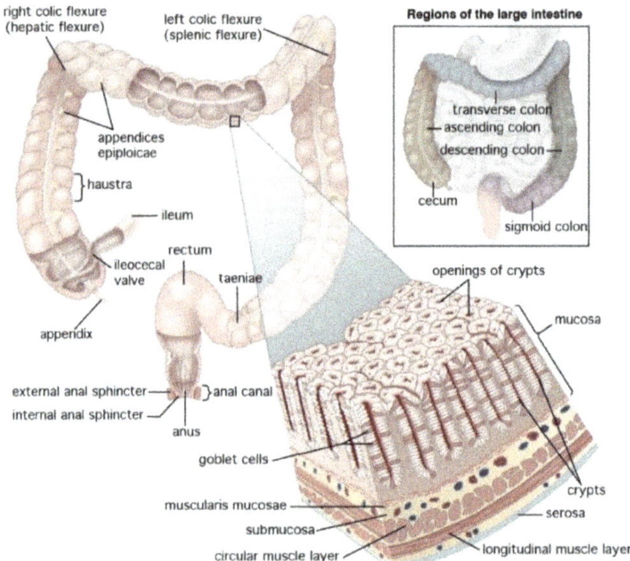

Histology of the large intestine (transverse cut)

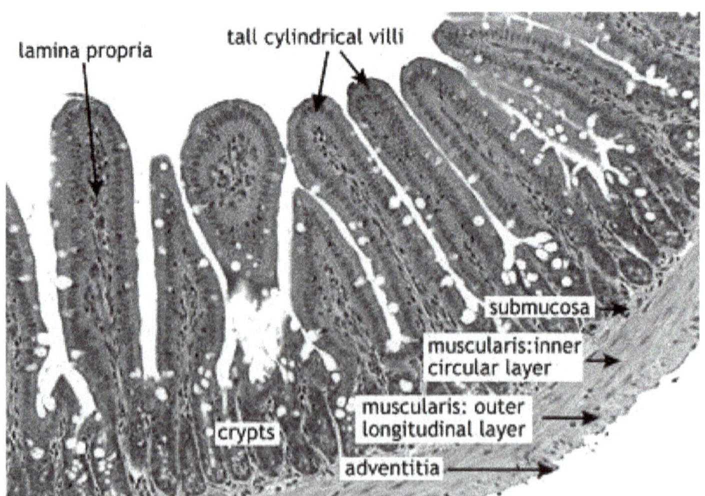

The **large intestine** is larger in diameter than the small intestine, but the small intestine is the longer of the two. In humans, the large intestine is about five feet long. The major functions of the large intestine are (1) water absorption and compaction of the remaining contents into feces, (2) absorption of vitamins, and (3) storage of feces before elimination.

The human large intestine is divided into three regions (use the space below for your response):

A. The **cecum** is a short, blind sac located where the small intestine joins the large intestine. The ileum opens to the cecum through the **ileocecal valve**. The **vermiform appendix** (or *appendix*) extends from the cecum.

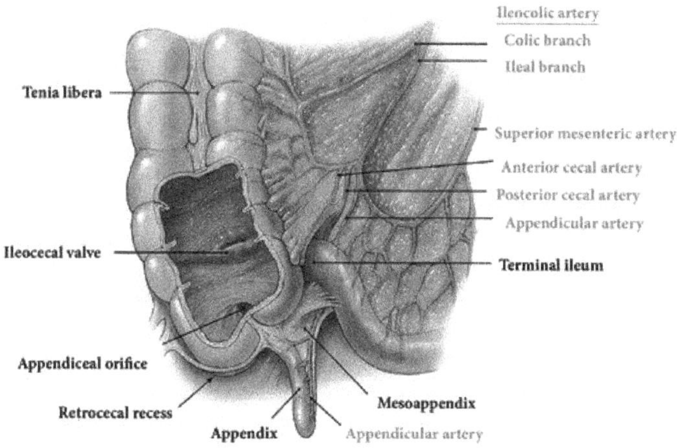

B. The **colon** can be subdivided into four regions: the **ascending colon**, the **transverse colon**, the **descending colon**, and the **sigmoid colon**. The colon's wall forms a series of folds called **haustra**, which expand and contract as food moves through the colon. C. The **rectum** is a straight tube extending to the anus.

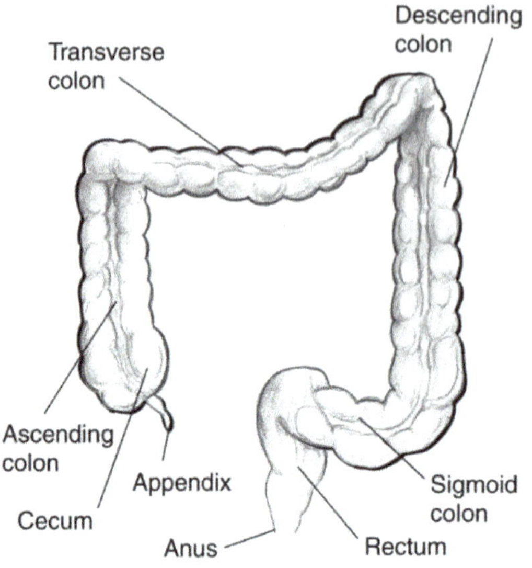

Colon cancers are one of the more common forms of cancer, with almost 100,000 cases diagnosed in the US each year. Mortality is high if it is not detected and treated early.

Identify the extrinsic glands in the illustration below.

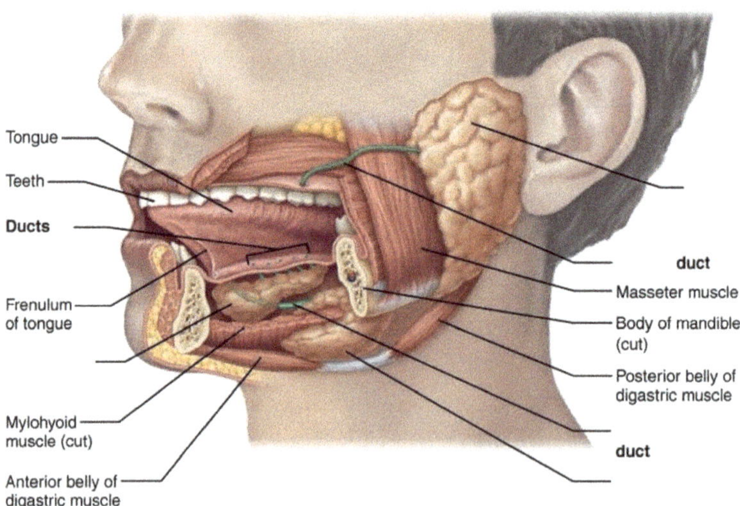

Chapter 9: Urinary System

I. Introduction to the Urinary System

The primary function of the urinary system is to remove waste materials from the body by filtering the blood. Every day, the kidneys filter the equivalent of about 200 liters of fluid from the bloodstream. When they are functioning normally, the kidneys eliminate toxins, metabolic wastes, and excess ions while retaining substances useful to the body.

The kidneys also provide additional functions for the body:

A. The kidneys produce an enzyme that is necessary for the formation of the hormone calcitriol.

Review the chapter on the integumentary system to recall the function of calcitriol if necessary.

B. The kidneys produce the hormone erythropoietin.

C. The kidneys participate in the regulation of the levels of various ions in the blood (*e.g.,* Na^+, K^+, and Ca^{2+}) and help maintain the proper pH of the blood.

D. The kidneys help regulate blood pressure by releasing renin and adjusting the amount of fluid the body retains.

Along with the kidneys, the urinary system includes the ureters, urinary bladder, and urethra.

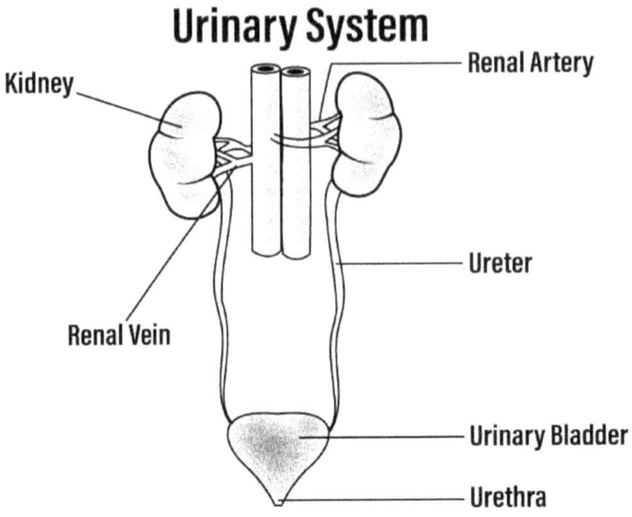

II. Gross Anatomy of the Kidney

<u>Location and support</u>

The kidneys are located in the upper lumbar region of the body. Each is bean-shaped and a bit smaller than the size of your hand. The **hilum** (or *hilus*) of the kidney is the point of entry or exit for the **renal artery**, **renal vein**, and **ureter**.

Gross anatomy of the kidney

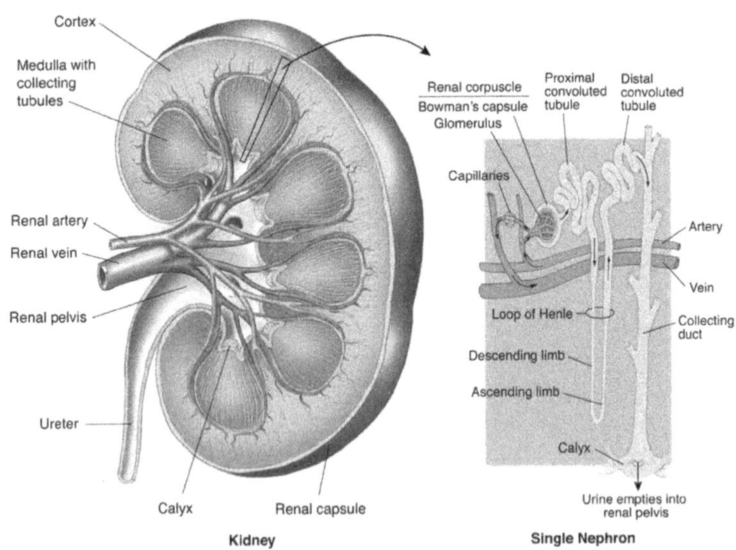

Posterior view of the position of the kidneys in the retroperitoneum. Note the position of the upper pole of the kidneys to T12 and L1.

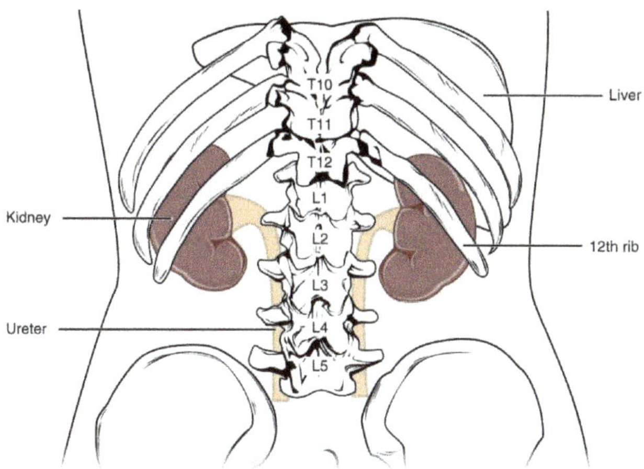

The kidneys are located dorsally in the body, outside the abdominopelvic cavity, between the parietal peritoneum and the muscle of the dorsal body wall.

A layer of tissue called the **fibrous capsule**, which consists mainly of collagen fibers, surrounds each kidney. A thick layer of adipose tissue, called the **perinephric fat**, surrounds the fibrous capsule. A layer of dense irregular connective tissue called the **renal fascia** encloses the perinephric fat and anchors the kidney to surrounding tissues. Finally, another layer of adipose tissue, the **paranephric fat**, lies outside the renal capsule.

Cross-sectional view of the kidney in the retroperitoneum below

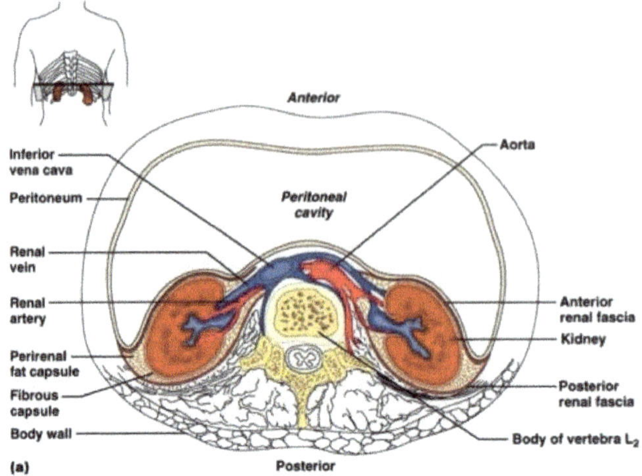

Retroperitoneal Position of the Kidneys

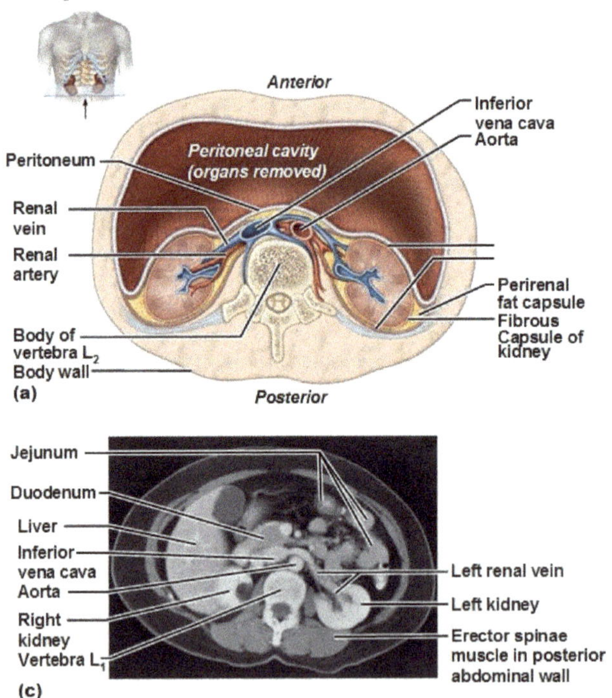

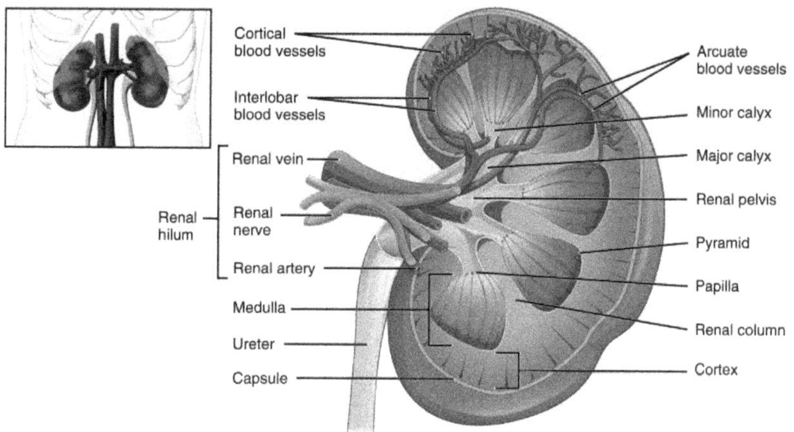

Sectional anatomy of the kidney

The outer portion of the kidney tissue is the **cortex**, and the deeper portions of tissue make up the **medulla**. The medulla is divided among eight to fifteen conical structures called **renal pyramids**. The base of each pyramid faces the cortex, and the apex of each pyramid (the **renal papilla**) projects inward. Pyramids are separated from each other by extensions of cortical tissue called **renal columns**. Each pyramid and its adjacent cortex region make up a **renal lobe**. An internal cavity, called the **renal sinus**, is located medially within each kidney. The sinus is a site for the renal artery branches, renal vein branches and ureter to enter and exit the kidney tissue respectively (see image below).

Urine production occurs in each renal lobe, and urine collects in ducts that lead through each pyramid to its renal papilla. Each papilla empties into a **minor calyx**. A few minor calyces join together to form a **major calyx**, and major calyces lead to the **renal pelvis**, which opens to the ureter.

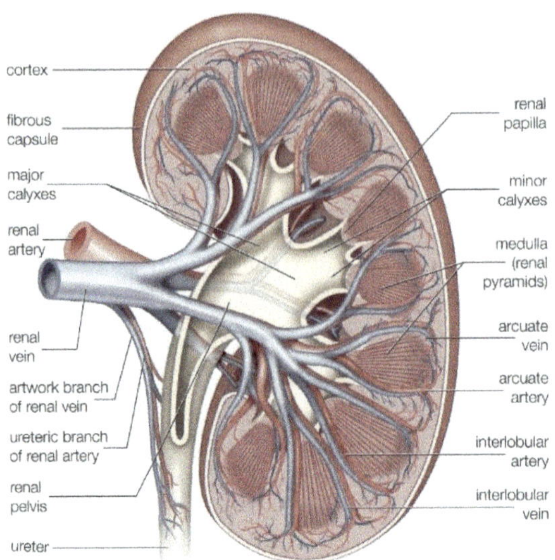

III. Functional Anatomy of the Kidney

Nephron

The functional unit of the kidney is the **nephron**. Each nephron consists of a **renal corpuscle** and a **renal tubule**. Most of each nephron resides in the cortex of the kidney, except for the nephron loop, which extends into the medulla.(see diagrams below).

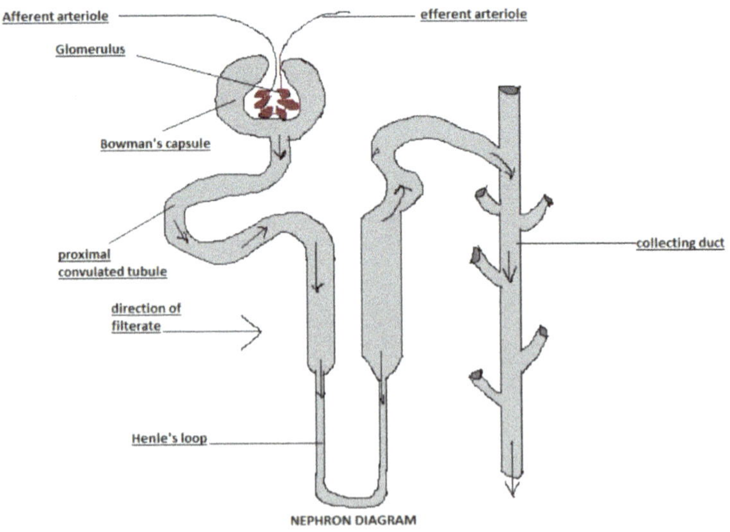

Cortical nephron and its blood supply below

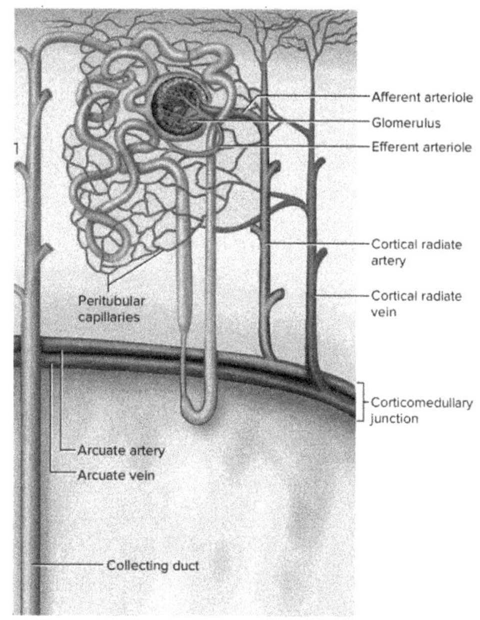
Afferent arteriole
Glomerulus
Efferent arteriole
Cortical radiate artery
Cortical radiate vein
Peritubular capillaries
Corticomedullary junction
Arcuate artery
Arcuate vein
Collecting duct

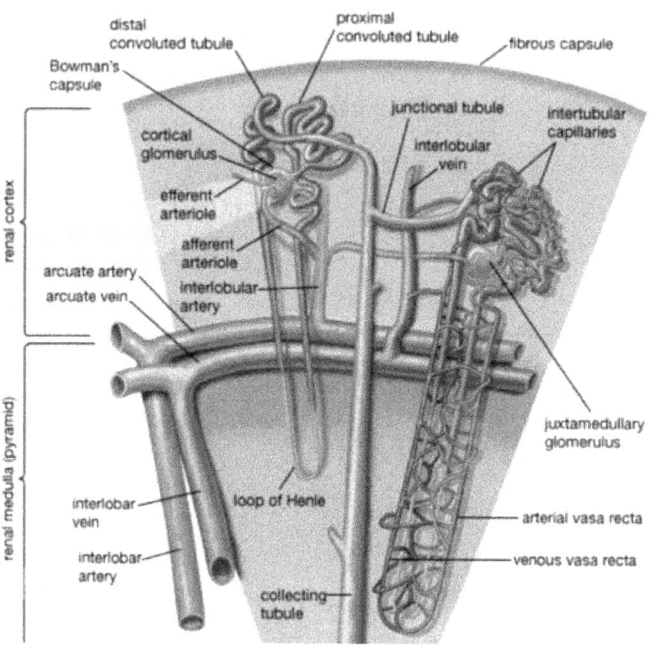
distal convoluted tubule
proximal convoluted tubule
fibrous capsule
Bowman's capsule
junctional tubule
intertubular capillaries
cortical glomerulus
interlobular vein
efferent arteriole
afferent arteriole
renal cortex
arcuate artery
arcuate vein
interlobular artery
juxtamedullary glomerulus
renal medulla (pyramid)
interlobar vein
loop of Henle
arterial vasa recta
interlobar artery
venous vasa recta
collecting tubule

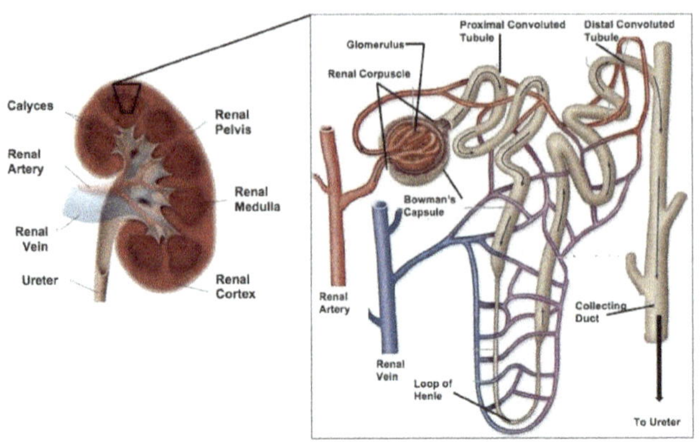

Renal corpuscle

The renal corpuscle forms one end of each nephron, and it is composed of two structures: the **glomerulus** and the **glomerular capsule** (or *Bowman's capsule*). The glomerulus is a network of fenestrated capillaries that filter blood that enters the kidney. Blood enters the glomerulus via an **afferent arteriole** and exits via an **efferent arteriole**. The glomerular capsule surrounds the glomerulus and collects fluid that is filtered out of the glomerulus. The **visceral layer** of the glomerular capsule is wrapped tightly around the capillaries of the glomerulus, and the **parietal layer** forms an outer covering. The space between these two layers is the **capsular space**, and this is where filtrate is initially collected.

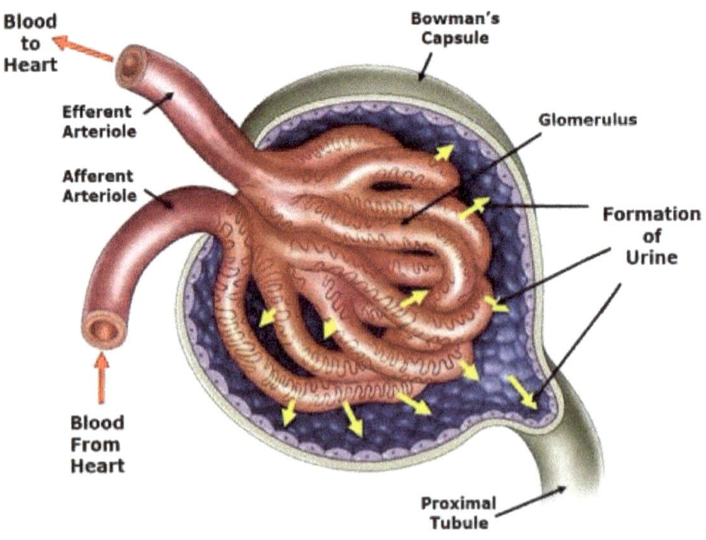

Renal tubule

The renal tubule is a long tube that can be divided into three parts. 1) The **proximal convoluted tubule (PCT)** carries fluid from the glomerular capsule. 2) Cuboidal epithelium with microvilli lines the lumen and absorbs desired molecules from the tubular fluid back into the tissues. The **nephron loop** (or *loop of Henle*) has both descending and ascending limbs, each with a thick and thin segment. The **distal convoluted tubule (DCT)** is similar in appearance to the PCT, yet it lacks microvilli.

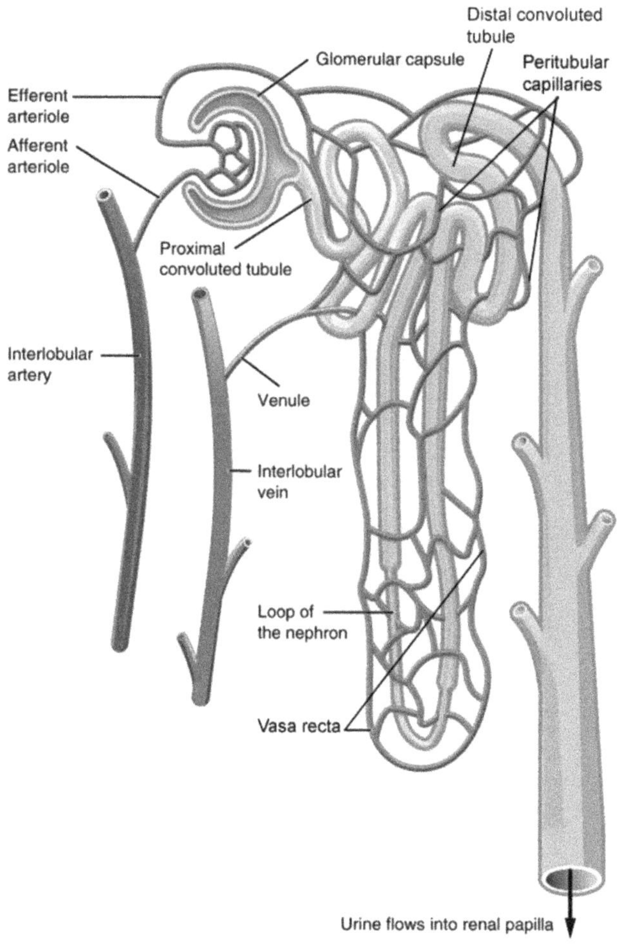

Sections of the renal tubule (you may fill in the blanks below)

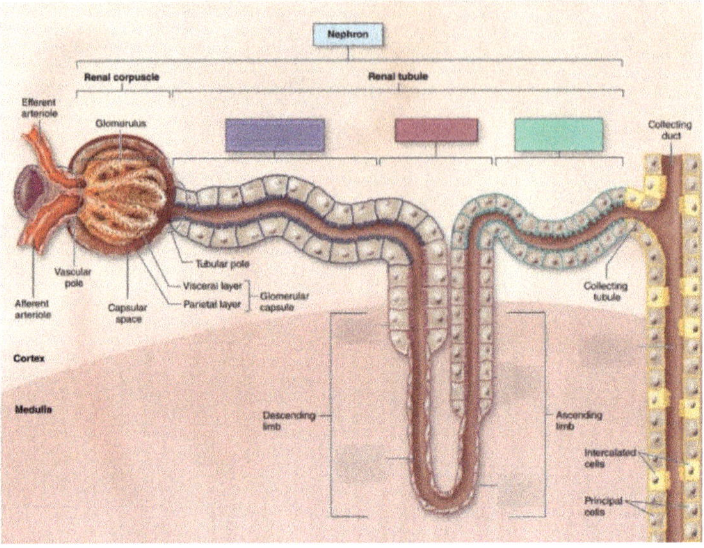

Collecting tubules and collecting ducts

From the DCT, urine travels to a **collecting duct**. From a collecting duct, urine will travel to a minor calyx, major calyx, etc. For simplicity, "collecting tubules" in with collecting ducts.

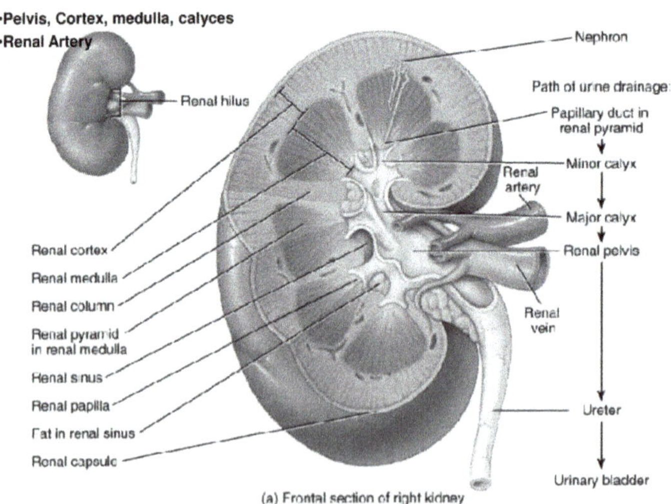

(a) Frontal section of right kidney

Juxtaglomerular apparatus

A portion of the DCT lies near the renal corpuscle and the afferent and efferent arterioles (see diagram below). As a result, the structure formed is called the **juxtaglomerular apparatus** (JGA).

Specialized cells of the juxtaglomerular apparatus called the **macula densa** will be discussed later.

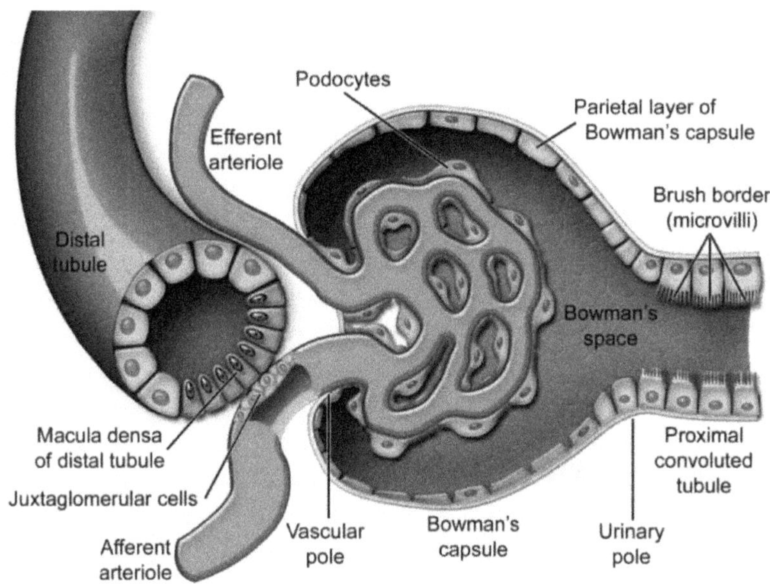

IV. Blood Flow and Filtered Fluid Flow

At rest, about 20-25% of cardiac output flows through the kidneys. This is on the order of 1200 ml per minute.

Blood flow through the kidney

Each **renal artery** enters the renal sinus and branches into **segmental arteries**. These branch into **interlobar arteries**, which extend through the renal columns toward the cortex. **Arcuate arteries** arch across the bases of the renal pyramids. **Interlobular arteries** extend into the cortex from the arcuate arteries.

Afferent arterioles branch off the interlobular arteries and lead to the glomeruli. Blood is filtered at the glomerulus, with some fluid leaving the glomerulus and entering the glomerular capsule. Blood that does not get filtered exits the glomerulus through an **efferent arteriole**. In part, because the afferent arteriole has a larger diameter than the efferent arteriole, blood pressure in the glomerulus is unusually high for a capillary bed. This pressure is the force that drives filtration.

From the efferent arteriole, blood travels to the **peritubular capillaries**, which surround the renal tubules. Blood also travels through the **vasa recta**, a network of capillaries surrounding the nephron loop.

Blood from the peritubular capillaries and vasa recta travels to the **cortical radiate**, **arcuate**, **interlobar**, and **renal vein**s.

Diagram of blood flow through the kidney below

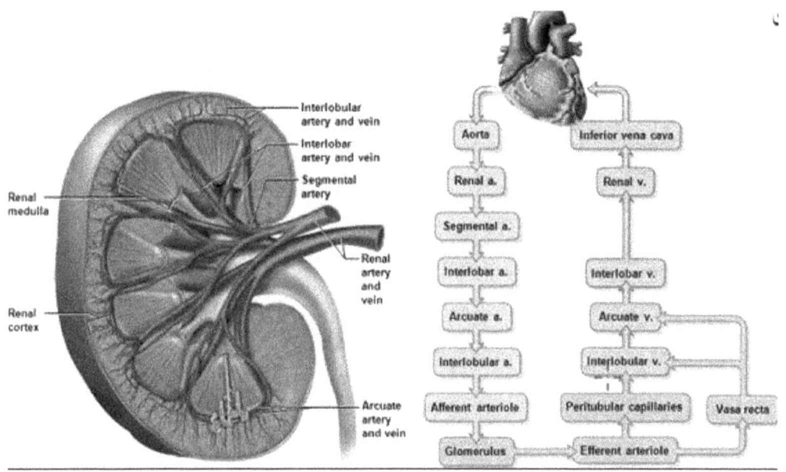

V. Production of Filtrate Within the Renal Corpuscle

Overview of urine formation

Daily, approximately 180 liters of fluid is filtered from the blood into the nephrons. This "filtrate" contains almost everything blood does except proteins and formed elements. By the time the filtrate reaches the collecting ducts, the body has absorbed much water, nutrients, and essential ions. Thus, the composition of filtrate is very different from the composition of urine that exits the body, and only about 1% of the filtrate leaves the body as urine.

One of the primary goals of urine formation is the elimination of organic wastes from the body. Three compounds in particular deserve mention:

A. Urea is the most abundant organic waste in urine. It is produced from ammonia, a by-product of protein metabolism.

B. Creatinine is a product of the breakdown of creatine in the muscle cells.

C. Uric acid is generated as nitrogenous bases in RNA are recycled.

There are three main processes involved in converting filtrate into urine: (1) filtration, (2) tubular reabsorption, and (3) tubular secretion. In this section, we will examine the process of filtration.

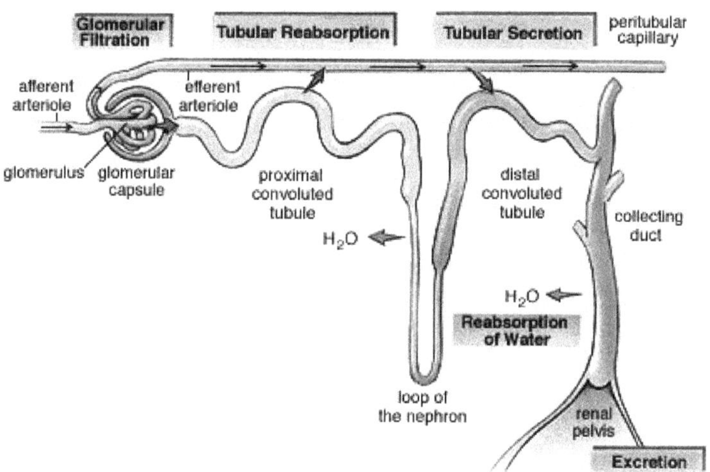

<u>Formation of filtrate and its composition</u>

Substances in the blood can be grouped into three categories according to how easily they pass through the filtration membrane of the glomerulus and glomerular capsule:

- Freely filtered–small substances such as water, glucose, other sugars, amino acids, ions, etc., pass freely through the filtration membrane.
- Not filtered–formed elements and large proteins normally do not pass through the filtration membrane.
- Partially filtered–medium-sized proteins normally do not get filtered to a significant degree, but they might, under certain circumstances, such as when blood pressure rises during exercise.

The large amount of filtrate formed contains many substances, including those shown in the table below:

Substance	Molecular weight	Relative filterability
water	18	1
sodium	23	1
glucose	180	1
myoglobin	1,700	0.75
albumin	69,000	0.005

This table illustrates the fact that filtration is selective on the basis of size. Small molecules are filtered freely into the nephrons, while larger particles are retained in the capillaries. The presence of large proteins or blood cells in the urine usually indicates damage to the kidneys.

Regulation of glomerular filtration rate

The rate of filtrate formation is known as the **glomerular filtration rate (GFR)**. A typical GFR for a healthy adult is about 125 ml/min, which is equivalent to about 180 liters/day. Maintaining an appropriate GFR is critical to kidney function, and GFR is generally regulated by intrinsic and extrinsic control mechanisms:

Intrinsic controls are as follows:

(1) The **myogenic mechanism** is when a rise in pressure on the afferent arterioles leads to reflexive constriction of the afferent arterioles. Likewise, a reduction in pressure on the afferent arterioles leads to reflexive dilation of the arterioles. Is this a negative feedback or a positive feedback system?

(2) The **tubuloglomerular feedback mechanism** is directed by the macula densa cells of the juxtaglomerular apparatus. When the flow of the filtrate is low, the macula densa cells promote dilation of the afferent arterioles. When the flow of the filtrate is high, the macula densa cells release chemicals that promote vasoconstriction.

Extrinsic controls include the following:

(1) The sympathetic nervous system. Moderate sympathetic activity has little effect on the GFR. However, higher levels of sympathetic activity tend to cause strong constriction of the afferent arterioles. What impact does this have on GFR?

(2) GFR is also affected by the hormone **atrial natriuretic peptide (ANP)**. ANP is released by atrial cardiac muscle cells in response to elevated blood pressure. It causes dilation of the afferent arterioles and an increase in GFR.

(3) Angiotensin II affects GFR by causing vasoconstriction. Angiotensin II is released when blood pressure is low and acts as a vasoconstrictor. In the kidneys, the vasoconstrictive effect of angiotensin II has a greater effect on the efferent arteriole than the afferent arteriole. This raises the pressure on the glomerulus and increases GFR. This likely helps to maintain GFR under conditions of low blood pressure.

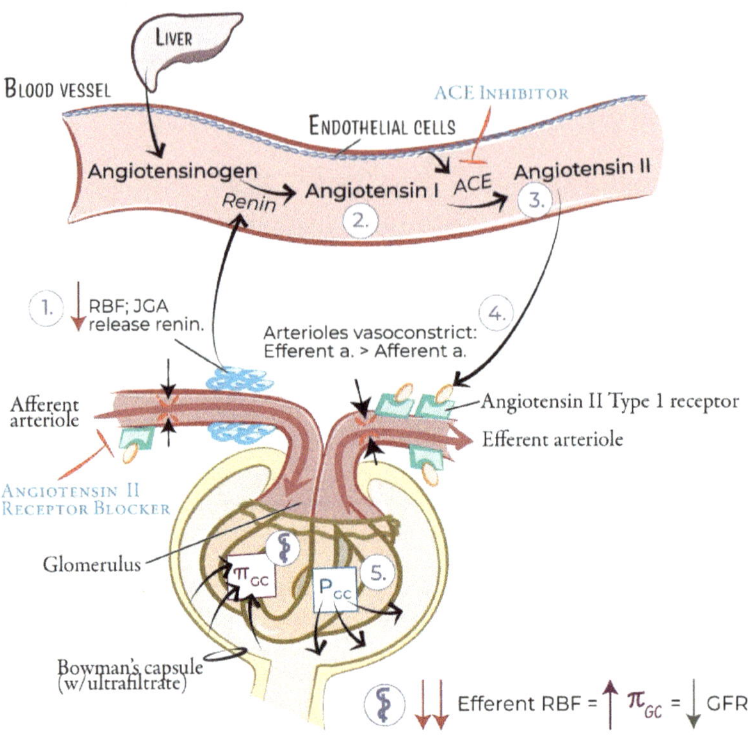

VI. Reabsorption and Secretion in Tubules and Collecting Ducts

A human cannot afford to lose 180 liters of fluid every day. Furthermore, you would not want to lose glucose, amino acids, etc. along with waste products. Whereas filtration is selective only by size, reabsorption is selective for particular molecules. For example, glucose is removed from the urine and returned to the body at the proximal tubule. All but about 1.5 liters of water (out of about 180 liters) per day is reabsorbed.

Substances reabsorbed completely

The PCT reabsorbs more or less 100% of filtered nutrients, such as glucose and amino acids. Likewise, any proteins that enter the filtrate are typically reabsorbed by the PCT.

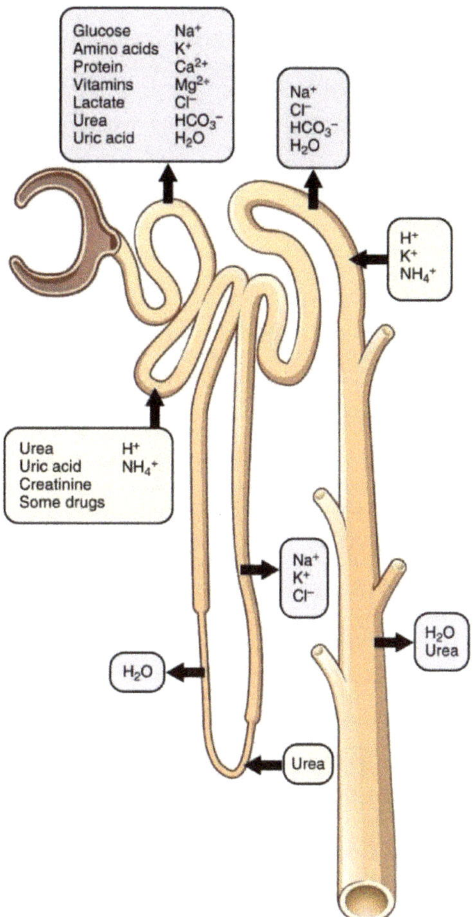

Substances with regulated reabsorption

Various substances in the filtrate are not reabsorbed completely, but the amount that is reabsorbed can be adjusted depending upon conditions. This allows the kidneys to regulate the amount of these substances in the blood.

Sodium reabsorption

The amount of sodium reabsorbed varies between 98% and almost 100%. While the difference between these two numbers may seem small, a large amount of sodium is filtered into the kidneys every day, so the difference between 98% and 100% is significant. Sodium is absorbed all along the renal tubule, with most being absorbed by the PCT and nephron loop. At the DCT and collecting ducts, aldosterone stimulates the synthesis of Na^+ channels and Na^+/K^+ pumps to increase sodium reabsorption. Water follows by osmosis. ANP inhibits the reabsorption of sodium. Under what conditions are aldosterone and ANP produced?

Water reabsorption

Reabsorption of water depends upon osmosis. Water can travel through cell membranes in two ways: (1) some water is able to move directly through the lipid bilayer by simple diffusion, and (2) water can move through channels in the cell membrane called **aquaporins**. The majority of water is reabsorbed by the PCT and nephron loop, largely as it follows the movement of sodium. At the DCT and collecting duct, the rate of water reabsorption is mainly determined by aldosterone and **antidiuretic hormone** (ADH). As salt reabsorption is regulated by aldosterone, the amount of water reabsorption also varies. Increased levels. Rising levels of ADH stimulate increased reabsorption of water by the collecting ducts. ADH is released in response to dehydration; thus, water is conserved in the body when needed. ADH makes the collecting ducts more permeable to water by increasing the number of aquaporins in the cells of the collecting ducts. Conversely, decreasing levels of ADH result in less water reabsorption.

Antidiuretic (ADH) role on kidney

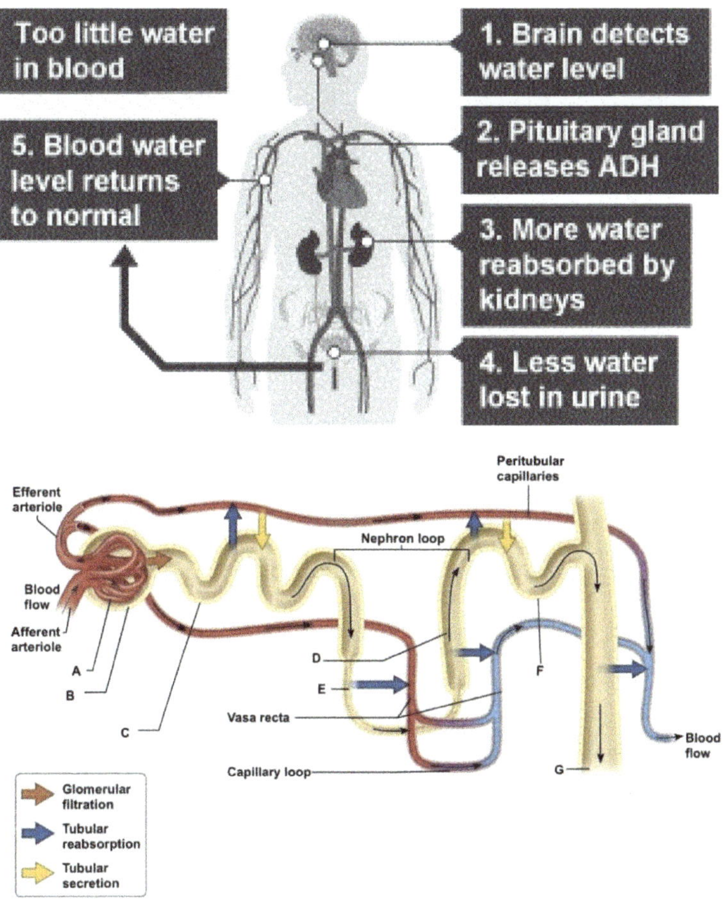

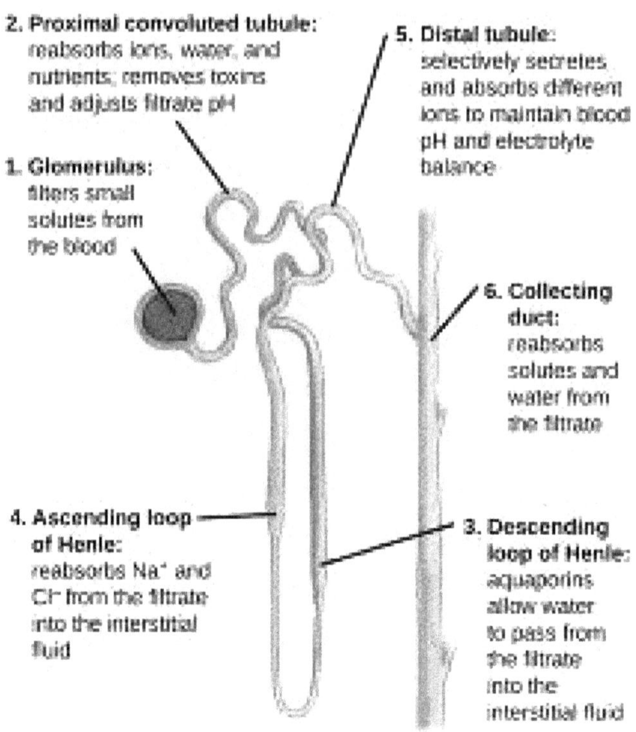

Aldosterone and Urine Formation

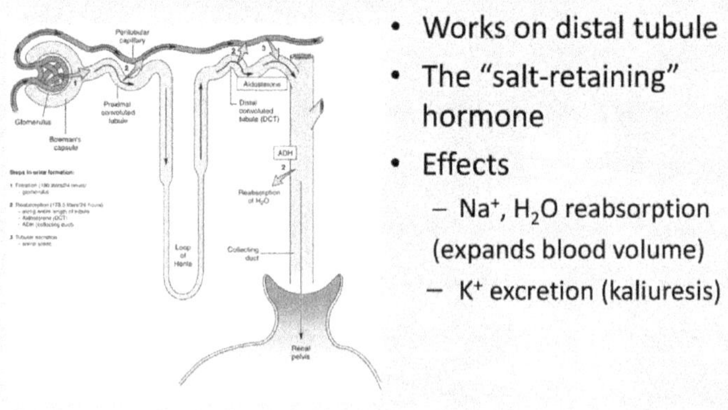

Movement of potassium

Potassium is both reabsorbed and secreted, and the net result depends on certain conditions. Most of the potassium in tubular fluid is reabsorbed by the PCT and nephron loop. A net reabsorption or secretion may occur in the collecting duct where some cells reabsorb potassium and others secrete it.

The secretion rate depends on aldosterone, with increasing levels of aldosterone stimulating secretion.

Bicarbonate ions, hydrogen ions, and pH

Nearly 100% of bicarbonate is reabsorbed by the PCT and nephron loop. Cells of the collecting ducts can secrete or reabsorb bicarbonate to help regulate the blood's pH. If the blood is more acidic, bicarbonate is reabsorbed, and H^+ is secreted, resulting in acidic urine. If the blod is more alkaline, bicarbonate is secreted into the filtrate, and H^+ is reabsorbed.

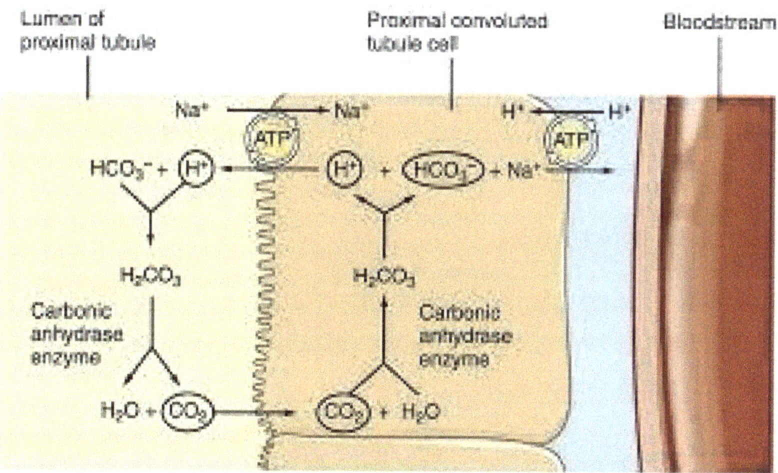

Substances eliminated as waste products

As mentioned previously, the main waste products eliminated by the blood are urea, uric acid, and creatinine. Urea and uric acid are both secreted and reabsorbed; creatinine is only secreted. In the case of urea, it is both reabsorbed and secreted as it passes through the PCT and loop, and the two processes roughly even out by the time filtrate enters the DCT. Reabsorption by the collecting duct removes about 50% of the urea from the filtrate, allowing the other half to be eliminated.

Establishing the concentration gradient

The nephron loop is responsible for creating a concentration gradient of salts in the kidney, which allows osmoconcentration of the tubular fluid. The concentration gradient is created by a system of **countercurrent exchange**. A positive feedback loop involving the descending and ascending limbs allows the concentration of salts in the interstitial fluid of the inner medulla to reach 1200 mOsm. This positive feedback loop is referred to as the "countercurrent multiplier."

The **vasa recta** is the portion of the peritubular capillaries that surrounds the loop of Henle. As blood travels through the descending portion of the vasa recta, it picks up solutes from the interstitial fluid of the kidney. Some water is lost, but proteins in the blood help retain water. As blood travels through the ascending portion, it absorbs water. The net result is that blood flowing through the vasa recta has a net gain of water. The blood eventually returns to the body via the renal vein.

Formation of dilute urine

Tubular fluid enters the DCT at an osmotic concentration of about 100 mOsm/l. How does this concentration compare to the normal concentration of body fluids?

When the body wishes to get rid of excess water, fluid from the DCT is simply allowed to pass through the collecting ducts and out of the body. When levels of ADH are low, the collecting ducts are impermeable to water, so as fluid travels through them, water is not reabsorbed into the tissues. Because cells of the DCT and collecting ducts can remove ions from the tubular fluid, the concentration of urine can be reduced to levels as low as about 50 mOsm/l.

Formation of concentrated urine

ADH regulates the permeability of the collecting duct and the DCT to water. If the body wishes to conserve water, ADH levels will increase, and the duct will become permeable to water. Osmosis leads to the reabsorption of water from the collecting ducts as urine passes through the medulla and increases interstitial osmolarity. Urine osmotic concentrations may rise as high as 1400 mOsm/l. How does the volume of urine relate to the concentration?

Caffeine and alcohol are both inhibitors of ADH. How do these chemicals affect the volume and concentration of urine?

VII. Urine Characteristics, Transport, Storage, and Elimination

Urinary tract (ureters, urinary bladder, urethra)

Ureters

After urine is produced in the kidneys, it travels through the **ureters** to the urinary bladder. A muscular layer of tissue in the ureter walls can produce peristaltic waves to propel urine toward the bladder.

Urinary bladder

The **urinary bladder** can typically hold up to about a liter of urine. Like the stomach, it has folds called **rugae**, which allow expansion. The bladder has three openings: one for each ureter and one for the **internal urethral orifice**. These three openings form a triangle at the bottom of the bladder, which is referred to as the **trigone**.

The lining of the bladder has three tissue layers:

A. The mucosa is the innermost layer. It is made of transitional epithelial tissue.

B. The **detrusor muscle** forms the middle layer. This consists of multiple layers of smooth muscle.

C. The peritoneum is the outermost layer on the superficial surface, and the adventitia is the outermost layer on the remainder of the bladder.

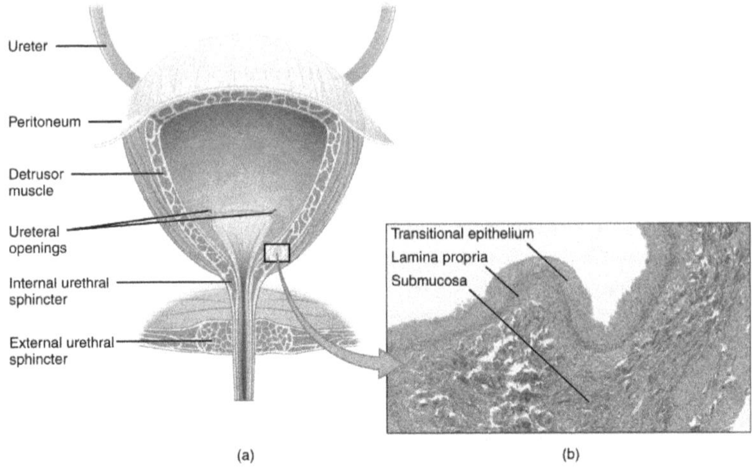

Urethra

Urine passes from the bladder to the outside of the body via the **urethra**. Passage of urine from the bladder into the urethra is controlled by two sphincters: (1) The **internal urethral sphincter** is made of smooth muscle, and it is under involuntary control. This sphincter is located at the base of the bladder. (2) The **external urethral sphincter** is made of skeletal muscle, and it is under voluntary control. This sphincter is located inferior to the internal urethral sphincter.

Micturition

Micturition is a term for urination. Micturition involves three basic processes that can be described as the "micturition reflex": **(1)** contraction of the detrusor muscle, **(2)** relaxation of the internal urethral sphincter, and **(3)** relaxation of the external urethral sphincter. The first two processes are stimulated by the parasympathetic nervous system and inhibited by the sympathetic. The external urethral sphincter, being a skeletal muscle, is controlled by the somatic nervous system. Stretching of the bladder by urine stimulates parasympathetic signals to the bladder. A person can then exercise choice over whether or not to urinate. If the person does not, the parasympathetic signals subside for a while.

Chapter 10: Reproductive System

This is the only system not required for the life of the individual. Instead, it is required for the continued existence of the species. The reproductive system in an individual produces, stores, nourishes, and transports either male or female **gametes**. The female reproductive system is also responsible for protecting, supporting, and nourishing a developing fetus.

I. Overview of Female and Male Reproductive Systems

Common elements of the two systems

Although the male and female reproductive systems have noticeable differences, you will find numerous similarities, too. Both reproductive systems include the following basic components:

A. **Gonads** are the primary reproductive organs–testes in males and ovaries in females. Gonads produce gametes, sperm, and eggs, respectively.

B. **Sex hormones** are produced by the gametes. They affect the body's sexual maturation and gamete development.

C. **Accessory reproductive organs** include (1) *ducts* that provide passages for the transport of the gametes, (2) various *glands* that secrete fluids required by the reproductive system, and (3) *perineal structures* that are important during copulation. (Perineal structures are also referred to as the **external genitalia**),

Human Reproductive System

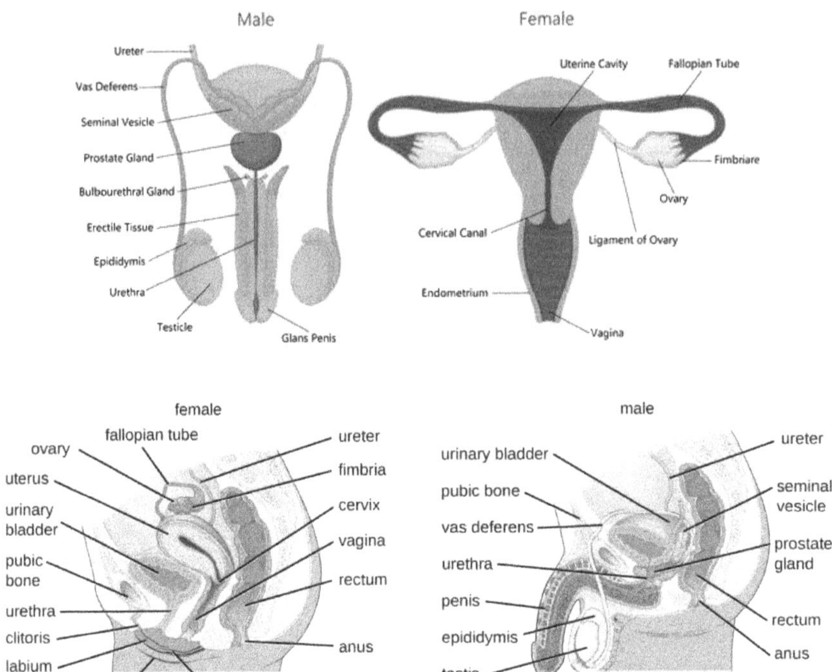

II. Gametogenesis

All cells in your body, except gametes, have two copies of all your genes. These cells are called **somatic cells**, and they may be referred to as **diploid** (or *2n*) to indicate the presence of two copies of the genes. Each gamete has only one copy of your genes, and gametes may be referred to as **haploid** (or *1n*) to indicate the presence of only a single copy of each gene. In order for a new individual to form, two 1n gametes must fuse to form a 2n **zygote**, a process called **fertilization**. The zygote then divides by **mitosis** to form an embryo, which may eventually grow into an adult human. Gametes are produced from 2n cells via the process of **meiosis**, which reduces the amount of genetic material by half (2n÷1n).

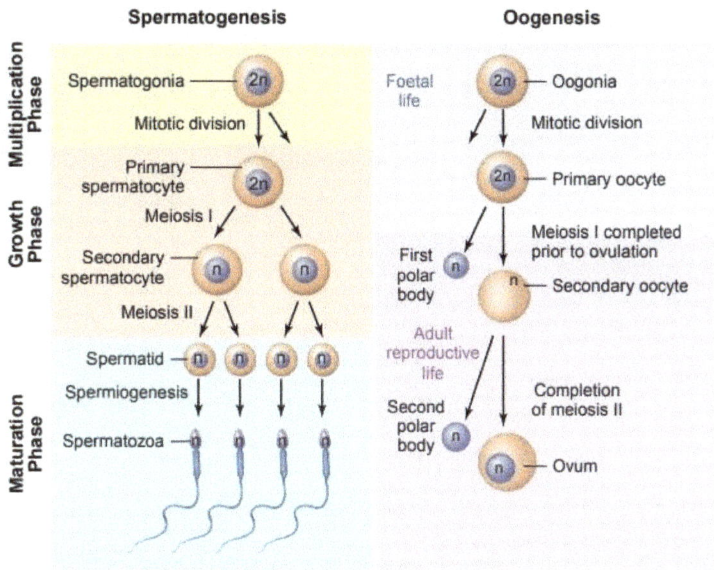

Gametogenesis

III. Female Reproductive System

The major structures of the female reproductive system are shown. The female gonads are the **ovaries**. The female duct system includes the **uterine (or *fallopian*) tubes**, the **uterus**, and the **vagina**. **Mammary glands** secrete milk to nourish a newborn child, and **greater vestibular glands** lubricate the vagina. Perineal structures include the **labia** and the **clitoris**.

<u>Ovaries</u>

Ovaries produce oocytes, secrete female sex hormones, and secrete inhibin.

Anatomy of ovaries

The ovaries are paired, oval-shaped organs about 2 cm by 3 cm in size. They are held in place on either side of the uterus by various structures made of connective tissue. From the laboratory, you should be familiar with the **broad ligament** and the **ovarian ligament**. A **suspensory ligament** attaches the lateral edge of the ovary to the pelvic wall.

Each ovary is encapsulated by a layer of dense connective tissue called the **tunica albuginea**. The tissue of the ovary itself is divided into an outer **cortex** and an inner **medulla**.

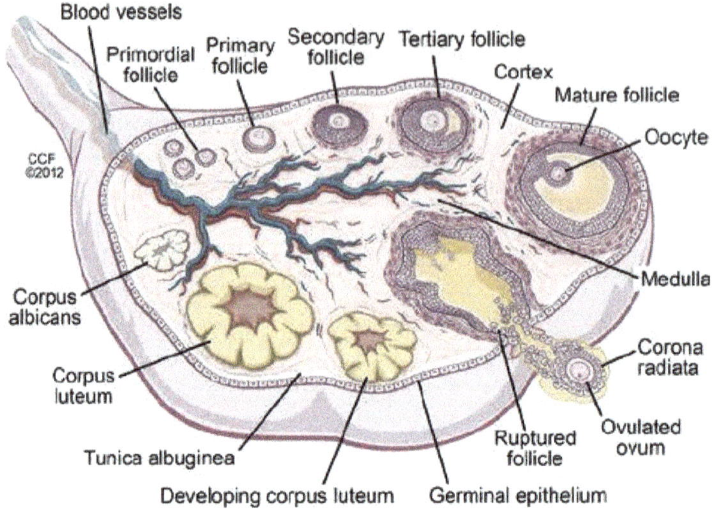

Ovarian follicles

Eggs are produced in the cortex of each ovary. The cortex contains many small, saclike structures called **ovarian follicles**. Each follicle contains an oocyte (the egg) and one or more layers of cells that surround the oocyte:

- A **primordial follicle** contains a **primary oocyte** surrounded by a single layer of flat, squamous-like **follicle cells**.
- A primordial follicle may develop into a **primary follicle**. The flat follicle cells develop into a layer of cuboidal cells, now called **granulosa cells**, which secrete **estrogen**. The granulosa cells will multiply and form multiple layers around the oocyte.
- A **secondary follicle** forms from a primary follicle as a fluid-filled cavity, called an **antrum**, develops between layers of granulosa cells. Structures called the **zona pellucida** and corona radiata surround the oocyte.
- As the oocyte completes meiosis I and becomes a **secondary oocyte**, the follicle becomes a large vesicular follicle.

Eventually, the vesicular follicle bulges from the surface of the ovary and releases the egg, a process called **ovulation**.

- After ovulation, the remainder of the follicle becomes a structure called the **corpus luteum**. The corpus luteum secretes estrogen and **progesterone** in anticipation of pregnancy.
- A corpus luteum will degenerate into a clump of scar tissue called the **corpus albicans**.

Oogenesis and the Ovarian Cycle

Oogenesis begins in women before birth. Female stem cells, called **oogonia**, complete mitotic divisions and differentiate into **primary oocytes**. The primary oocytes begin meiosis I before birth, but this process is not completed until after puberty.

Some primary oocytes continue to develop each month after the onset of puberty. One of the primary oocytes completes meiosis I and becomes a **secondary oocyte**. The secondary oocyte begins meiosis II, but this process stops prior to ovulation.

If a sperm does not penetrate the secondary oocyte, it deteriorates. If a sperm penetrates it, then it rapidly completes meiosis II to become an **ovum**. The nuclei of the ovum and sperm then fuse as **fertilization** occurs.

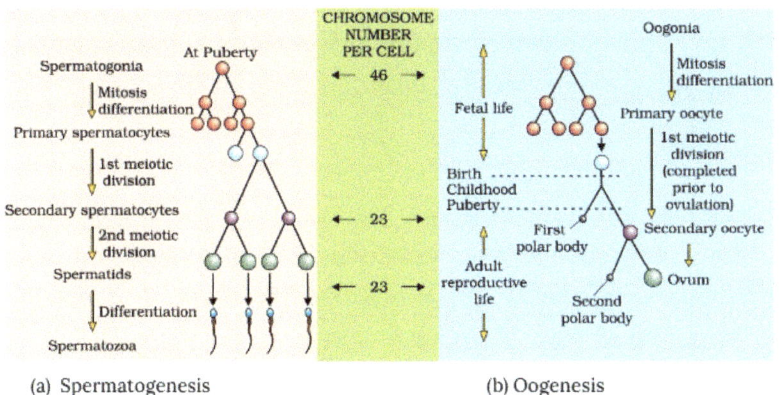

(a) Spermatogenesis (b) Oogenesis

From puberty to menopause

Oogenesis continues from puberty until menopause. Oocyte growth occurs in the cortex of the ovary in a process known as the **ovarian cycle**. Prior to puberty, the release of **gonadotropin-releasing hormone (GnRH)** is inhibited by low levels of estrogen released by the ovaries. With the onset of puberty, the hypothalamus becomes less sensitive to estrogen, and GnRH is released in pulses. GnRH stimulates the anterior pituitary to release **luteinizing hormone (LH)** and **follicle-stimulating hormone (FSH)**. The ovarian cycle begins, and it consists of three phases:

1. The follicular phase

At puberty, a woman's ovaries contain about 400,000 primordial follicles, each of which contains a primary oocyte. Each month a small number of primordial follicles are activated to continue development. FSH triggers follicular cells to divide and form several layers of granulosa cells, creating primary follicles. The oocyte rapidly enlarges, and the granulosa cells produce estrogen. A few primary follicles develop into secondary follicles as granulosa cells begin secreting follicular fluid, which makes the antrum within the follicle.

Usually, a single secondary follicle makes it to the point of maturing into a vesicular follicle, which bulges from the surface of the ovary. Rising levels of LH prompt the primary oocyte to complete meiosis I, become a secondary oocyte, and enter meiosis II.

2. Ovulation

The vesicular follicle releases the secondary oocyte in response to LH. If any other secondary or vesicular follicles have developed, they degenerate. Occasionally, more than one oocyte is ovulated. Does ovulation of more than one oocyte have the potential to produce fraternal twins or identical twins?

3. The luteal phase

After ovulation, the antrum fills with clotted blood. The blood is absorbed, and granulosa cells of the ruptured follicle proliferate and form a structure called the corpus luteum. The corpus luteum produces progesterone and estrogen. In the absence of fertilization, levels of progesterone and estrogen drop. The corpus luteum is destroyed by fibroblasts and becomes the corpus albicans. The ovarian cycle comes to an end. A new cycle will begin with the activation of another cluster of primordial follicles.

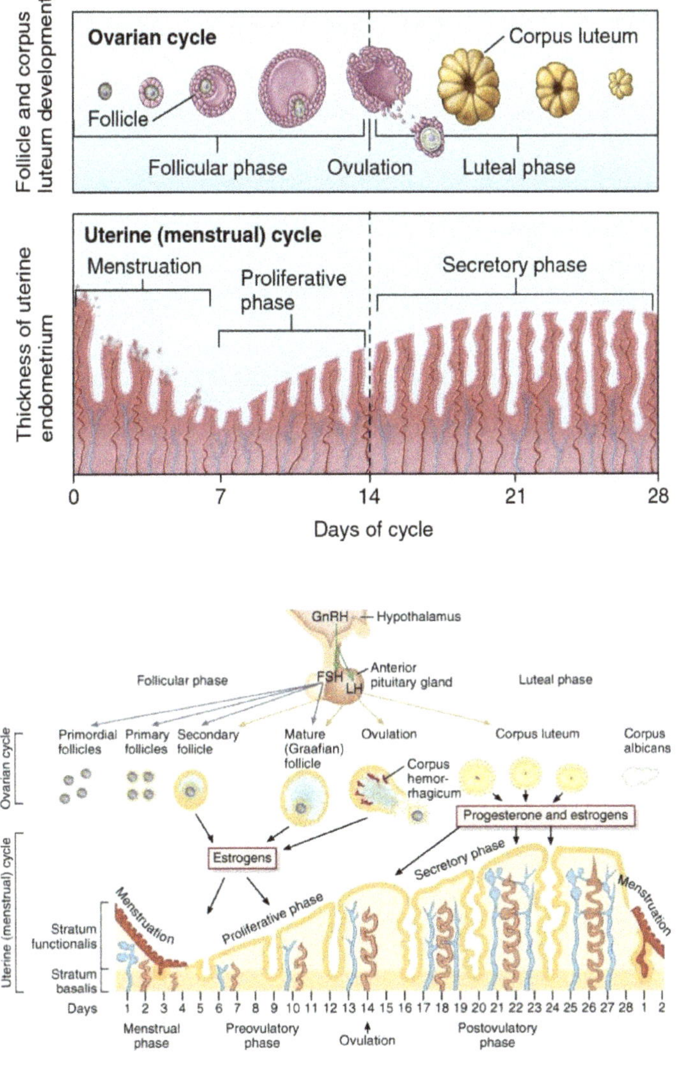

Uterine tubes, uterus, and vagina

Uterine tubes

After an oocyte leaves the ovary, it usually enters one of two uterine tubes, also called "fallopian tubes" or "oviducts". The entry into the uterine tube is a funnel-shaped structure called the **infundibulum**. The infundibulum is draped over the ovary and has finger-like projections called **fimbriae**. The uterine tube narrows as it joins the uterus; the narrow portion is called the **isthmus**. Fertilization of an egg by a sperm often occurs in the uterine tube.

What is an ectopic pregnancy?

Uterus

The uterus provides an environment for the developing embryo and fetus. The uterus can be divided into three regions: The **body** is the major portion of the uterus, the **fundus** is the rounded part of the uterus superior to the attachment of the uterine tubes, and the **cervix** is the inferior portion of the uterus that extends into the vagina. The lumen of the uterus connects to the vagina via the **cervical canal**. The **internal os** of the canal opens to the uterus, and the **external os** of the canal opens to the vagina.

The uterine wall has a thick layer of smooth muscle called the **myometrium**. The inner lining of the uterus is called the **endometrium**. If an embryo implants into the endometrium, glandular and vascular tissues within the endometrium help nourish the developing embryo. The outer layer of the uterus is called the **perimetrium**; it is the visceral peritoneum of the uterus.

The endometrium has two distinct layers: (1) The **functional layer** changes in response to the monthly pattern of hormones. (2) The **basal layer** is the underlying layer from which a new functional layer forms each month.

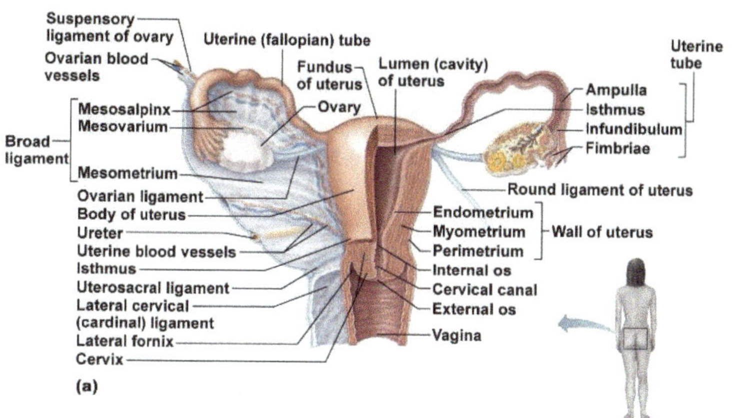

The vagina

The vagina is a tube that leads from the uterus to the outside of the body. The wall of the vagina has three layers: an outer **adventitia**, a middle layer of smooth muscle called the **muscularis**, and an inner **mucosa**. The external opening of the vagina is called the **vaginal orifice**. Epithelial cells of the vagina secrete glycogen, which is converted to lactic acid by bacteria. The acid lowers the pH of the vagina and helps prevent infection.

Why is it that taking antibiotics can make a woman more susceptible to yeast infections?

Uterine (menstrual) cycle and menstruation

The **uterine cycle** (or *menstrual cycle*) is a series of changes in the endometrium that occurs in response to monthly hormonal changes. The cycle can be divided into three phases:

The menstrual phase: The cycle begins on the first day of **menses**, which involves the degeneration of the functional layer of the endometrium. Weakened arterial vessels rupture, and blood and degenerated tissue break away from the inner wall of the uterus. This process generally lasts from one to five days and should be relatively painless. Painful menstruation (called **dysmenorrhea**) is typically a sign of inflammation or some other abnormal condition.

The proliferative phase: This phase involves the proliferation of cells in the basal layer of the endometrium, which remains after menses. This results in the restoration of the functional layer of the endometrium, and it is stimulated by estrogen secreted by the developing follicles.

The secretory phase: This phase begins as ovulation occurs. Glands in the endometrium secrete mucus rich in glycogen, which may provide a source of energy for a developing fetus. In the absence of fertilization, secretory activities decline and the cycle is repeated.

Menstruation Summary

Days	Ovary	Uterus	Hormones
1-5	Menstruation (Menstrual Flow)	Shedding of the endometrium	-No particular hormonal surges
6-13	Follicles develop in ovaries	Endometrium is restored	-Estrogen rises to maximum -FSH rises a bit
14	Egg bursts from the ovary and follicular cells differentiate into the corpus luteum	Endometrium gradually thickening	-Estrogen drops sharply -LH surges
15-28	Corpus Luteum is maintained and slowly degenerates if fertilization does not occur	Endometrium thickens to maximum	Estrogen/Progesterone rises then falls if fertilization does not occur -FSH/LH inhibited

Hormones and the female reproductive cycle

Use the illustrations to help you understand the relationships between hormone secretion and ovarian and uterine cycles.

GnRH release in females varies markedly over the course of a month. Levels of GnRH increase from day zero to a peak at around day 13, and then levels fall rapidly. GnRH targets the anterior pituitary, which promotes the secretion of LH and FSH, which, in turn, act on the ovaries. Both hormones, though primarily FSH, stimulate the enlargement of follicles. FSH stimulates granulosa cells in the follicle to multiply and produce estrogen. LH also stimulates the production of estrogen.

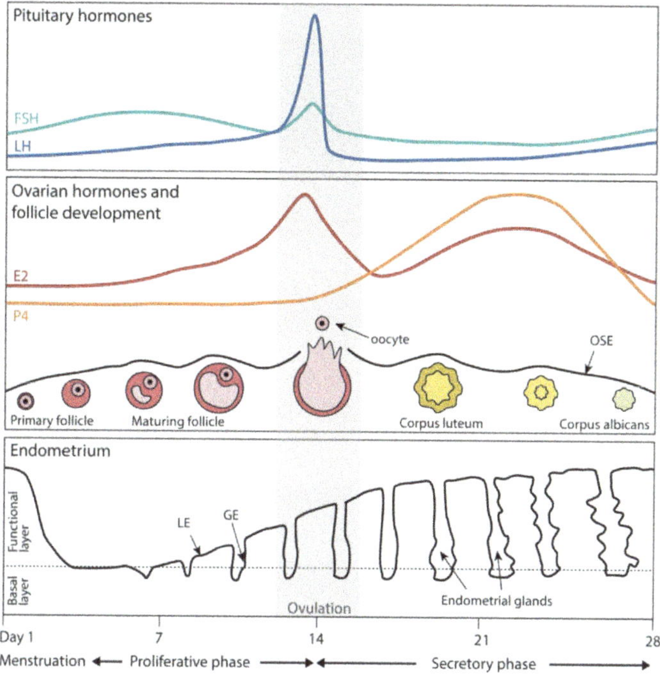

The timed release of these hormones coordinates the ovarian and uterine cycles in the following manner:

*During menses, FSH levels rise, and primary follicles develop.

*Developing follicles release estrogen. The formation of vesicular follicles causes a steep rise in estrogen secretion roughly a week into the cycle.

*GnRH release increases in response to estrogen, stimulating LH secretion. GnRH also causes a rise in FSH, although it does not rise as high as LH.

*At about day 14, estrogen, LH, and FSH levels peak.

*High levels of LH trigger ovulation and promote progesterone secretion by the corpus luteum.

Progesterone prepares and maintains the lining of the uterus.

*Levels of LH and FSH drop rapidly after ovulation. In the absence of fertilization, estrogen and progesterone levels will drop as well.

*Declines in progesterone and estrogen levels result in menses.

*If an egg is fertilized, the corpus luteum maintains estrogen and progesterone secretion. However, after 5 or 6 weeks, the placenta (the interface between the uterus and fetus) begins to secrete estrogen and progesterone to maintain the pregnancy.

External genitalia

The external genitalia are often collectively referred to as the **vulva**. Overlying the pubic symphysis is a layer of fatty tissue called the **mons pubis**. Extending posteriorly from the mons pubis are the **labia majora**. Medial to the labia majora are the **labia minora**. The labia minora surround a region called the **vestibule**, which contains the external urethral orifice and the vaginal orifice. **Greater vestibular glands** surround the vaginal orifice and release mucus to help lubricate the vagina during intercourse.

Anterior to the vestibule is the **clitoris**, which is the female equivalent of the penis. The clitoris is partially covered by a fold of skin called the **prepuce of the clitoris**. Similar to the penis, the clitoris contains corpora cavernosa, which become engorged with blood during sexual arousal.

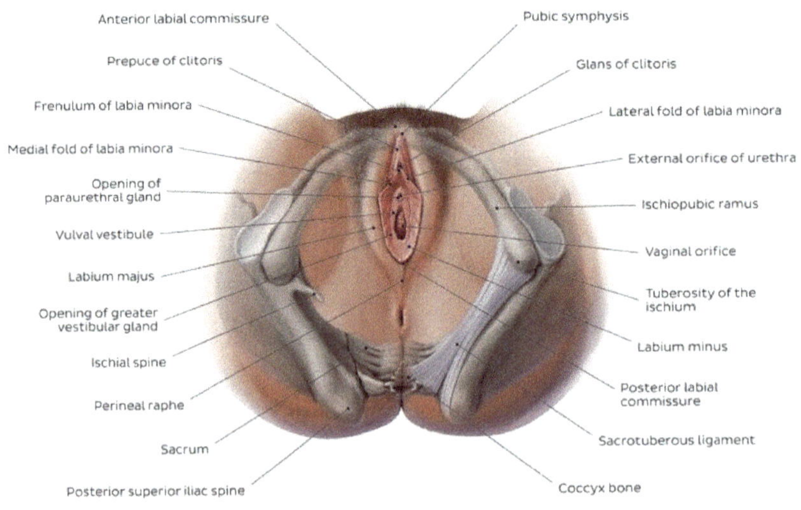

Midsagittal view of female urogenital tract

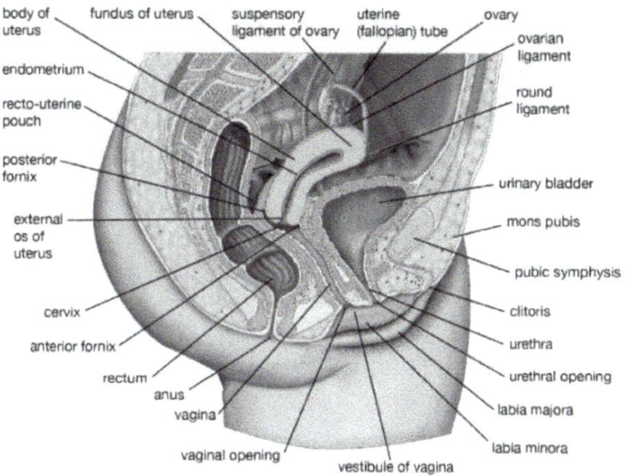

Mammary glands

Mammary glands are present in both males and females but normally function only in females. Each mammary gland lies anterior to the pectoral muscles. Slightly below the center of each is a pigmented **areola**; at the center of each areola is a **nipple**.

Each mammary gland is made of 15 to 25 **lobes** that radiate from the nipple. Cells within each lobe are arranged into clusters called **lobules**. The lobules contain smaller structures called **alveoli**, which secrete milk when a woman is **lactating**. Milk travels from the alveoli to the nipple via **lactiferous ducts.**

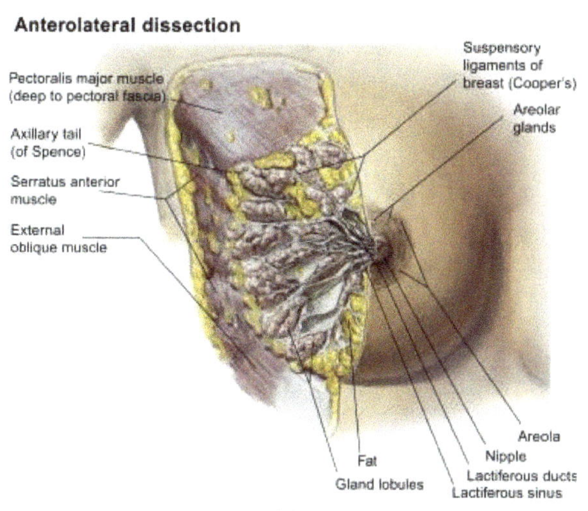

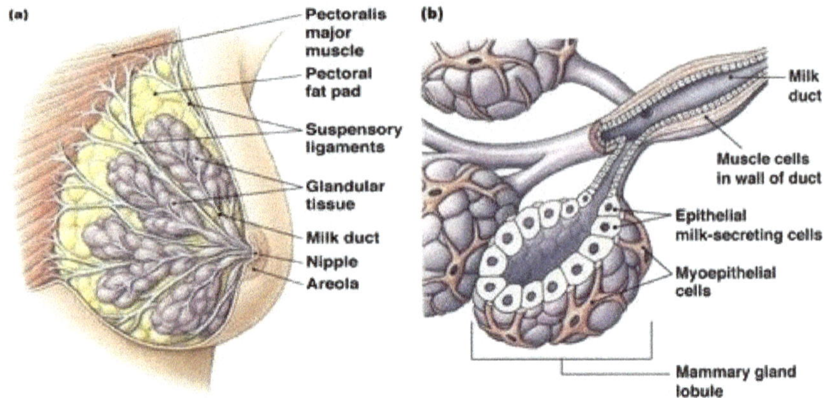

Female sexual response

During sexual excitement, the clitoris becomes engorged with blood in a manner similar to that of the penis. Vestibular glands secrete mucus to lubricate the vagina. These processes are controlled by the parasympathetic. Orgasm in the female is accompanied by rhythmic contractions of the uterus and vagina. The events of orgasm are linked to the sympathetic nervous system.

IV. Male Reproductive System

The basic structures of the male reproductive system are shown. The male gonads are the **testes**. The male duct system begins with the **epididymis**, which leads to the **ductus deferens** (or *vas deferens*), the **ejaculatory duct**, and finally, the **urethra**. Accessory glands include the **seminal vesicles**, the **prostate gland**, and the **bulbourethral glands**. Perineal structures include the **scrotum** and the **penis**.

Midsagittal section of the male urogenital tract

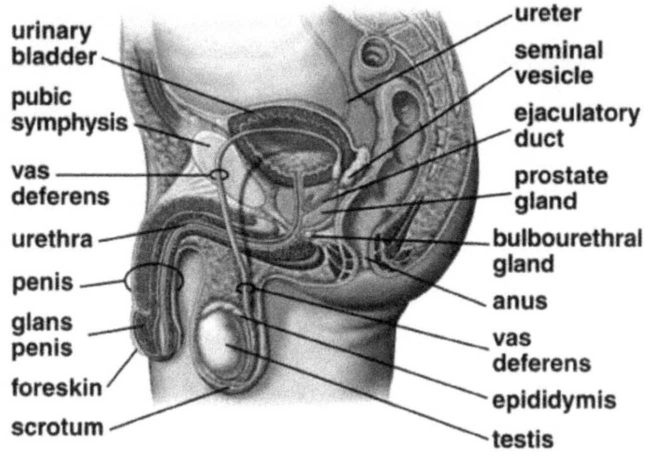

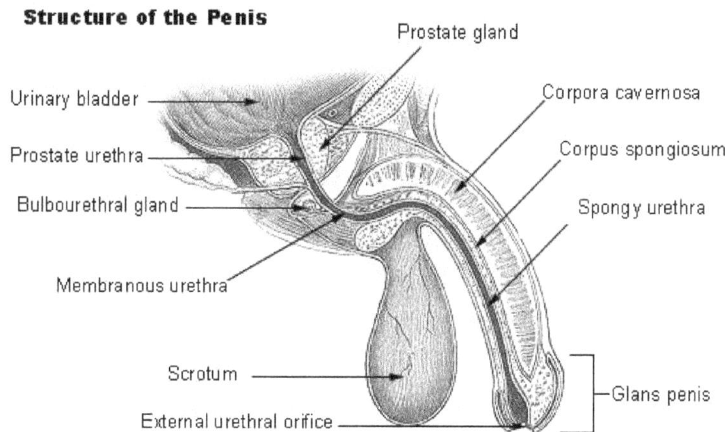

Scrotum

During early development, testes form inside the body near the kidneys. Prior to birth, the testes typically descend into the scrotum, a sac made of a layer of skin and an underlying layer of tissue called the **superficial fascia**.

Bands of skeletal muscle called the **cremaster muscle**, lie deep in the dermis of the scrotum. Contraction of the cremaster muscle pulls the testis close to the body wall. Contraction or relaxation allows regulation of the testes' temperature, which should be about 1°C cooler than the core body temperature for normal sperm development.

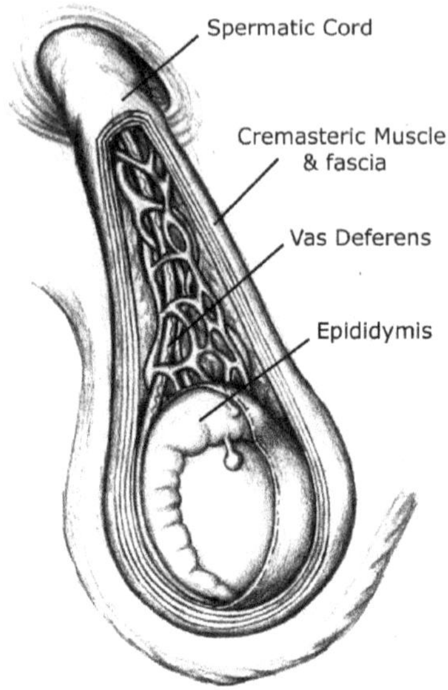

Spermatic cords extend from the abdominopelvic cavity to the testes. Each cord consists of fascia and muscle that enclose the ductus deferens, blood vessels, lymphatic vessels, and nerves, which lead to and from the testes. Each cord passes through the abdominal musculature through the **inguinal canal**, which links the scrotal cavity to the abdominopelvic cavity. Excessive force on the abdomen may push part of the intestine into the inguinal canal, resulting in an **inguinal hernia** (see illustration below).

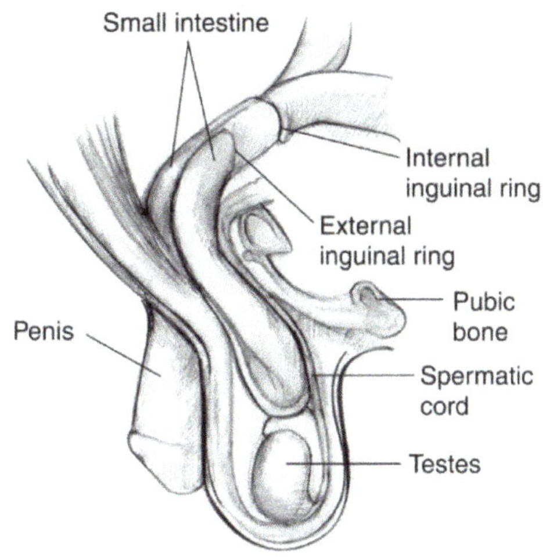

Testes and spermatogenesis

Each testis is surrounded by an outer **tunica vaginalis** and an inner **tunica albuginea**. Each testis is divided into compartments called **lobules**. Each lobule contains a few **seminiferous tubules**; a single testis contains about one-half mile of seminiferous tubules. Surrounding the seminiferous tubules are **interstitial cells** (or *Leydig cells*), which produce androgens. The primary androgen secreted by the interstitial cells is **testosterone**.

Longitudinal cut of testis showing internal structures

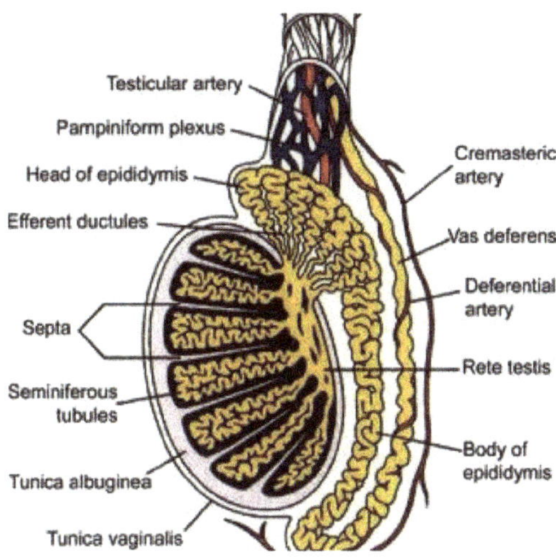

Hormonal regulation of androgen production and sperm development

Males are sterile until puberty, which is marked by the secretion of GnRH by the hypothalamus. GnRH's target is the anterior pituitary, which is stimulated to release FSH and LH, similar to the case in women. FSH stimulates spermatogenesis. LH stimulates interstitial cells to produce testosterone and other androgens. Testosterone also stimulates spermatogenesis, and it has other functions as well: It causes the development of secondary sex characteristics such as deepening of the voice, reproductive organ growth, hair growth, and growth of musculature (which is the effect desired by people who abuse anabolic steroids). As levels of FSH rise, the testes secrete **inhibin**, a hormone that eventually inhibits FSH secretion by the pituitary.

Flowchart showing hormonal regulation of spermatogenesis

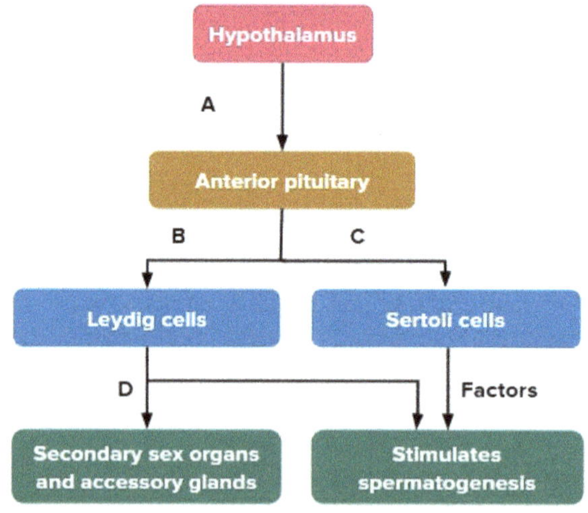

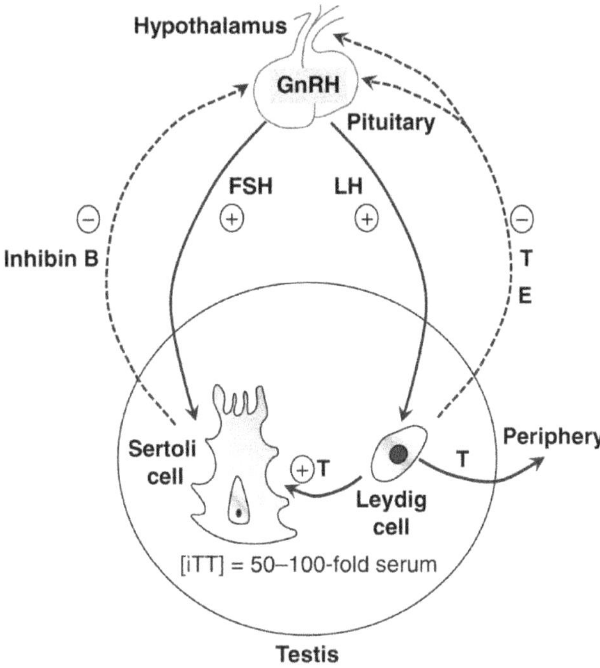

Development of sperm: spermatogenesis and spermiogenesis

Sperm are produced in the testes via a process known as **spermatogenesis**.

Spermatogenesis begins in the seminiferous tubules with stem cells called **spermatogonia**. Spermatogonia divide by mitosis to keep replenishing themselves. Some spermatogonia move on to the next stage of development by differentiating into **primary spermatocytes**. The primary spermatocyte undergoes the first division of meiosis (meiosis I), which leads to the production of **secondary spermatocytes**. Secondary spermatocytes undergo the second division of meiosis to become **spermatids**. Spermatids undergo differentiation to produce **spermatozoa**, which are fully-formed sperm. This last process of differentiation is known as **spermiogenesis**. Spermatozoa travel to the epididymis, where they are stored. Know the various stages in spermatogenesis and the processes (underlined) that lead to each stage.

Below is the diagram of the human spermatozoon. Note the head containing the DNA molecule is the actual portion of the sperm that separates from the whole sperm and penetrates the egg (ovum).

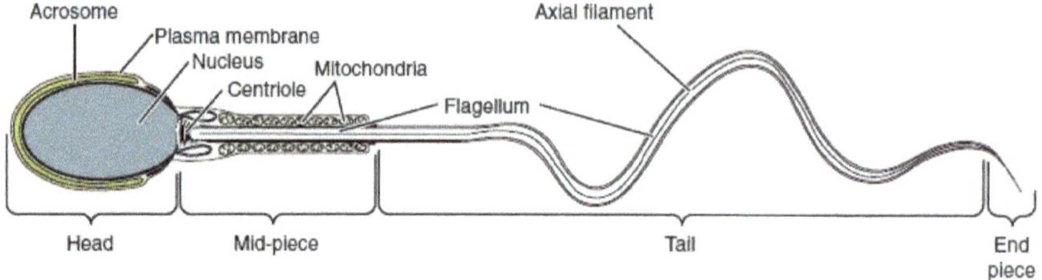

Transverse section of the seminiferous tubules with spermatogonia undergoing spermatogenesis.

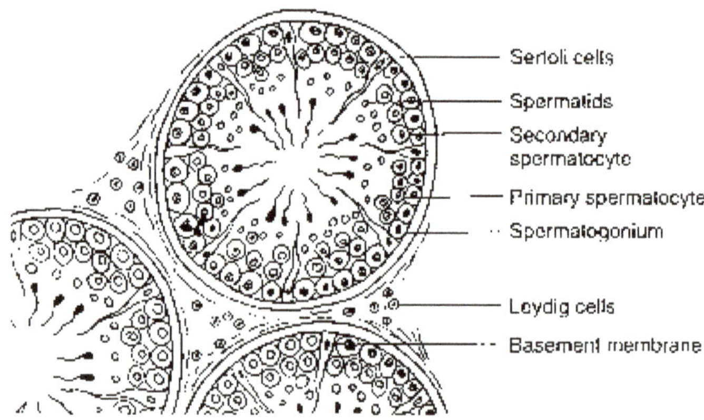

Sagittal section of a testis and Epididymis

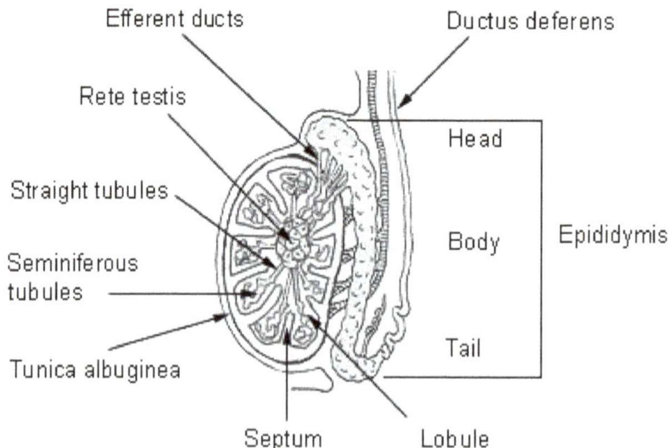

Duct system in the male reproductive tract

The epididymis leads to the ductus deferens, which leads to the ejaculatory duct and then the urethra. Be able to trace the path of sperm from the seminiferous tubules to the external urethral orifice.

Accessory glands and semen production

Sperm and fluid from the epididymis account for only about 5% of semen volume. The remainder is contributed by various accessory glands:

A. *Seminal vesicles* produce about 60% of semen volume. The fluid secreted by the seminal vesicles has a high concentration of fructose. Why? It also contains **prostaglandins**, which stimulate contractions of smooth muscle along the reproductive tract. Seminal fluid is alkaline, which neutralizes the acidic environments of the urethra and the vagina. Finally, secretions of the seminal vesicles stimulate the beating of sperm flagella.

B. The *prostate gland* produces about 20-30% of the semen volume. This gland encircles the urethra as it exits the urinary bladder. The prostate fluid contains **seminal plasmin**, an antibiotic that may help prevent urinary tract infections (UTIs). Inflammation or tumor growth of the prostate is fairly common in older men. Symptoms include pain in the lower back and painful urination.

C. The *bulbourethral glands* secrete thick, alkaline mucus, which also helps to neutralize acids in the urinary tract and vagina. Semen typically contains 20-100 million sperm per milliliter of fluid, and a typical ejaculation contains 3-5 ml of semen. For good fertility, greater than 60% of sperm should be swimming.

Penis

Refer to the diagram of the structure of the penis below. The penis and the scrotum make up the male external genitalia, and the penis is used for **copulation**.

The distal portion of the penis is the **glans**. In a circumcised male, the glans is exposed; in an uncircumcised male, the glans is covered by the **foreskin** (also called the **prepuce**). The body of the penis contains three long cylinders of **erectile tissue**. This tissue is porous, and as blood fills the tissue, the penis becomes erect. The two dorsal portions of erectile tissue are called the **corpora cavernosa**, and the ventral portion is called the **corpus spongiosum**.

Longitudinal section of prostate gland, penis and other structures of male reproductive organ

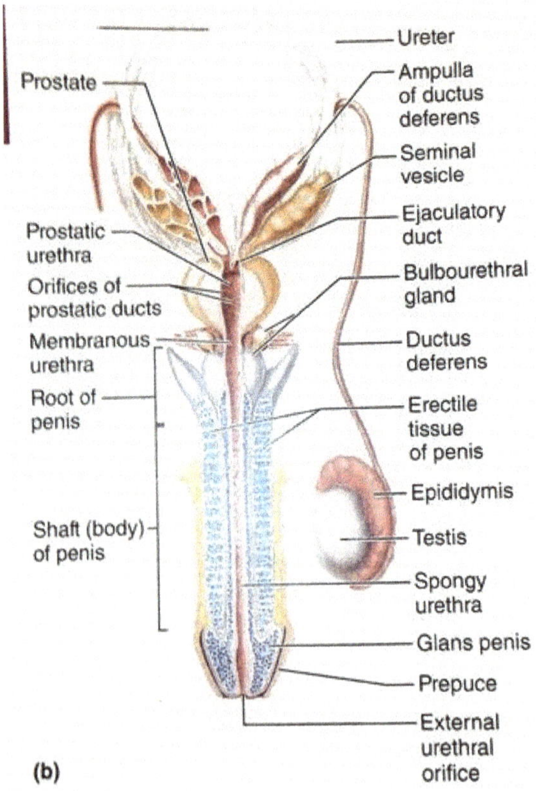

Penile erectile structures compared to clitoral erectile tissue on left and cross-section of penis on right

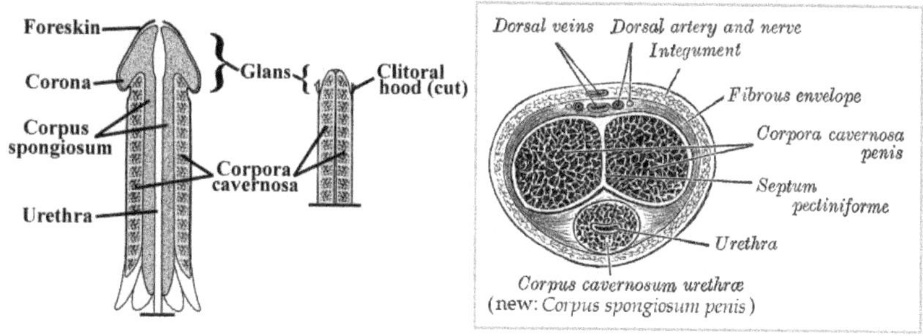

Male sexual response

Erection is caused by the dilation of blood vessels that lead to the erectile tissue. Dilation of the blood vessels is caused by the release of nitric oxide, which is stimulated by the parasympathetic nervous system.

Ejaculation is caused by the sympathetic nervous system, which causes contractions of various muscles along the male reproductive tract.

Remember: *point (erection by parasympathetic)* and *shoot (ejaculation by sympathetic).*

Differentiation of external genitalia in the human embryo and fetus

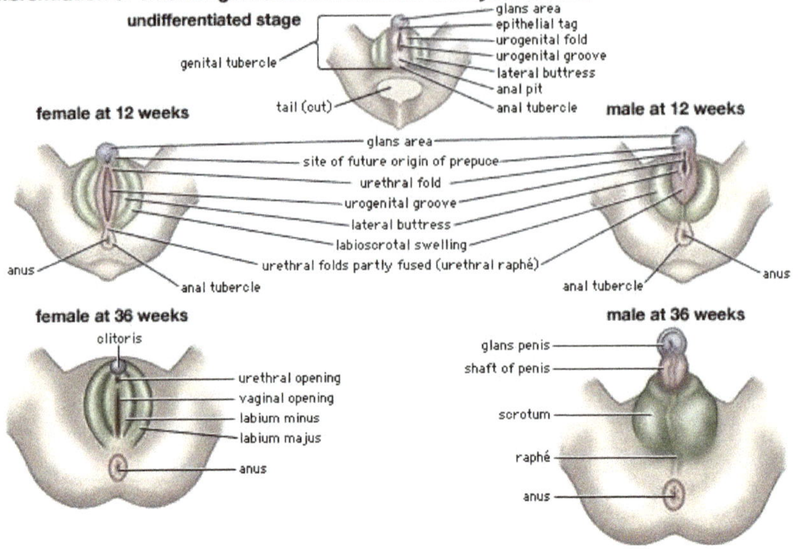

Answers to cardiac (quantitative analysis) questions:

1. First convert 6.4 L/min to milliliters per minute (recall that 1 L equals 1,000 mL). So, 6.4 L/min X 1,000 is now 6,400 mL/min then, HR = CO / SV; 6,400 mL / 74 mL/beat = 88 bpm.
2. CO must again be converted from L/min to mL/min since we want to solve for SV (whose unit is mL/beat), so CO = 5,200 mL/min. Using the formula for SV, we get: 5,200 mL/min / 110beats/min = 47mL/beat.
3. PP = SBP – DBP; or PP = SV / 2; DBP = SBP – PP; so, PP is SV divided by 2 (PP = 94mL/2 = 47 mmHg). DBP = 133 mmHg – 47 mmHg = 86 mmHg;
   MBP = DBP + 1/3PP = 86 mmHg + 1/3 47 = 101.7, or 102 mmHg.
4. Convert HR to beats/minute (so 12 beats/10 seconds X 60 seconds = 72 bpm)
   PP = SBP – DBP = 110 – 70 = 40 mmHg
   SV = PP x 2 = 40 x 2 = 80 mL/beat
   CO = SV x HR = 80 x 72 / 1,000 = 5.76 L/min
   a. HR = 75 beats/45 sec x 60 sec = 100 bpm; PP = 150 – 95 = 55 mmHg

SV = 55 x 2 = 110 mL/beat
CO = 110 x 100 / 1000 = 11 L/min

b. % change in HR = [(HR_A – HR_B)] / (HR_B)] X 100
(100 – 72)/72 = 38.9 %

5. SV = 5,000 / 49 = 102 mL/beat; PP = SBB – DBP (or PP = SV / 2) and DBP = SBP – PP, so 105 – 51 = 54 mmHg

Answers to respiratory physiology questions:

1. TV = VC – (ERV + IRV) = 4,800 mL – 4,300 mL = 500 mL
2. COPD (or any obstructive lung disease) decreases vital capacity
3. Since 1microliter equals 0.001 mL then 450,000 microliters equate to 450 mL. So, ERV = VC – (TV + IRV); and the answer is 4,050 mL.
4. COPD increases residual volume

Index

www.ingramcontent.com/pod-product-compliance
Ingram Content Group UK Ltd.
Pitfield, Milton Keynes, MK11 3LW, UK
UKHW062303290726
14090UKWH00017B/858